THEORY OF
STRUCTURAL
GEOLOGY

W0230668

THEORY OF STRUCTURAL GEOLOGY

BY

DR. N.W. GOKHALE
RETIRED PROFESSOR OF GEOLOGY,
SHREYAS, 82 HUBLI ROAD,
DHARWAD 580007
KARNATAKA

CBSPD

CBS Publishers & Distributors Pvt Ltd

New Delhi • Bengaluru • Chennai • Kochi • Kolkata • Lucknow • Mumbai
Hyderabad • Jharkhand • Nagpur • Patna • Pune • Uttarakhand

Theory of Structural
GEOLOGY

ISBN-13: 978-81-239-0453-5

Copyright © Author & Publisher

First Edition: 1996

Reprint: 2001, 2003, 2006, 2008, 2010, 2012, 2013, 2014, 2017, 2019, **2024**

All rights reserved. No part of this book may be reproduced or transmitted in any form or by any means, electronic or mechanical, including photocopying, recording or any information storage and retrieval system without permission, in writing from the author and publisher.

Published by **Satish Kumar Jain** and produced by **Varun Jain** for
CBS Publishers & Distributors Pvt Ltd
4819/XI Prahlad Street, 24 Ansari Road, Daryaganj, New Delhi 110 002, India
Ph: 011-23289259, 23266861   Website: www.cbspd.com
                                    e-mail: delhi@cbspd.com

Corporate Office: 204 FIE, Industrial Area, Patparganj, Delhi 110 092, India
Ph: 011-4934 4934   Fax: 011-4934 4935   e-mail: publishing@cbspd.com;
                                             publicity@cbspd.com

Branches

- **Bengaluru:** Seema House 2975, 17th Cross, KR Road, Banasankari 2nd Stage, Bengaluru 560 070, Karnataka, India
  Ph: +91-80-26771678/79   Fax: +91-80-26771680   e-mail: bangalore@cbspd.com
- **Chennai:** 7, Subbaraya Street, Shenoy Nagar, Chennai 600 030, Tamil Nadu, India
  Ph: +91-44-26680620, 26681266   Fax: +91-44-42032115   e-mail: chennai@cbspd.com
- **Kochi:** 42/1325, 1326, Power House Road, Opp KSEB, Power House, Ernakulum Kochi 682 018, Kerala, India
  Ph: +91-484-4059061-65, 67   Fax: +91-484-4059065   e-mail: kochi@cbspd.com
- **Kolkata:** 147, Hind Ceramics Compound, 1st Floor, Nilgunj Road, Belghoria, Kolkata-700056, West Bengal, India
  Ph: +91-33-25633055, 033-25633056   e-mail: kolkata@cbspd.com
- **Lucknow:** Basement, Khushnuma Complex, 7 Meerabai Marg (Behind Jawahar Bhawan), Lucknow-226001, UP, India
  Ph: +91-522-4000032   e-mail: tiwari.lucknow@cbspd.com
- **Mumbai:** PWD Shed, Gala no 25/26, Ramchandra Bhatt Marg, Next to JJ Hospital Gate no. 2, Opp. Union Bank of India
  Noorbaug, Mumbai-400009, Maharashtra, India
  Ph: 022-66661880/89   e-mail: mumbai@cbspd.com

Representatives

Hyderabad	0-9885175004	Jharkhand	0-9811541605	Nagpur	0-8692091830
Patna	0-9334159340	Pune	0-9664372571	Uttarakhand	0-9716462459

Printed at SRK Graphics, Delhi (India)

PREFACE

Matter described pertaining to structural geology in any book nearly remains the same. Much depends upon how it has been presented. Akin to the CLAN CONCEPT in igneous petrology, it is felt necessary to consider the genetically related structures together into one group. This has led to the formation of only two groups namely, Rupture structures, and Plastic structures. Emphasis has been laid on the recognition of structures in the field without which no further studies at all are possible. Structures go unnoticed because of ignorance about their style of appearance in outcrops. Realising this lacuna, several field photos and 3 dimensional block diagrams have been drawn. This will help to locate more and new structures from different parts of India by those geoscientists who desire to do so. This book has described several unusual structures hitherto unknown to the geoscientists of India. Deformative forces being instrumental in the development of the various structures, this aspect has been described in one chapter, elaborating the Modus Operendi of individual structures. Concept of "paired geological events" has been introduced for the first time in the geological literature. Chapters entitled "Some concepts, Importance of structures, Classification of the deformative forces "are again not described separately by many authors. The matter described in the said chapters in this book will be found very useful by the researchers and the thinkers. Examples of structures from India as available in the literature have been enlisted. Seeing is knowing them. Therefore these examples will enable the budding geoscientists to visit and study these places of structural importance.

The coloured and the black/white field photographs are the outcome of the precious and conscientious field work of my doctoral researchers. These have enhanced the value of the text because the structures captured are very clear, illustrative and self explanatory ones. Some ary very rare and unique indeed like "brick work like joints, rotational fault of Jamkhandi, bedding plane fault of Basidoni, and mud crack like columnar joints of Salgaon" and so on.

I place on record my appreciative thanks to the contributors of the field photographs, Drs. V.C. Chavadi, N.N. Gothe, V.B. Koppad, S.C. Puranik, D.I. Deendar, A.H. Kouhsari, K.Bhimsen, G.S. Pujar, H.D. Desai, B.C. Prabhakar, B.P. Waghamare, Shriyuts S.H. Paramshetti, M.R. Shinde (teachers from University and College Departments of Geology, Karnataka and Maharashtra states, and Iran), and Dr. V.N. Hegde (O.N.G.C. India), Dr. G.V. Hegde (Department of Mines and Geology, Karnataka state), Dr. D. Muralidharan (N.G.R.I. Hyderabad) who are working in various organisations.

The account given in this book is such that it is useful to a beginner, as well as to those who are desirous of breaking new ground in the field of structural geology. The text is free from constraints of curricular syllabus. It will be therefore found utilisable by students of any Indian university. The book is the outcome of my 36 years of experience in the field of research.

Dr. N.W. GOKHALE
Retired Professor of Geology,
Karnataka University,
Dharwad.

CONTENTS

1

STRUCTURAL EQUATION

Introduction, Objectives of structural studies, Structural equation, Components of equation, Role of kinds of rocks, Role of deformative forces, Additional components of structural equation, Classification of structures, Basis of size of structure, Basis of mechanism of deformation, Basis of deformative forces, Varieties of deformative forces, Mechanical properties of rocks, Environment of deformation, Mechanism of deformation, Mechanisms of elastic, plastic and ruptural deformations, Square prism experiment and its role in deformation of rocks.

Figures 1-20
Photo 1

INTRODUCTION

The subject of structural geology describes aspects of building up of the crust of the earth. Changes are taking place continuously, some times slow, some times fast. The word "structure" conveys various meanings; to some it may be folds, joints, faults, others may include schistosity, foliation, unconformity etc. All these are structures, but the way these were developed, those ways differ. Some rocks are seen to be folded, others are seen to be faulted. The question therefore arises as to why some are folded and others are faulted.

The different structures developed in the rocks, are these of any use, at least academically, if not economically to the geoscientists? The experience is that the structures are useful both from the economic and the academic points of view. These also need to be described in the book. Then comes the stage of explaining the genesis of the different structures. Of late, the age of the development of the different structures, is also being considered in the structural studies.

Geology fundamentally is a field science, and structural geology is probably the only branch which is to be understood in the field. There is a minimum amount of imagination required on the part of the geologist to appreciate the structure, since it is directly observable in the field. Minimum theorising is required. However any structure being a three dimensional one, it can be understood the best only in the field. Seeing is believing. As such, good examples of structures from India will be of great help. Therefore, wherever possible, such instances are cited so that the structure could be seen and understood better, because the places cited could be visited by those who desire to do so. Further it will also help to recognise similar structures elsewhere in the country by the budding geoscientists.

"The Dharwarian and the Eastern Ghat trends meet in southern Mysore and in the Nilgiris; the Satpura and the Aravalli trends meet in northern Bombay and southernmost Rajasthan. Three different strike directions meet in a triangle in Bhandara in Madhya Pradesh. These and other regions where the major trends meet each other require to be carefully studied" (Krishnan 1968 p. 49). Krishnan thus underscores the importance of structural features either in the understanding of the evolution of the crust of the earth, or in the context of (economic) mineralisation.

The most debatable part in structural geology is to envisage the deformative forces that had brought about the development of the observed structures. This is an abstract aspect and there could arise difference of opinion between the geoscientists. Each one tries to present a model from the available data by considering certain assumptions. The geological data are seldom complete, the assumptions could therefore vary from person to person. Therefore more than one version can prevail regarding the origin of a structure.

OBJECTIVES OF STRUCTURAL STUDIES

The role of a geologist and hence that of structural geologist is to recognise the structure, and try to reconstruct the stages through which it was developed. The earth is atleast 4000 m.y. old. The crustal rocks are formed throughout that period. As such there will be superimposition of the structures formed during the geological past. Academically speaking, each structural event needs to be isolated, and its time of formation established. Further, the several structural events require to be arranged in a chronological order.

Earlier it was said that the deformative forces are needed to be considered. But what caused the development of these forces also should be discussed. The crustal rocks constitute a contiguous body. Therefore, if at one place compressive forces are produced, it becomes necessary to look for a place in the crust where tensional forces were produced. Thus the geological events must be taking place in pairs. If a landmass were to rise, elsewhere a landmass has to sink. If a gravity fault were to occur, the base of such a column of rocks has to impound upon the rocks below it, and these latter rocks must get folded and or crushed, otherwise additional space for the sinking column of the rocks due to the gravity fault, cannot be created.

STRUCTURAL EQUATION

The main theme, or the central idea, or the main pursuit of structural geology can be reduced to a simple equation namely,

$$\left(\begin{array}{c} \text{observed deformation} \\ \text{in the crustal rocks} \end{array} \right) = \left(\begin{array}{c} \text{kinds of} \\ \text{rocks} \end{array} \right) \times \left(\begin{array}{c} \text{deformative} \\ \text{forces} \end{array} \right)$$

The above given equation may be written in the form of a sentense. *The different structures noticed in the crustal rocks are the outcome of the interaction taking place between the different kinds of rocks and the different types of the deformative forces.* The geologist first identifies the different structures and then tries to explain their mode of formation. It is observed that the different rocks develop different structures. Hence it is inferred that the different varieties of structures may be due to the variation in the rock types. But it is also observed that though the rock is same, the style and the intesity of structures, is different. Obviously the cause for this should be found in the kind of the deformative forces and not in the kind of rock. Thus it is clear that the different structures are a function of the different kinds of the rocks, as well as the different kinds of the deformative forces that had acted on them.

COMPONENTS OF EQUATION

Each component of the above given equation needs to be expanded further. The starting point is the existence of a structure or structures. Most of the studies are concentrated on this aspect, and very little is said about the kinds of the rocks, and the kinds of the deformative forces. Structures need to be classified, and this formulates the stem of structural geology. Depending upon the basis chosen, varieties of structures multiply. Thus a fold gets further split into a syncline, an anticline, a reclining fold etc. This aspect will be elaborated in a separate secfion. Many times it goes difficult to recognise the structure owing to the paucity of data. In such cases only a broad name is given. When more data are available, further distinctions are then made to classify it in a more precise manner. This study according to some structural geologists formulates "Descriptive structural geology". Structure being a three dimensional object, its mere outcrop on the surface of the earth or in a section (these tentamount to two dimensions), is not sufficient. If the three dimensions of the structure be not available, then the name of "Surficial structural geology" is applied. When all the three dimensions are known, then it is called as "Subsurface structural eology". These considerations increase the recognition of more varieties of structures.

ROLE OF KINDS OF ROCKS

Likewise the factor of "kinds of rocks" needs to be analysed. A correlation between the three main kinds of rocks namely, the sedimentary, the igneous and the metamorphic, and the styles of structures, is warranted. Are these major groups of rocks deformed alike? The answer is "no". Folds are not found in the igneous rocks. Faults are not easily noticeable in these rocks. The pattern of joints in the sedimentary and the igneous rocks, are different. The conclusion therefore is inescapable that the rocks have a "say" in the development of the structures. Actually a structural geologist does not consider the rock in their petrographical sense, but in their reactions to the deformative forces. This is spoken off as the "mechanical properties of rocks". Consideration of these properties is necessary because though a shale and a sandstone are sedimentary rocks, these do not yield or react to the deformative forces in a similar manner. A shale flows and produces a fold, while a sandstone ruptures and produces a fault or a shear. This feature of "kinds of rocks" will be further described in a separate section.

ROLE OF DEFORMATIVE FORCES

The deformative forces are the main stay in the development of the structures. Without them, no structures will be produced. Therefore depending upon the kind and the intensity of the deformative forces, the resultant structures will differ. The details about the deformative forces will be elaborated in a separate section, but here it is felt necessary to give information about the mode of generation of the deformative forces and their transmission in the crustal rocks because without this knowledge, the evolution of the structures cannot be understood in completeness.

ADDITIONAL COMPONENTS OF STRUCTURAL EQUATION

In the equation given above, only three main components have been noted namely :

   (i) structure,
   (ii) kinds of rocks, and
   (iii) the deformative forces.

But there are two more controlling factors that contribute towards the development of the structures. Argument is as follows. A sandstone is a brittle rock, and therefore it must yield in a "ruptural manner". But this rock is also observed to be folded. Fold is a plastic style of deformation. Therefore the inference is inescapable that the sandstone behaved in a "plastic manner". Therefore its mechanical property is observed to be changed. That means, there is yet one more control over the development of the structures. That factor is the "depth of formation of the structure". A rock located at a shallow depth behaves differently than the same rock when situated at a greater depth. One more controlling factor is "time". A small force acting over a greater length of time can bring about the same type of structure like that produced by a large force acting over a short period of time. The geological processes operate over periods of millions of years. Hence while deriving the magnitude of the deformative force, the factor of "period or time" is not to be lost sight of. This is called as the "rheidity" of the rock, and this property varies from rock to rock.

Thus in the structural equation given above, two more controls are to be included which are "depth and time". The factor of depth is called by the term "environment of structural deformation". The time factor is called the "rheidity" of the rocks. The complete analysis of the "structures" developed in the crustal rocks, should give information in respect of all the components of the structural equation as described above.

CLASSIFICATION OF STRUCTURES

Structures can be classified in several ways depending upon the nature of the bases. The different bases of classification and the resultant structures recognised therefrom will be described in the following paragraphs.

   1. **The basis of size of the structure:** Here the visibility and size of the structures are combined, and the varieties recognised are

(i) mega structures, and

(ii) micro structures. Further distinctions are possible within mega structures namely,

    (a) large, and

    (b) giant.

Thus there could be an anticline whose width of limbs is measurable in kilometer (giant anticline), a few meters (large anticline), or a few centimeters (minor anticline), and so on. Again a fault where the displacement may be of the order of a few kilometers (giant fault or a nappe structure), a few tens of meters (large fault), or a few centimeters (minor fault), and so on.

2. **The basis of mechanism of deformation**: Different structures like the fold, the fault, the joint the slaty cleavage, the lineation, the foliation etc., are observed in the crustal rocks. Externally and by appearace these do differ, but precisely, these structures basically differ in their modes of formation. The terms "rupture" and "plastic" structures are therefore used. Thus the joints, the faults, the shears, the fractures etc., are called as the "rupture structures", while the folds, the schistocity, the foliation, the lineation etc., are called as the "plastic structures".

3. **The basis of deformative forces**: A deformed rock is also looked at from two more angles. In the *first category*, the *entire rock* is considered as *one unit*, and which is subjected to the deformative forces. Thus the rock gets folded, faulted, ruptured and so on. Here the deformative force or forces had acted on the *entire rock as one unit*. In the *second category*, the *unit* of the rock is observed to be affected by the deformative forces. The unit of a rock is a mineral, and in order to affect the minerals, the deformative force should penetrate deep into the rock. The minerals thus get rearranged to suite the compelling deformative forces. There need not be any external or visible deformation like the development of a fold, joints., but such structures like the orientation of the grains, the laths, or the flattening and or the elongation of the grains etc., might take place. Slaty cleavage, schistocity, lineation etc., belong to such a type of deformation. These rocks are structurally called as the "tectonites", and others where the indidual minerals are not deformed or rotated by the deformative forces, are called as the "non-tectonites".

From the foregoing account, a schematic classification of the structures can be written as given below.

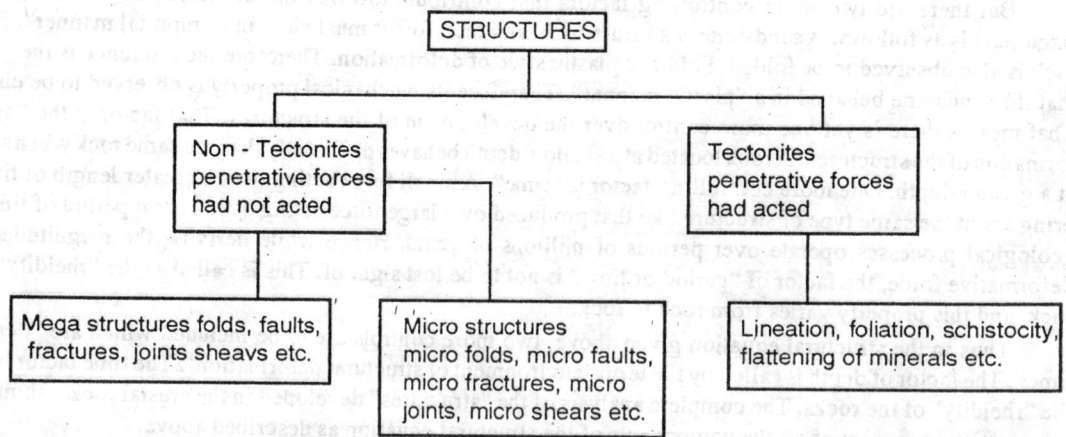

The individual structures are further classifiable into plunging and non-plunging folds (in the case of folds), reverse or normal faults (in the case of fault), B ⊥ B and B ^ B tectonites (in the case of tectonites).

VARIETIES OF DEFORMATIVE FORCES

The deformative forces are classified into several kinds. However their distinction in the field is very difficult or many times not possible at all. For example, joints are noticed in almost every rock, but it will not be possible to attribute them to the shearing, the tensile or the compressive forces, when data are meagre. Even then different kinds of forces are recognised, namely the compression, the tension, the torsion, the couple and so on. The general characters of these forces are described below.

COMPRESSION

In this case the forces act towards a point, and these forces act *in one plane and in one line* (Fig. 1A). If these were not to act in one line, these can become shearing forces. as is shown in Fig. 1B. Further, if a plane is not stipulated, then the surface may be curved one, and instead of shear, rotatinonal forces may develop as is evident in Fig. 1C. Further it is equally necessary to consider the intensity of the force on the two sides, as well as in the vertical direction. If the intensity be not constant, the resultant structures are different. This aspect will be discussed later.

TENSION

In this case, the forces act away from a point, but these act in the *same plane and in one line* (Fig. 2A). If these do not act in one line, then shear or rotation may develop (Figs. 2B, 2C).

SHEAR

In this case the forces *initially itself do not act in the same line* though these act in the *same plane.* If the surface be curved, this will lead to torsion (Figs. 3A, 3B, 3C). Originally these forces could be compressional or tensional ones.

TORSION

These forces are created when the surface is curved one. In fact the forces have the characters of a shear, but as these are acting on a curved surface, these therefore bring about torsion or twist (Figs. 4A, 4B). If the surface be plane, then it may culminate into a shear or a rotation of the specimen may take place.

COUPLE

This situation is created when two sets of shearing forces act on a body as shwon in Fig. 5. When the shears A_1 - A_1 and B_1 - B_1 act on the square, it will rotate in the direction of the arrows. In order to prevent this movement, another set of shears A_2 - A_2 and B_2 - B_2 acts on the body, and this prevents the movement of the square.

It has been simplified that the intensity of the deformative forces, is constant (in the vertical sense), and that these forces are acting on a plane or a curved surface. In the crust of the earth, neither the intensity of the forces is constant, nor the surfaces are plane or curved ones. The surface of contact between the two rock units is likely to be irregular. Therefore the theoretically expected behaviour (style of deformation) is not realised in the deformed crustal rocks. The conclusion or the inference drawn therefore should not be that unusually strong forces had acted, but there were variations in the intensity of the deformative forces (in the vertical sense) and the surface of contact of the rock units was far from being planar or curviplanar one in attitude. The other implications and the actual mode of deformation will be described in a separate section

MECHANICAL PROPERTIES OF ROCKS

These properties are two in the main namely,

   (i) competency, and

   (ii) incompetency.

This is spoken off as the strength of the rocks. The other terms used while describing the mechanical perperties of rocks are

     (a) brittle,

     (b) ductile, and

     (c) melleable.

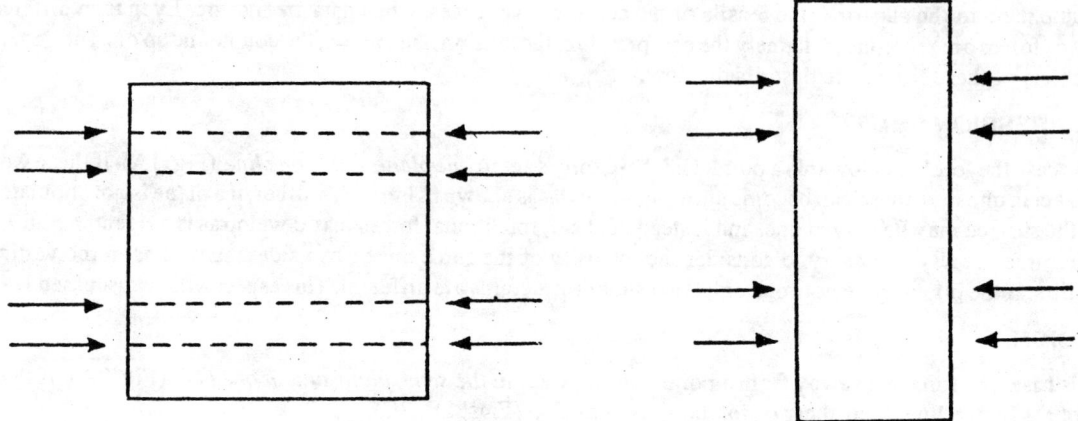

Fig. 1A. Forces act in same plane, same line, and towards each other. A square changes to a rectangle.

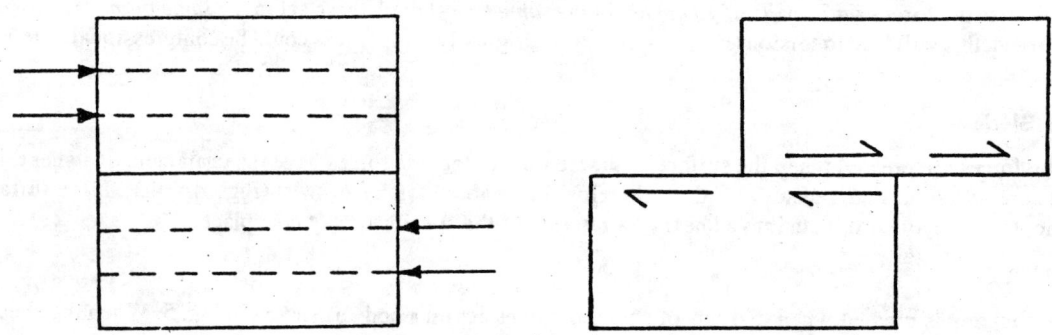

Fig. 1B Forces act in the same plane, towards each other, but in different lines. This may lead to shear and a break may occur along which displacement can take place.

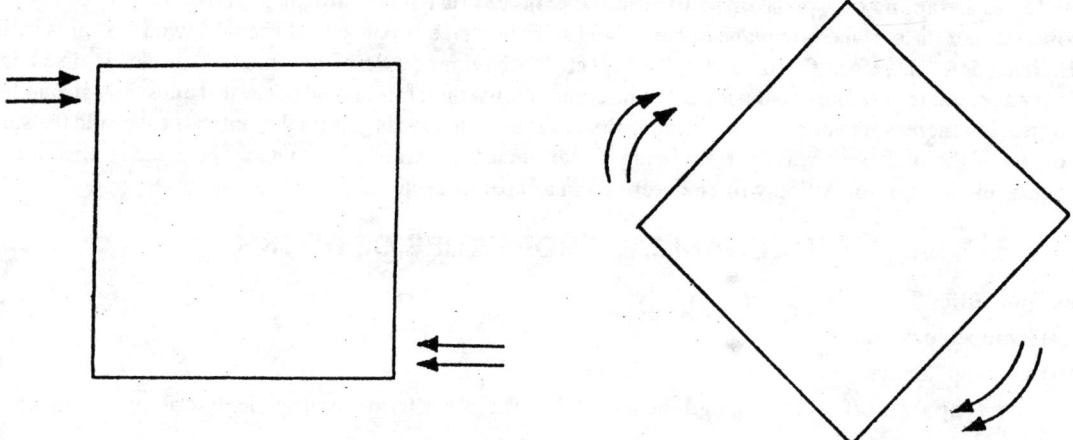

Fig. 1C. Forces act in the same plane, towards each other, but at the diametrically opposite ends of a body. Rotation of the body may take place.

Fig. 1 A,B,C. Behaviour of compressive forces

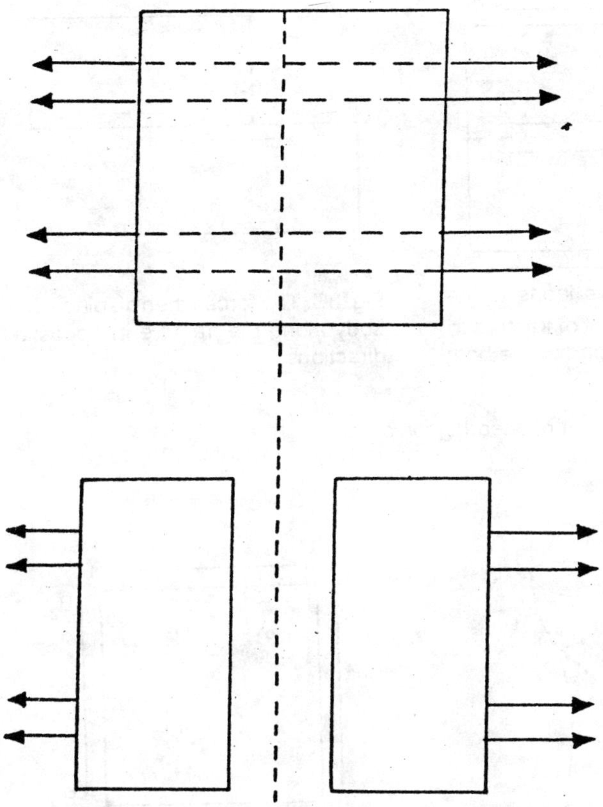

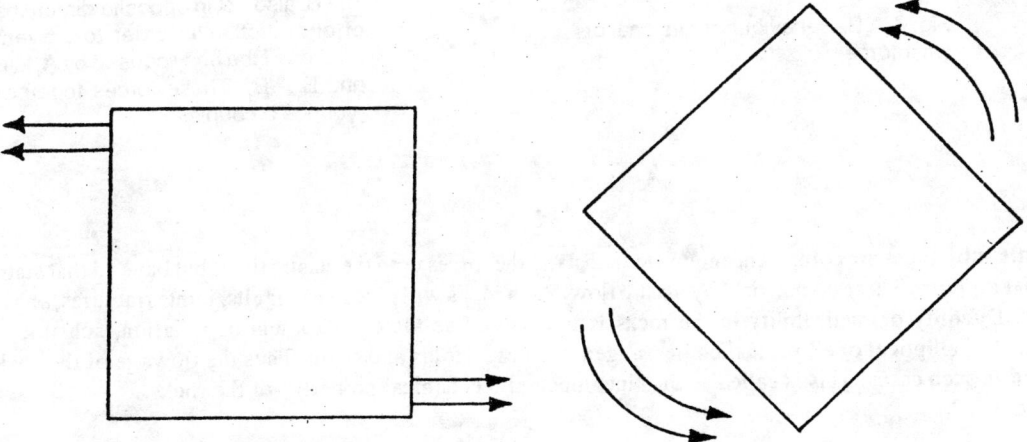

Fig. 2A. Forces act in the same plane, same line, but in opposite directions. Fracture is developed which is perpendicular to the tensional forces.

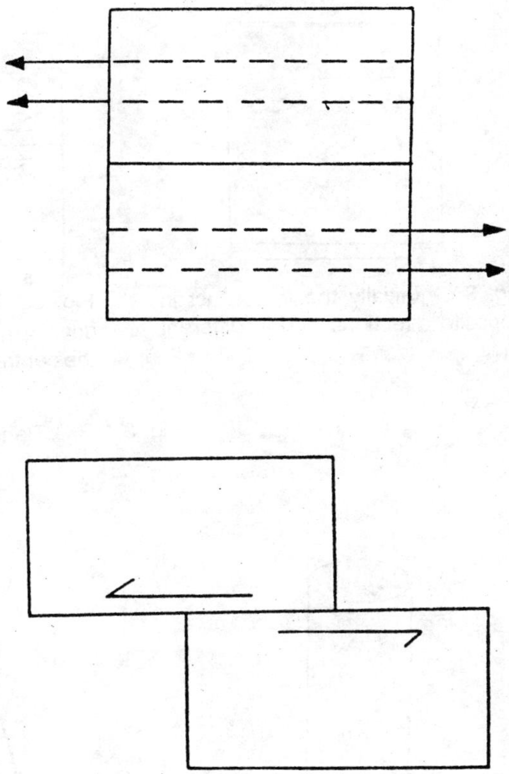

Fig. 2B. Forces act in same plane but in opposite directions, and in different lines. This may lead to shear, and a break may occur along which displacement can take place.

Fig. 2C. Forces act at the diametrically opposite ends of the body, and in opposite directions. This may lead to the rotation of the body.

Fig. 3A. Initially the forces act in opposite directions and in different lines.

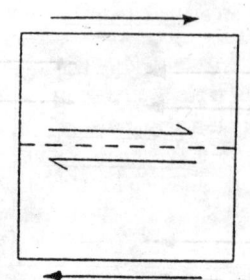

Fig. 3B This leads to the development of fracture in the central part of the body.

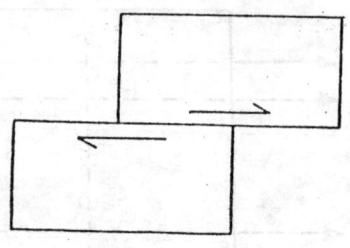

Fig. 3C. The forces then displace the body along the fracture in opposite directions.

Fig. 3 A,B,C. Behaviour of shearing forces.

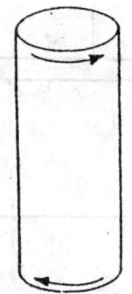

Fig. 4A. Forces act in a curvilinear fashion, in opposite directions and at the opposite ends of the body.

Fig. 4B. A curved fracture plane is produced.

Fig. 4 A,B. Behaviour of torsional or twist force.

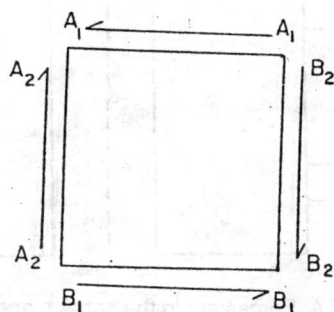

Fig. 5. Couple

A_1 - A_1 and B_1 - B_1 act in opposite direction of one another. A_2 - A_2 and B_2 - B_2 also act in opposite directions of one another in order to prevent rotation of the body caused by A_1 - A_1 and B_1 - B_1. These forces together produce a couple.

Brittle substances are competent enough to withstand the forces upto the elastic limit, but beyond that stage, the material ruptures. These do not yield by plastic flow. Such rocks will give rise to faults, joints fractures, shears and so on. Ductility or melleability in the rocks is understood as the development of foliation, schistocity, development of elliptical or ellipsoidal bodies, augen structures, folds and so on. Thus the flowage of the rocks is observed in such cases. This is called as the "incompetent mechanical property" of the rocks.

The mechanical preperties can be also classified as:
1. absolute, and
2. collective ones.

Sandstone, shale, granite, gabbro etc., possess certain mechanical property, when considered individually. This is called as the absolute or "in born" intrinsic mechanical property. But when a group of rocks are affected by the deformative forces (and this is the common situation that exists in the crust of the earth), then the mean or the average or the leading mechanical property of the unit becomes effective or operative. Such a behaviour is called as the "collective or the dominating mechanical property". Thus though a sandstone which should have yielded rupturally, it gets folded along with the other rocks of the group, because the "collective" behaviour is of "plastic or the flowage" type. Hence the kind of the structure produced in the rocks will depend not upon the individual mechanical properties, but on the collective mechanical property of the rock unit.

ENVIRONMENT OF DEFORMATION

So far the three components of the "structural equation" have been elaborated. However the absolute and the collective mechanical properties of the rocks in turn are not constant in respect of the depth at which the actual deformation is taking place. These properties change, because with depth, the temperature increases so also the confining pressure. Slowly the brittleness decreases and becomes negligible. On the other hand, the rocks become more and more ductile or melleable (plastic). There is yet another factor to be taken into account. It is the availability of the fluids by way of the circulating gorundwater, or the gases and vapours emanating from the consolidating magmas and so on. These fluids act as the lubricants and the style of deformation may depart considerably. All these components constitute the environment of structural deformation.

Therefore for a proper understanding of the varieties of the structures, the mechanical properties, the kind of the deformative forces and the environment under which these are acting, should be critically analysed. As already said, the information in respect of the several variables, is incomplete, and hence the structural geologist finds it difficult to correctly and conclusively account for the observed variations in the structures developed in the crustal rocks.

MECHANISM OF DEFORMATION

When subjected to the deformative forces, the rocks react in a progressive manner. The first reaction is the resistance to any change that might occur within the body. This however is not registered externally in the form of any permanent deformation. It is the competency of the rock and this property varies from one rock to another. The reaction created within the rock is spoken off as the "stress". This technical term is very useful in the structural analysis, and it will be elaborated in a separate section.

When the deformative forces increase, the rock reacts by a visible change like bending, stretching, twisting and so on. No sooner the deformative forces are withdrawn, the rock regains its original shape and form. This style of deformation which is a temporary feature, is called as the elasticity of the rocks. The passage of the earthquake waves through the crustal rocks, is a proof that these are perfectly elastic ones. The degree of elasticity, however, varies from rock to rock.

The second stage of deformation is the plastic type or the flow type. This is the first permanent change that takes place in the rocks unlike the elastic bending or stretching of the rocks, which occurs as long as the forces are operative. However, the rocks are not broken or fractured, but are folded, bent or stretched. The elongated minerals in a foliated quartzite is a good example of this kind of deformation. Rocks differ in their property of ductility or melleability, and accordingly, the sizes and the shape of the folded beds, vary.

The third and ultimate stage of deformation is the breaking up of the original rock into two or more parts. This is the rupture stage and it gives rise to such structures as the joints, the fractures, the shears, the faults and so on. The time spent in the three stages of deformation namely, the elastic, the plastic and the rupture, varies from rock to rock, and from environment to environment. Accordingly a combination of structures could be expected in the deformed crustal rocks.

So far only the different stages of deformation have been discussed. The structural geologist is also interested in ascertaining as to what happens internally within the rock, which manifests externally as a fracture, a fold, a schist and so on. This is technically called as the mechanism of deformation, and it will be elaborated in the following paragraphs.

MECHANISM OF ELASTIC DEFORMATION

This is a temporary phase and it is argued that some sort of a "stop gap" arrangement must be taking place within the rock. It may be by way of reducing the distance between the units of the rock, or elongation of the units of the rock, or some other sort of arrangement of the units (Figs. 6A to 6I).

MECHANISM OF PLASTIC DEFORMATION

This is a permanent change and it is therefore observable even after the deformative forces are removed, or cease to act on the rocks. There are three ways in which the plastic deformation is operative namely,

1. rearrangement of the platy, flaky minerals, if present,
2. recrystallisation under the effect of stress (Rieckie's principle), and
3. formation of new platy and or flaky minerals and their rearrangement.

For the plastic deformation , it is necessary that the rocks should be plastic ones, as well as these must be deformed at considerable depth. This latter situation produces considerable confining pressure which induces plasticity in the rocks, if the rocks be not already plastic ones. The individual grains are stretched in the direction of the least stress or perpendicular to the direction of the maximum stress. This stretching occurs according to the Rieckie's principle. It states that melting occurs at the point of maximum stress and deposition of the melted material takes place at the point of minimum stress (Figs. 7A, B). This will result in the development of elliptical (two dimensions) or ellipsoidal (three dimensions) bodies and the rocks will be flattened giving rise to elongated grains. Thus a sandstone gets converted into a foliated quartzite (Figs. 8A, B and Photo 1) with the development of mosaic texture.

In some other cases, stretching of the grains will lead to the bending of the rocks into folded forms (Figs. 9 A, B). At the same time, if there are platy, flaky or accicular minerals in the rocks, these get rotated into parallel positions and may be further confined to parallel planes. This results in the development of slaty cleavage, schistocity, foliation and lineation in the rocks (Figs. 10 A to E). It is not necessary that the platy or accicular minerals alone should be present in the rocks. If the rocks were to have ingredients such that their two dimensions are longer than the third one, then such grains may get oriented into parallel positions and produce deformation (Figs. 11 A,B).

So it may said that the plastic deformation manifests itself by the rearrangement of the pre-existing minerals such that a maximum economy of space is brought about without fracturing the original rocks. Analogy may be given of the cylinders or rods thrown or dumped at random. In such cases, only a few cylinders or rods may be packed into the space with a lot of space wasted. Under the presence of compelling stress, the rods or the cylinders are rearranged and oriented so that even more cylinders or rods could be accomodated in the same initial volume. This results in the orientation or an orderly arrangement of the rods or the cylinders. This has been shown in Figs. 12A, B.

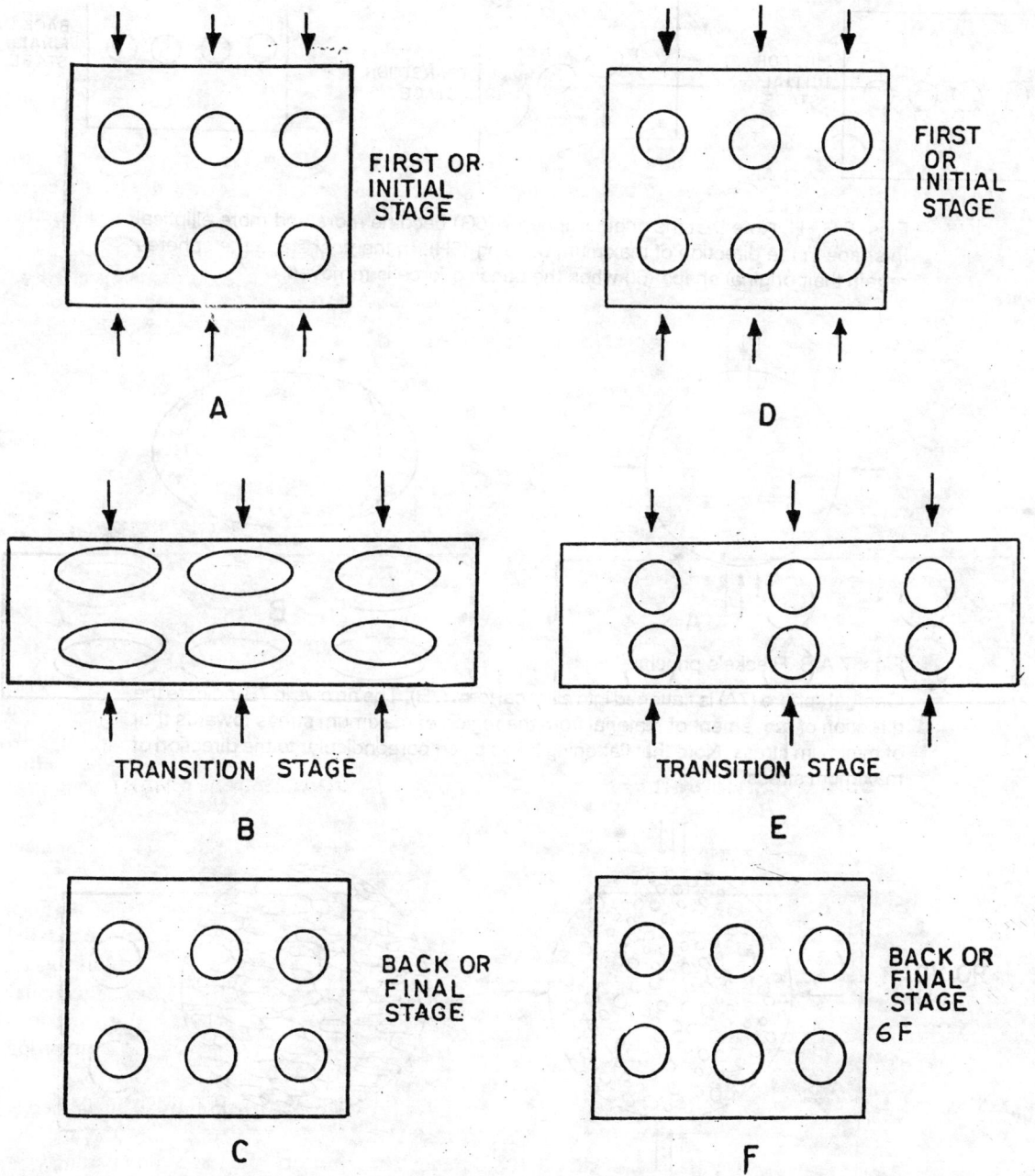

FIRST OR INITIAL STAGE

A

FIRST OR INITIAL STAGE

D

TRANSITION STAGE

B

TRANSITION STAGE

E

BACK OR FINAL STAGE

C

BACK OR FINAL STAGE
6 F

F

Figs. 6 A,B,C. Note that the original spheres are flattened only in the transition stage because of the application of the pressure. 6C Original shape is regained when the force is removed.

Fig. 6 D,E,F. Note that spheres come closer to each other in transition stage (6E). When the force is removed, the original distance between the spheres is regained (F).

Fig. 6A to I. Mechanism of elastic deformation.

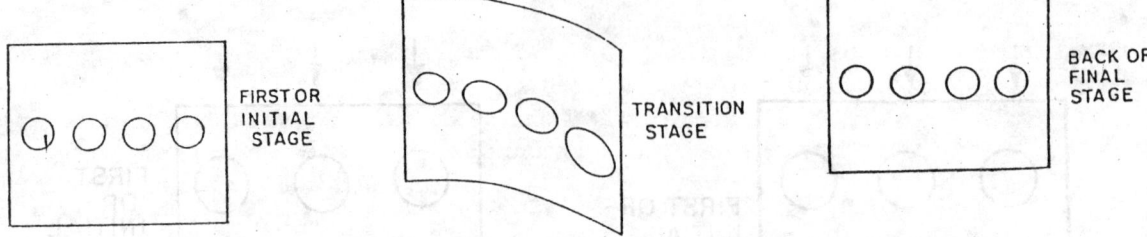

Figs. 6 G,H,I. Note that the original spheres (6G) become more and more elliptical in shape in the direction of maximum bending (6H). In the final stage the spheres regain their original shape (6I) when the bending force is removed.

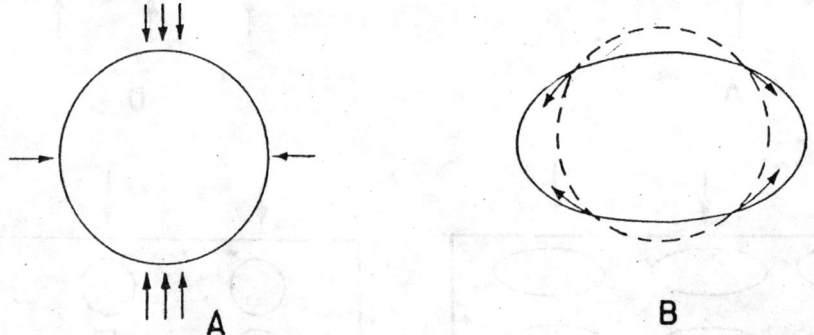

Figs. 7 A,B. Riecke's principle

Original sphere (7A) is flattened into elliptical form (7B). The arrows in 7B indicate the direction of movement of material from the region of maximum stress towards that of minimum stress. Note that flattening takes place perpendicular to the direction of maximum stress.

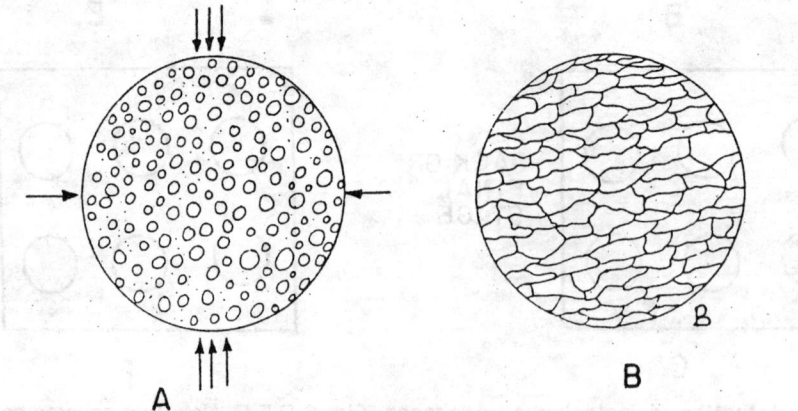

Figs. 8 A,B. Mechanism of plastic deformation

Note that the original spherical, rounded to subrounded grains (8A), yield plastically and get permanently flattened out producing a mosaic texture (8B). Note that the flattening (foliation) is perpendicular to the direction of maximum stress.

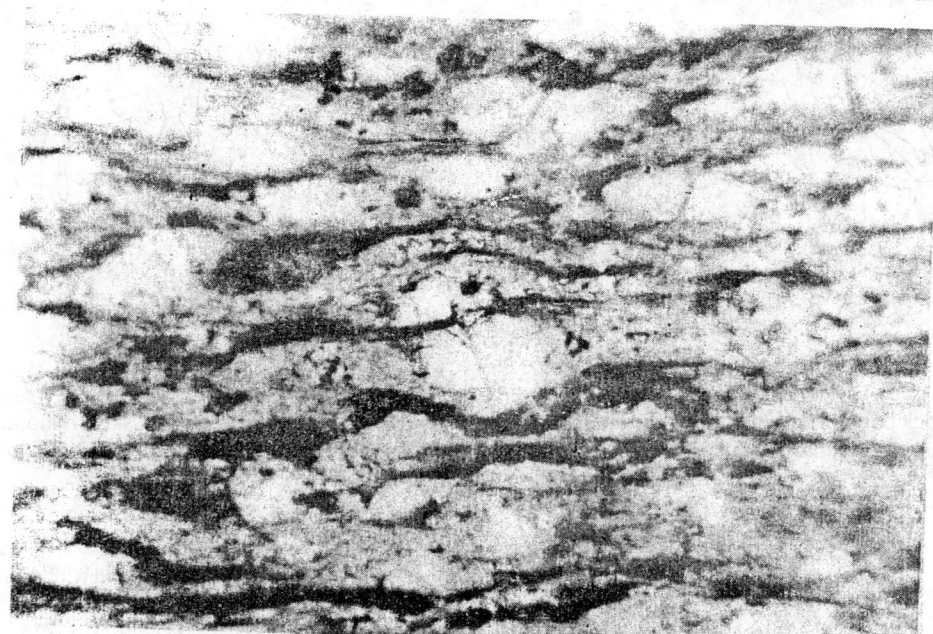

Photo 1. Microphotograph of quartzite showing elliptical/lensoidal augen shaped pieces of quartzite. The groundmass also is of smaller pieces of quartzite. The rock specimen is from N.E.B. range, Bellary district, Karanataka state. Courtesy Dr. H.D. Desai.

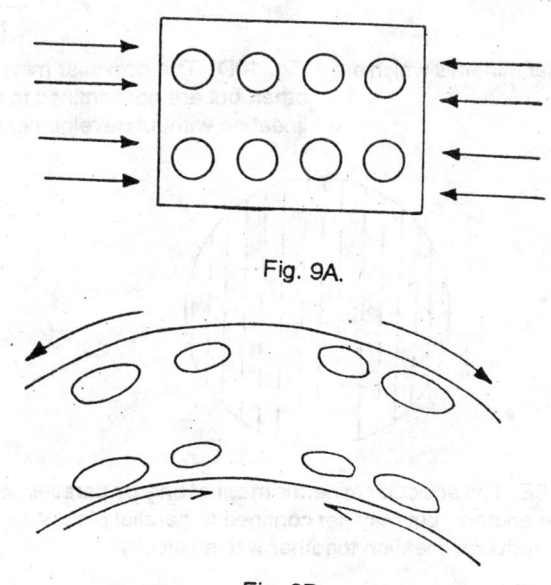

Fig. 9A.

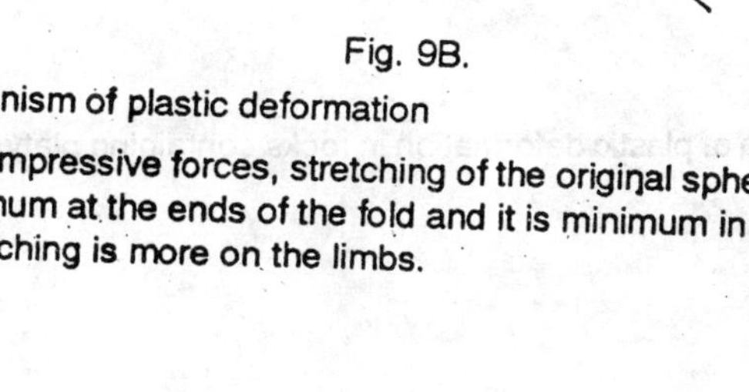

Fig. 9B.

Figs. 9 A,B. Mechanism of plastic deformation

Note that due to compressive forces, stretching of the original spheres (9A) takes place. Stretching is maximum at the ends of the fold and it is minimum in the more central part (9B), because stretching is more on the limbs.

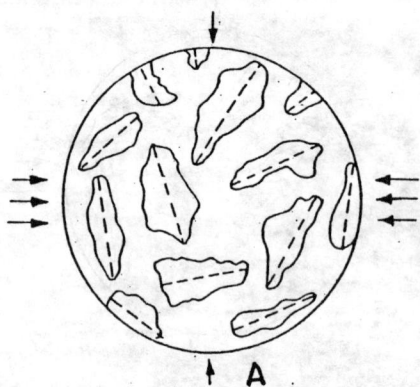

Fig. 10A. A rock containing platy, flaky minerals whose longer axes are not oriented. It is subjected to stress.

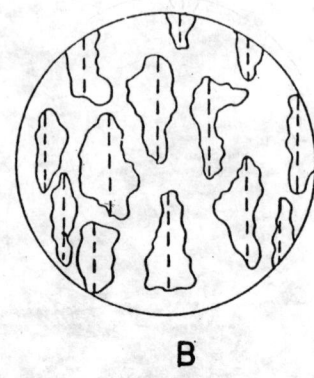

Fig. 10B Longer axes of the platy, flaky minerals display parallelism between themselves. This leads to development of schistocity or foliation.

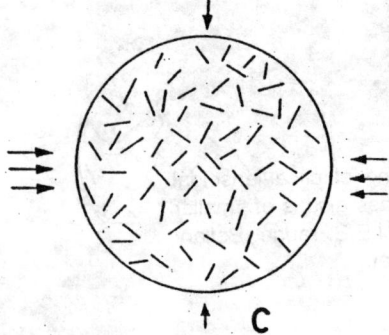

Fig. 10C. A rock containing accicular minerals which are unoriented, is subjected to stress.

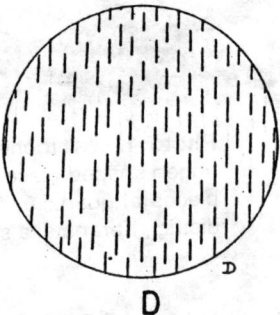

Fig. 10D. The accicular minerals orient parallel to each other, but are not confined to any planes. This produces lineation without development of schistocity.

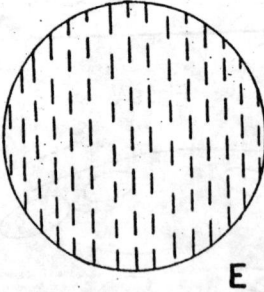

Fig. 10E. The accicular minerals may not only be parallel to one another, but may get confined to parallel planes. This produces lineation together with shistocity.

Figs. 10 A to E. Mechanism of plastic deformation in rocks containing platy/flaky/accicular minerals.

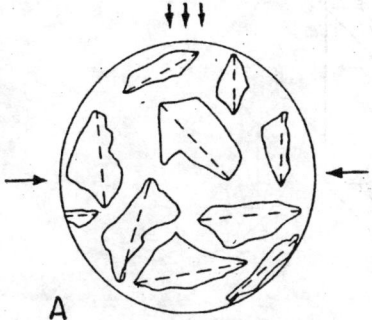

Fig. 11A. A rock containing xenoliths or fragments whose 2 dimensions are greater than the 3rd one, is shown.

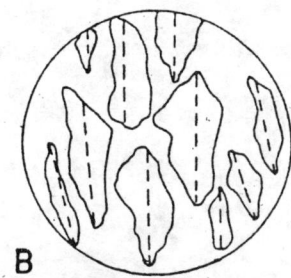

Fig. 11B. Due to stress the xenoliths or the fragments get oriented such that their longer axes are parallel to one another, and are perpendicular to the direction of maximum stress.

Figs. 11 A,B. Mechanism of plastic deformation.

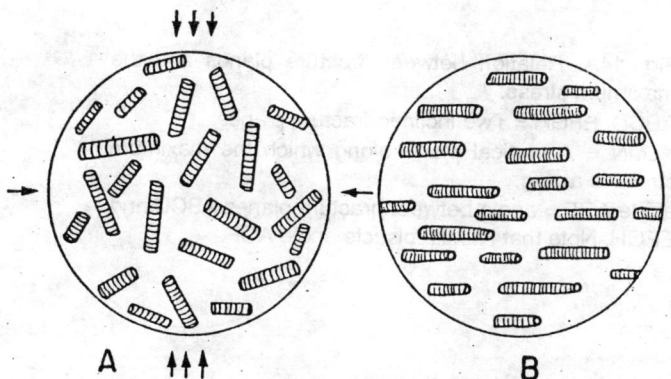

Figs. 12 A,B. Cylindrical bodies lying pell - mell or in an unoriented state are shown in Fig. 12 A.

Due to the action of stress, cylindrical bodies get rearranged into parallel dispositions. Fig. 12 B.

Note that more cylinders could be accomo- dated in the same space, thus bringing about economy of space.

MECHANISM OF RUPTURAL DEFORMATION

There is a systematic way in which the rupturing takes place and this depends upon the *direction of the maximum compressive stress,* as well as the presence of confining pressure. Experiments conducted on a squate prism in the laboratory are very informative and useful, and the same are described below.

CASE I SQUARE PRISM IS VERTICAL RIGHT AND LEFT SIDES UNCONFINED

In Fig. 13, faces MNOP and QRST are unconfined, which in other words it means that the faces QNOR and TMPS are confined ones. Further, the compressive forces are acting vertically on the faces MNQT and PSRO. Under this situation, fractures ABCD and EFGH are produced, each plane sloping with an inclination of 60° towards the concerned unconfined side of the prism (either TQRS or MNOP). These two intersecting fractures bear an angle of 60° between themselves, and the direction of the maximum stress bisects such an angle. These features are shown in Fig. 14A. Angle AOF = 60° (Fig. 14A), and it is bisected by the vertically acting maximum stress. The important observation to be made is that the fracture plane ABCD or EFGH is dipping 90 degrees. This inclination of 60° is controlled by the direction of the maximum compressive (stress) force. It is further interesting to note that as per the mathematical expectations, the angle subtended between the fracture planes should be of 60° only (Fig. 14C). This discrepancy of 30° is attributed by the geologists to the heterogeneous nature of the rock unit as is normally existing in the crust of the earth.

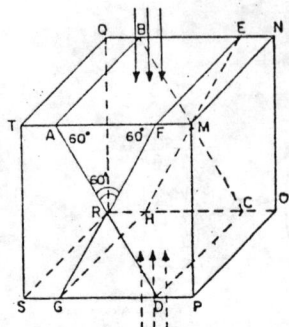

Fig. 13. Square prism experiment. Arrows indicate direction of compressive forces which act vertically. MNOP and QRST are the unconfined sides. MPST and QNOR are the confined sides. Fracture plane ABCD is inclined at an angle of 60° towards MNOP, and fracture plane EFGH is inclined at an angle of 60° towards QRST.

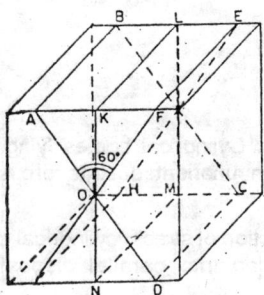

Fig. 14A. Relation between fracture planes and the maximum stress.

ABCD, EFGH = Two inclined fracture planes.

KLMN = A vertical plane along which the maximum stress is acting.

angle AOF = angle between fracture planes ABCD and EFGH. Note that KLMN bisects angle AOF.

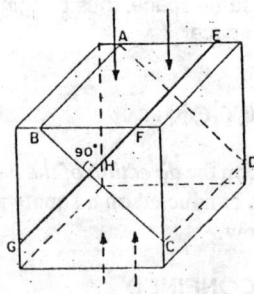

Fig. 14B. Mathematically expected relation between fractures and the stress. Angle between ABCD and EFGH is 90 degrees.

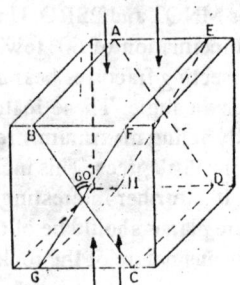

Fig. 14C. Reality in nature. The angle between ABCD and EFGH is 60° and not 90 degrees.

CASE II SQUARE PRISM IS VERTICAL. FRONT AND BACK SIDES ARE UNCONFINED

In this case, the front and the back vertical faces of the square prism are unconfined. Here also two sets of fractures ABCD and EFGH are produced, which again dip towards the concerned unconfined sides of the square prism. The angle subtended between these fractures is also of 60° (Fig. 15). In this case also the maximum vertical compressive stress bisects the angle of 60 degrees.

Thus from the cases I and II described above, it is clear that if all the four sides of the square prism were to be unconfined, then *Four Sets of Fractures* will be produced, each set of two fractures dipping towards the concerned unconfined side of the square prism. *This is the systematic relation existing between the compressive forces and the sides of the square prism or a body of rocks undergoing deformation.* It is possible that only one or 2 sides are unconfined, and not all the four. The fractures developed under such a situation has been presented in Figs. 16 A,B,C. It is very necessary to remember that in nature the rocks do not occur in the form of square prism and with vertical sides. The actual meaning of it is that the surface of contact of one rock with the other, and its nature (whether vertical, inclined or curved and so on) controls the kind of ruptures produced in the crustal rocks. The geologist therefore should give the proper latitude to these actual and real situations prevailing in nature, before inferring and interpreting the origin of the fractures observed in the crustal rocks.

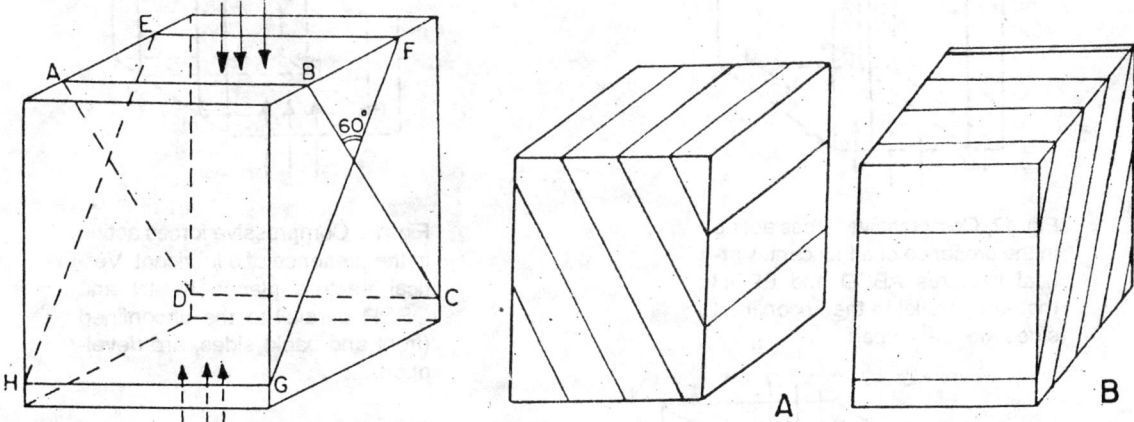

Fig. 15. Fracture planes ABCD and EFGH are inclined towards the unconfined front and the back sides of the square prism. The angle between the fracture planes is of 60 degrees.

Figs. 16 A,B. Only one set of parallel fracture planes may be developed, if only one side of the square prism be unconfined.

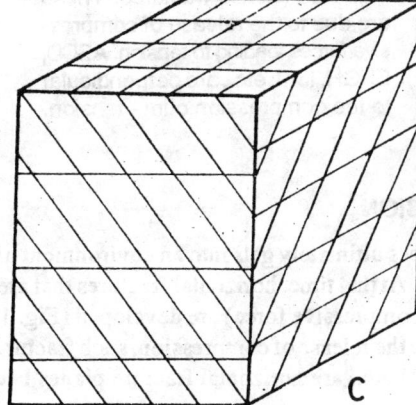

Fig. 16C. Two sets of fracture planes may be developed the trace of which on the horizontal surface results in an intersecting pattern. The angle of intersection will be observed to be of 90 degrees. (top of block).

CASE III SQUARE PRISM IS VERTICAL AND LUBRICANT USED

The square prism experiment when conducted by placing a lubricant while applying the pressure with the piston, fractures resulting under this situation are quite different from those described in Cases I and II above. Vertical fracture planes parallel to the unconfined sides of the prism, are now produced instead of inclined ones as described in the earlier cases. Fracture planes ABCD and EFGH are produced considering that the side surfaces of the square prism are unconfined (Fig. 17). If the front and the back sides of the square prism be unconfined, then the fracture planes KLMN and OPQR are produced (Fig. 18). Fractures developed in Case III are called as the "Extension fractures". Thus the development of the vertical or the inclined fractures is due to the presence or the absence of a lubricant (fluids in nature) during the application of the deformative forces.

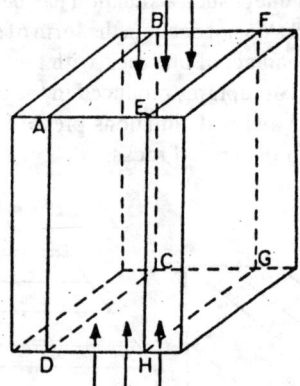

Fig. 17. Compressive forces acting in the presence of a lubricant. Vertical fractures ABCD and EFGH that are parallel to the unconfined sides are developed.

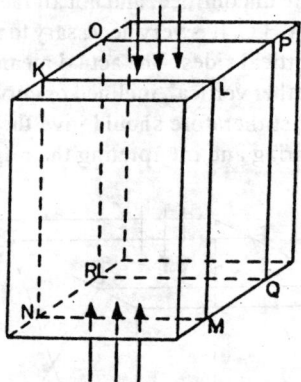

Fig. 18. Compressive forces acting in the presence of a lubricant. Vertical fracture planes KLMN and OPQR parallel to the unconfined (front and back) sides, are developed.

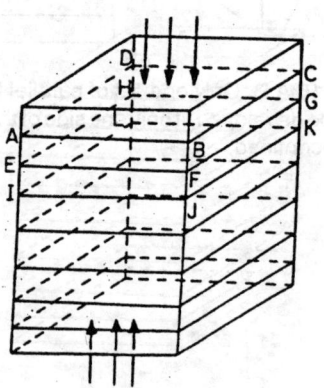

Fig. 19. Release fractures. These are due to the release of compressive forces leading to tension. ABCD, EFGH, IJKL etc., are perpendicular to the compression cum - tension.

CASE IV SQUARE PRISM SUBJECTED TO TENSION

A square prism which is subjected to the compressive forces ultimately gets into an environment of tension, when the earlier compressive forces cease to act or these die out. At that time, horizontal fractures that are perpendicular to the direction of tension (earlier it was the direction of compressive force) are developed (Fig. 19). Since these are due to the dying out of the compressive forces or due to the release of compression, such fractures are therefore called as the "release fractures". Fractures ABCD, EFGH etc., are horizontal fracture planes because these are perpendicular to the vertical tensional forces.

CASE V FRACTURES PRODUCED UNDER COUPLE

When a rock is brought under the influence of couple, three different types of fractures are possible, namely, shear, tension and thrust. The shear fractures will be parallel to the direction of couple, thrusting is possible along the longer diagonal of the rhomb, while tension fractures will be parallel to the shorter diagonal of the rhomb (Figs. 20 A to F).

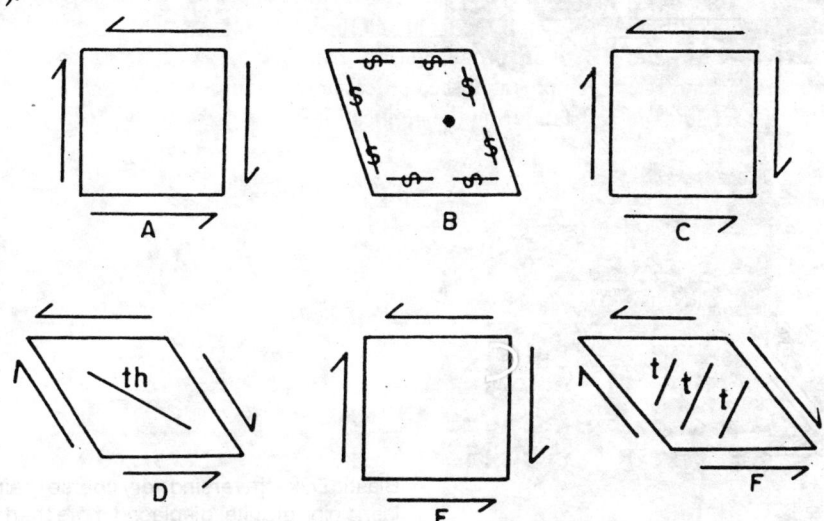

Fig. 20 A,B,C,D,E,F. Mechanism of rupturing under the action of coupling forces.

s = shear fractures parallel to the sides of the rhomb.

th = thrust fault whose trend is along the longer diagonal of the rhomb.

t = tension fractures which are perpendicular to the longer diagonal of the rhomb.

Note that a square is steadily changed to a rhomb under the action of coupling forces.

So far the mechanisms of plastic and ruptural deformation have been described. The purpose of doing so was to bring out the regularity or a definite method noticeable in the development of the structures. Ultimately the structural geologist strives hard to establish the direction in which the deformative forces had acted. This is possible only through the analysis of the structures actually developed in the crustal rocks. Therefore the orientation of the fractures, the axes of the folds, the trend of schistocity etc., are to be utilised to derive the direction of the forces. Thus if the fractures are inclined at high angle of 60°, then it is evident that the deformative forces had acted vertically. Later while dealing with the faults, it will be shown that the trough faults or the tear faults can develop only if the maximum compressive force had acted vertically and horizontally, respectively. So from the orientation of the faults, or the fractures, the direction of the forces can be derived.

Blastic Dyke traversing very coarse grained porphyroblasric pink granite, displaced more than once across its length. In the photo two faults are observabled. The locality is Akutothpalli, Aurepalle, Mehboobnagar District, Andhra Pradesh. (Courtesy Shri; D. Muralidharan).

Rhombic pattern of joints developed in fuchsite quartites of Delapuran Village, Bellary District, Karnataka State. One set trends in N 75°E- S75°W (Frequent), Other set Trends in N 45° W-S45° E direction. (Courtesy Dr. V.N. Hegde).

2

RUPTURE STRUCTURES

Classification of ruptures into fracture, fault, joint and shear. Rocks and corresponding joints, Classification of joints, Non-Genetic, Basis of dip amount, Shapes of joint blocks, Trend of joints, Attitude of joints w.r.t. that of rocks, Genetic classification, Tension and shear joints, Portraying of joints, Symbolic representation, Histograms, Ray or rossette diagrams, Stereographic method, Schmidt or Grid method, Free counter method, Mellis or circle method, New approach towards processing of stereograms.

Faults, Technical features, Kinds of movements along fault plane, Heave and throw of faults, Classification of faults, Schematic diagram of classificatory bases of faults, Rotational faults, Faults of translatory movement, Classification based on attitude of fault plane, Movement along the fault plane, Dextral, sinistral, wrench, tear, normal, reverse, thrust and gravity faults, Schematic classification of thrusts, Basis of attitude of fault plane w.r.t. that of rocks. Patterns of faults. Basis of deformative forces.

Recognition of faults. Schematic representation of bases used in recognising faults in the field. Displacement, Evidences from fault plane, Morphological features, Mineralisation.

Complicated faulting and establishment of relative ages of displacement. Examples of different kinds of faults from India.

Figures 21 to 83

Photos 2 to 49

In keeping with the three stages of deformation, structures corresponding to them namely, the elastic (but this is not a permanent feature), the plastic, and the rupture structures, are produced. The structures classified as the joints, and the faults, are designated as the "rupture structures". Any structure wherein a break or a fracture takes place, and the original rock unit is broken into two or more parts, is first designated as a rupture or fracture. Later depending upon the other characters, these will get specific names of a fault, a joint, a shear and so on. The characteristics of these different ruptures will now be discussed.

CLASSIFICATION OF STRUCTURES

The rupture structures are classified into the following varieties

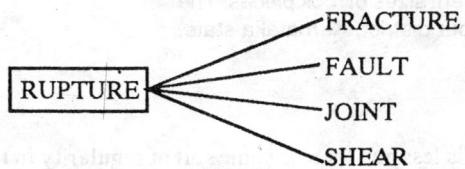

Description of each category noted above is given below.

RUPTURE

When the rock is broken along numerous directions so that it becomes a mess of broken pieces, it is to be called as "rupture". Each rupture line is short in respect of its length (Fig. 21). It does not produce any definite pattern, because there is no regularity in respect of the directions, dip amount, distance between two rupture planes and so on. In a cataclasite such a structure is often noticeable, like the one developed in the quartzarenitic rocks exposed at Timmapur village (Photo 2).

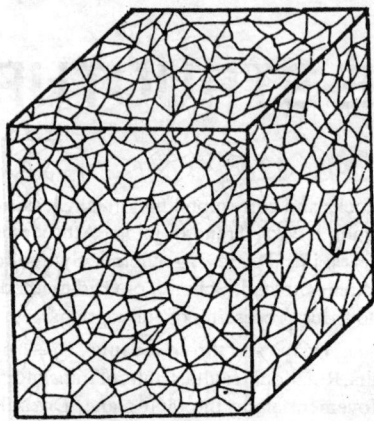
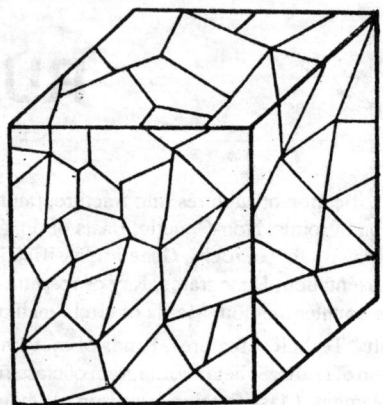

Fig. 21. Rupture structure. Note the irregularity of rupturing, and small and irregular sizes of ruptured blocks.

Fig. 22. Fracture structure. Note the reduction in the number of fractures as well as increase in the size of the individual fractured blocks.

Photo. 2. Field photo of quartzarenitic rocks exposed 500 meters WNW of Timmapur village (75° 32' 56" E, 15° 54' 10" N) showing shattered rocks due to development of shears trending in different directions. The frequent directions of shearing are indicated by the 2 hammers (central part of the photo). Note angularity and different sizes of rock pieces. The structure is exposed near Timmapur, Bijapur district, Karnataka state. Courtesy Dr. G.V. Hedge.

FRACTURE

Here also the rock is broken in several planes, but the intensity is less and there is some sort of regularity in terms of the inclination of the planes, direction etc. The size of the each "fracture block" is also bigger as compared to that noticed in the ruptured rocks. Further, akin to rupture, here also there is no parallelism between the different rupture planes. As a result of this an intersecting arrangement of the fracture planes is observable (Fig. 22).

JOINTS

These generally share the properties of the rupture, the fractures or the shears. However one feature possessed by this structure is not seen in the other varieties. That feature is that the "joint block" and the "joint planes" are considerably bigger than that noticeable in the ruptures or the shears. The number of rupture planes is considerably less than that developed in the fractured or the ruptured rocks. The rupture surfaces are more planar in nature. Some regular pattern is discernible like the columnar, the rhombic, the rectangular, the cubic, the mural and so on. Further, there is a systematic relation between the joint planes and the bedding planes (in the case of sedimentary rocks), joint planes and the surface of contact of the rock (igneous body) with the country rocks, and so on. Some patterns of joints are shown in Figs. 23 and 24.

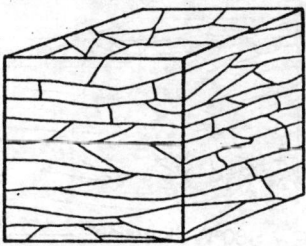

Fig. 23. Joints. Note the regularity in the development of joint blocks due to fracturing. Also note the reduction in the number of fractured blocks.

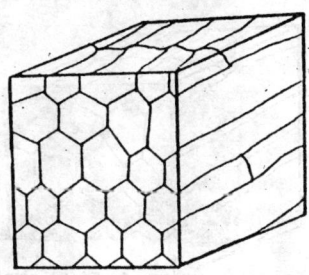

Fig. 24. Columnar joints. Note that the fractured blocks are regular in shape and are hexagonal in form. Some columns are long while some are cut along the length.

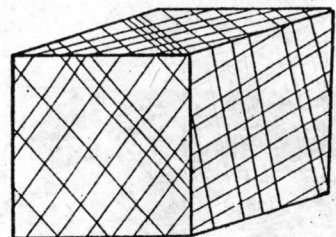

Fig. 25. Rhombic patterns of joints. Note that two sets of intersecting joint planes produce angles of 60 and 120 degrees between them. A regular distance is maintained between the intersecting joint planes.

SHEAR

In this case, the ruptures are reduced to 3, 4, 5 or so in number. These are generally parallel to each other, or these may intersect one another to produce rhombic or similar patterns (Fig. 25, Photos 3, 4). The most important feature of the shears is that the rocks are traversed by such ruptures, only at some places, while at other places, these are not at all developed. Therefore the shears give rise to zones (Fig. 26). The shear planes are generally vertical, or have high angle of inclination. When the shears are of the intersecting type, one set may be closely spaced while the other is widely spaced (Fig. 27). Another important feature is that the shears are longer than those classified as fractures or the ruptures.

FAULT

In this case, the rupture planes are very much reduced in number, generally 2 or 3, but most commonly only one plane is found to be developed. These ruptures are considerably long or it should also be said that these ruptures are the longest in extent. The distance between the two or more fault planes is quite wide as compared to that noticed in the ruptures, the fractures, the shears or even the joints.

Photo. 3. Field photo of shears trending N60°E - S60°W direction. Hammer is parallel to the trend of the shears. A fault trending in similar direction is suspected at the place. This structure is exposed in granites, 1 km. N70°W of Venkatapuram village, Hospet taluka, Bellary district, Karnataka state. Note the closeness between the several shear planes and their total absence in the other parts of the exposure. Courtesy Dr. V.N.Hegde.

Photo. 4. Field photo showing shears running in 3 directions, N31°W - S31°E, N36°W - S36°E, and N40°E - S40°W. Note that one set is more frequent and is closely spaced. Rocks are hornblende schists of Desaiwada area, Chendvanvadi, Vengurla, Maharashtra state. Courtesy Dr. D.I. Deendar.

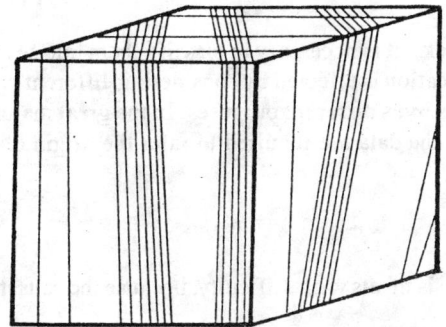

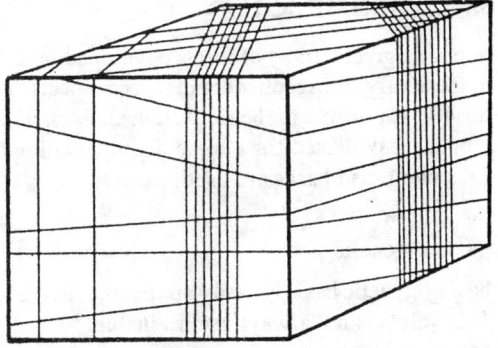

Fig. 26. Shear joints. Note that the shears are closely spaced and there are areas where no shearing is produced.

Fig. 27. Two sets shears are developed. One is closely spaced and parallel to each other. The other is irregularly developed.

JOINTS

CHARACTERISTICS

Joints are the fractures in rocks and along these fracture planes very little movement might have taken place, either in the horizontal, vertical or oblique directions. Actually the term joint conveys a wrong meaning as though the rocks are joined along the fracture planes. On the contrary, the rocks are broken and are not joined along the planes. The joint surfaces are by and large "plane surfaces", however these as well could be curved ones. Generally the joint plane is not very extensive, excepting the sheeting structures noticed in the granites, and the bedding joints developed in the bedded formations. This creates problems in the field, when the rocks are bedded ones, as well as jointed. The dip and strike of a planar surface encountered in the field has therefore to be distinguished as to whether it is an extensively developed joint plane or it is a bedding plane, since both look alike. The experience of the geologists is that the bedding planes are normally more extensive than the joint planes. However, considerable precaution is to be exercised while making the distinction between them while collecting data.

As said earlier, the joint planes are more extensive than the fractures or the ruptures, but less so as compared to the faults or the shear planes. Joint planes do not extend much in depth. These are cut by the other joint planes and so on. Thus the several joint planes join together to produce a "joint block". Such a block may be triangular, cuboidal, rectangular, columnar or generally irregular in shape.

ROCKS AND CORRESPONDING JOINTS

Joints are developed in all the three varieties of rocks i.e., the igneous, the sedimentary and the metamorphic varieties. However it should be endeavoured to find out whether the pattern of joints in these three classes of rocks, is different. This would be an additional criterion to make the distinction between a metamorphosed igneous rock, and a metamorphosed sedimentary rock. It has been observed that the cuboidal, the rectangular and the columnar joints are more common in the igneous rocks. Further, there are more number of joints trending in several directions, and these do not bear any systematic angular relation with each other. In the sedimentary rocks, dip and strike joints are very common, and these are nearly perpendicular to each other. Bedding joints are also common. Thus a set of three joint planes nearly perpendicular to each other, are to be expected in the sedimentary rocks than in the other two varieties of rocks. Dyke rocks show cuboidal joints, some lava flows exhibit columnar joints. Thus to some extent it is possible to infer the nature of the rock whether igneous, metamorphic or sedimentary, from the style of the joints.

CLASSIFICATION OF JOINTS

Since there are a great variety of joints developed in the crustal rocks, it is necessary to classify them into several groups, if these vary from each other in some respects. The classification is effected by considering different bases that are readily applicable in the field studies. Classification also serves different purposes. In the great majority of cases, data are collected for a mere documentation. In others, the data are required to infer the origin of the joints. Thus two broad bases are used, namely,

   (i) the genetic, and

  (ii) the non-genetic.

     The non-genetic basis is easily applicable, but the genetic basis meets with difficulty, because the causative factor of the joints is not always ascertainable.

NON - GENETIC VARIETIES

Here the external appearance, relation with respect to the rocks etc., are considered to effect a classification. The different bases and the kinds of joints recognised under each category, are given below.

  1. **Amount of dip of joint plane/surface:** The joint planes/surfaces could be horizontal or dipping ones. The amount of dip naturally has to vary from low through medium to high. Accordingly three types of joints are recognised, namely,

     (a) horizontal or low angle joints (0° to 30°),

     (b) medium angle joints (dip around 45°), and

     (c) high angle or vertical joints (60°, 80° or more).

  2. **Shapes of joint-block:** Here the geometrical shapes of the joint blocks are considered as the basis. Accordingly the following varieties are distinguishable.

     (a) triangular,

     (b) cubic,

     (c) rhombic,

     (d) rectangular,

     (e) columnar, and

     (f) irregular.

     These varieties are shown in Fig. 28. Sometimes, joints are developed which do not belong to the categories noted above. Thus Gokhale et al. (1983) have rpeorted "peculiar jointing" pattern from the Gadag schist belt, wherein the pattern is really unusual (Photo 5).

  3. **Trend of joints:** Here the strike of the joint planes is considered, and the groups are created on the basis of the frequency of the actual directions of strike. Thus joints may be trending N - S, E - W, NE - SW, NW

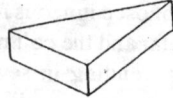
A. Triangular joint block

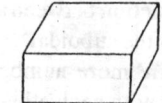
B. Cuboidal joint block

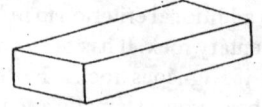
C. Rectangular joint block

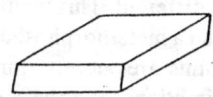
D. Rhombic joint block

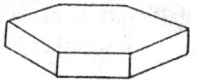
E. Columnar joint block

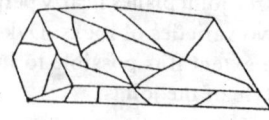
F,G. Irregular joint block

Fig. 28. A to G. Different shapes of joint blocks.

Photo. 5. Field photo showing peculiar (brick work like) jointing in banded iron quartzites of Nagavi village, Dharwad district, Karnataka state. Refolding, thrust faulting and a sudden change in the strike direction has resulted in the development of rather regularly sized and spaced rectangular blocks of joints. Elsewhere in the vicinity, this structure is not registered. Courtesy Dr. S.C. Puranik.

- SE, N10°W - S10°E, N60°E - S60°W and so on. These trends later may constitute groups or sets of joints, depending upon the value of frequency of joints. Sometimes the trends vary within 30° to 50°, and these fan out from a centre, producing a radial pattern. Hegde (1984) and Pujar (1989) have reported such joints which are presented in Photos 6 and 7, respectively.

Photo. 6. Field photo of radial joints produced in the quartzarenites located about 1 km. NE of Kabbalgeri, Badami taluka, Bijapur district, Karnataka state. Note that the joints converge at a point where some persons are standing. As many as 7 directions of radial joints starting from N 60°W - S 60°E to N5°W - S5°E have been produced. Courtesy Dr. G.V. Hegde.

Photo. 7. Field photo showing radial joints resembling "Percussion Structure". The several directions of joints are, N 40° E, N45° E ,N 85° E, N 10° W, N 58° ,W, N 60° W and N85° W. The photo also shows cross joints in between 2 adjacent radial joints (left hand central part of the photo). Two such cross joints are prominently seen, one trending N 40° E and another trending N60° E. Note that at the intersection of the radial joints, the rock is broken and hence it is called as "Percussion joint - structure". The structure is developed in pink coloured quartz arenites exposed 2 km. NE of Gorvankolla village, Belgaum district, Karnataka state. Corutesy Dr. G.S. Pujar.

4. **Attitude of joints with respect to that of rocks:** Here dip and the strike directions of the rocks (if these are bedded formations) are compared with that of the joint planes. However this basis cannot be applied to the igneous rocks, and also commonly to the metamorphic rocks. Following categories are recognisable under this basis.

(a) **Dip joints:** When the trend of the joints is parallel to the dip direction of the rocks, it is called as a dip joint. In Fig. 29. the rock dips to the right hand side of the observer, and the joints ABC, DEF, GHI and JKL trend parallel to the direction of dip of the rocks.

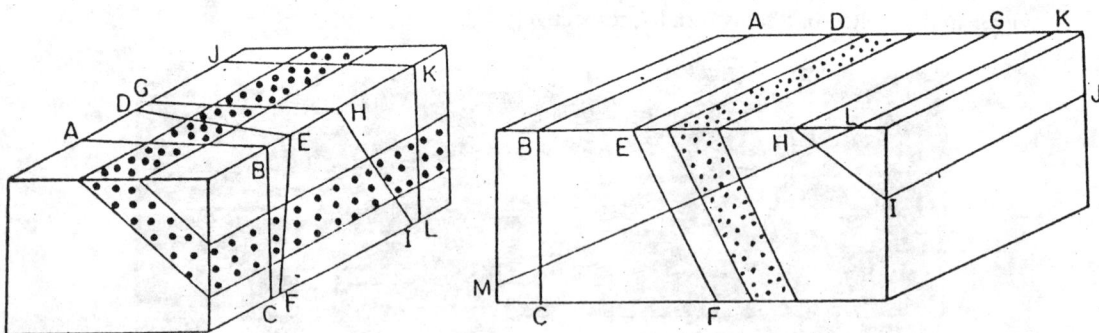

ABC, DEF, GHI and JKL are all dip joints as their trends (lines AB, DE, GH and JK) are parallel to the direction of the dip of the country rocks (dotted bed). Note that ABC and JKL are vertical joints, while DEF and GHI are high angle dip joints. In the absence of bedding plane, all the joints will get classified merely as vertical or high angle joints and not as dip joints.

Fig. 29. Dip joints

ABC = vertical strike joint

DEF = bedding joint

GHIJ = inclined strike joint, but dipping in the same direction as that of the bedding plane

KLM = inclined strike joint, dipping in opposite direction as that of the bedding plane

Dotted part dipping bed.

Fig. 30. Strike joints

(b) **Strike joints:** In this case, the trend of the joint is parallel to the strike direction (trend) of the rocks. Thus if the rocks be trending N - S and the trend of the joint planes also be N - S, then such joints are called as strike joints. In such cases, the dip direction of such joint planes need not coincide with the dip direction of the rocks in which the joint planes are developed. In Fig. 30, the disposition of the joints and that of the bedding plane is shown. ABC, DEF, GHIJ and KLM are all strike joints.

(c) **Bedding joints:** These are special types of strike joints wherein their dip direction and the amount of dip are nearly the same as that of the rocks in which the joints are developed. Thus if the rocks be trending N - S and dipping 50° due west, and if the joint planes also be trending N-S or nearly so, and also dipping 50° due west or nearly so, then such joints are designated as the bedding joints. In Fig. 30, DEF is a bedding joint, but ABC, KLM or GHIJ are not bedding joints though these latter joints are strike joints. Thus bedding joints are necessarily strike joints, but vice - versa need not be true.

(d) **Oblique or diagonal joints:** In this case, the trend of the joint plane is at an angle to that of the rocks in which the joints are developed. Thus if the joints be trending N - S when the rocks are trending NW - SW, then such joints are called as oblique or diagonal joints. In Fig. 31. the disposition of such joints is shown.

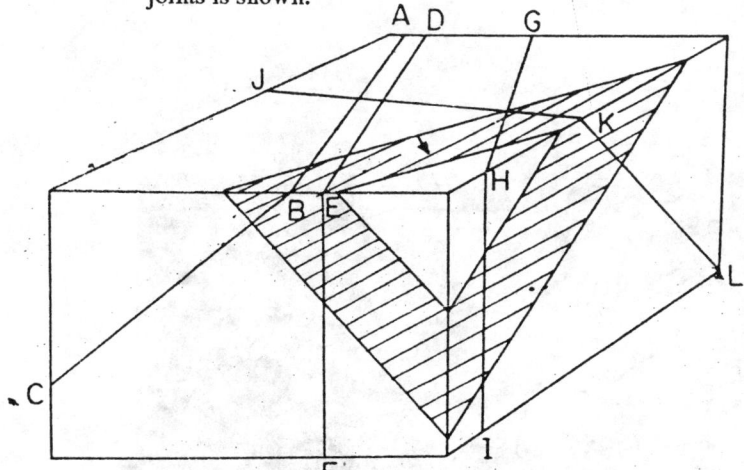

DIPPING
BED

ABC, JKL = inclined joints

DEF, GHI = vertical joints.

Fig. 31. Oblique/diagonal joints.

GENETIC VARIETIES

Here the cause of the development of the joints is considered. The kind of the deformative forces are taken into account, and accordingly the following varieties are recognised.

(a) **Tension joints:** These joints are very commonly encountered wherever the rocks are folded. Tension developes around the axial portion of the folds. A cooling magma also experiences tension and results in the development of joints. Analogy is given of the mud cracks which are produced through the drying up of the lakes. Columnar joints noticed in the volcanic rocks again are good examples of this kind of joints. Cocoanut Island, St. Mary' group of islands near Malpe, south Kanara, Karnataka, exposes well developed columnar joints (Photo 8). Jotiba hills (Wadiratangiri, Panhala taluka, Kolhapur district, Maharashtra) in the vicinity of Kolhapur, Naldurg near Aurangabad provide excellent examples of the columnar joints developed in the Decan Basalts. All joints are developed on the western side of the Jotiba hill. Vertical columns have height of about 15 meters with a diameter of 1.5 meters (Photo 9). Oblique columnar joints are quite rare, but these are excellently developed in some parts of Jotiba hill. These columns are 8 meters long and are almost tapering producing a radiating pattern (Photo 10). As oblique columns are not common, these therefore have been considered as unusual ones.

Photo. 8. Columnar joints in the lava flows of Cocoanut Islands, St. Mary's Group of islands Dakshina Kannada district, Karnataká. The columns are vertical, 2.3 meters long and about 30 cm. wide. The perfect hexagonal shape developed by the columns is clearly observable in the lower photo. Courtesy Dr. B.P. Waghamare.

The exposure of the columnar joints has been declared as a National Geological Monument by the Geological Survey of India on 16th November 1979.

Photo. 9. Field photo of very well developed columnar joints in basaltic flows exposed on the western side of Jotiba hills, Wadiratnagiri, Panhala taluka, Kolhapur district, Maharashtra state. Height of each column is about 40 feet, with a diameter of about 4 - 5 feet. Courtesy Prof. M.R. Shinde, Department of Geology, Rajaram College, Kolhapur.

Photo. 10. Field photo of unusually tapering columnar joints exposed on the western side of Jotiba hills, wadiratnagiri, Panhala taluka, Kolhapur district, Maharashtra State. The length of each column is about 25 feet. Courtesy. Prof. M.R. Shinde, Department of geology, Rajaram Science College, Kolhapur.

Unusual columnar joints (?) have been described by Bhimsen (1989) in the exposures of Deccan basalts in the vicinity of Ajra. These almost look like "mud cracks", and are nearly horizontal in attitude (Photos 11, 12). As they appear like mud cracks, these have been called as unusual columnar joints. Horizontal columnar joints have been reported by Hegde (1984) in a vertical dyke, and these have been documented in Photo 13.

Joints extending from wall to wall of vertical dykes are also examples of tension joints. Such joints are many times mineralised and give rise to ladder vein deposits. In the case of horizontal sills and sheets of igneous rocks, the joints are perpendicular to the top and the bottom of the sills and the sheets. Some tension joints have been shown in Figs. 32 A to E. In larger igneous intrusions like the batholiths, the stocks and the bosses, the joint planes that are almost perpendicular to the surface of the contact with the country rocks, are produced. These are called as the "fan joints" and are shown in Fig. 33.

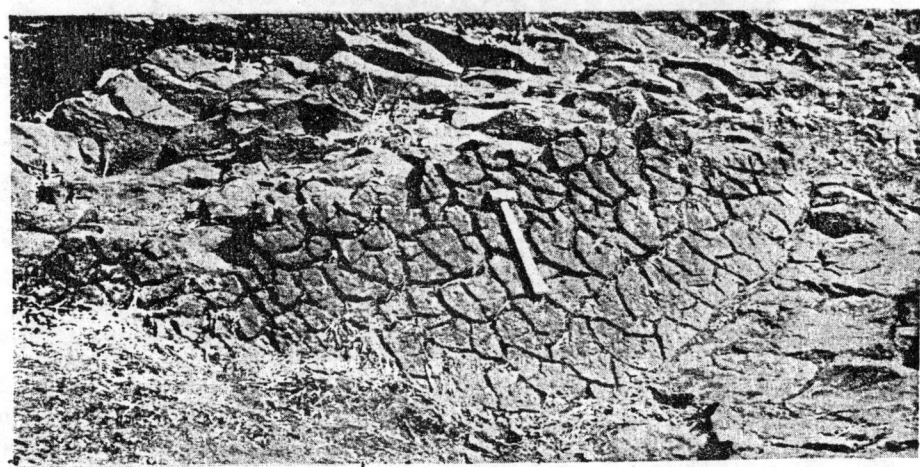

Photo. 11. Field photo of unusual columnar joints developed in the basaltic flows exposed 2 km. N20°W of Salgaon village, Ajra taluka, Kolhapur district Maharashtra state. These absolutely look like "mud cracks" and are hence called as unusual. Yet another horizon of columns is seen to the immediate backside of the hammer. Courtesy, Dr. K. Bhimsen.

Photo 12. Field photo of unusual columnar joints developed in basaltic flows exposed 2 km. N20°W of Salgeon village, Ajra taluka, Kolhapur district, Maharashtra state. The columns are horizontal in attitude as seen in the foreground. Vesicular structure can be observed in the right hand (bottom) corner of the photo. In the farthest background, a portion of the slickensided surface, is visible which marks an E - W trending fault. Courtesy Dr. K. Bhimsen.

Photo 13. Field photo of horizontal columns developed in a vertical basaltic dyke traversing granitic rocks. Hammer is on one such horizontal column. The columnar joints are exposed 5 km. S40°W of Gangavati town, Bellary district, Karnataka state. Courtesy Dr. V.N. Hegde.

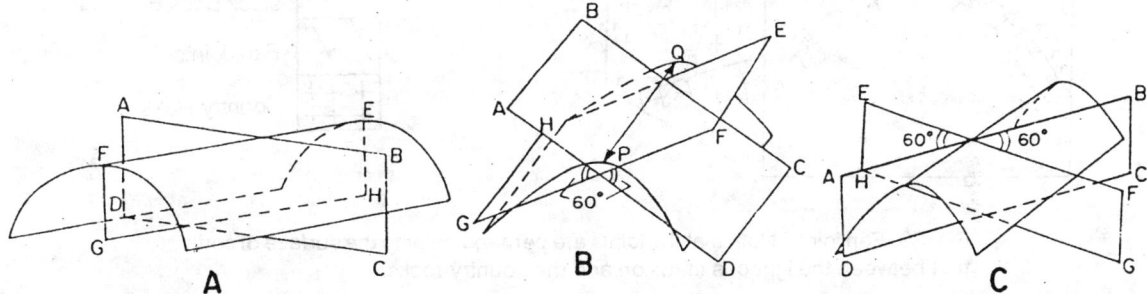

Fig. 32A. ABCD = perpendicular to axial plane,

EFGH = parallel (along) axial plane. Note that both joints are vertical.

Fig. 32B. ABCD and EFGH are inclined joints, and these bear 60° angle between them.

Fig. 32C. ABCD and EFGH are vertical joints, but between themselves these bear an angle of 60 degrees.

Figs. 32 A,B,C. Tension joints associated with folds.

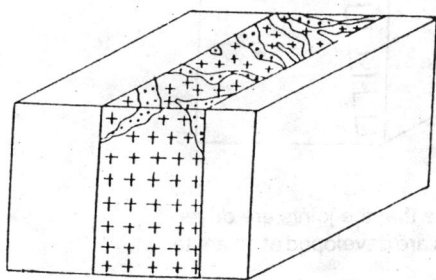

Fig. 32D. Ladder veins. Note that the veins (dotted parts) are perpendicular to the walls of the dyke (+++). The dyke is vertical and the veins fill up tension joints created in the dyke.

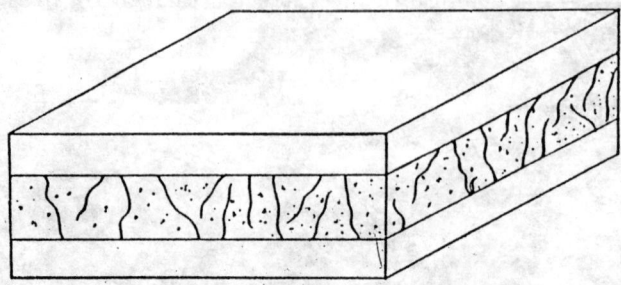

Fig. 32E. Joints associated with horizontal sheets and sills. Note that the joints exttend from the top to the bottom of the igneous body. These therefore are vertical in attitude. These are tension joints.

Figs. 32 A,B,C,D,E,. Tension Fractures/Joints.

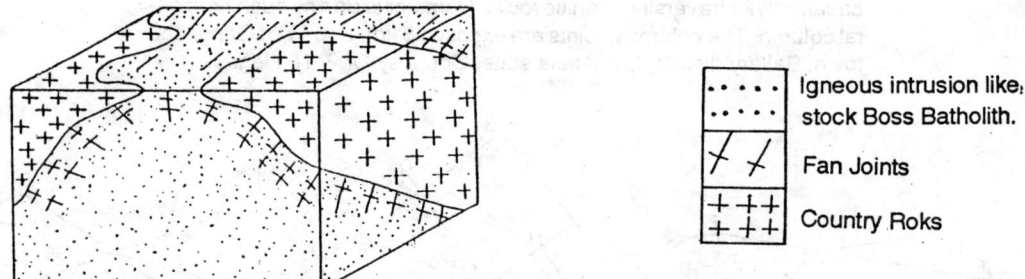

	Igneous intrusion like, stock Boss Batholith.
╱ ╱	Fan Joints
+ + + + + +	Country Roks

Fig. 33. Fan joints. Note that the joints are perpendicular to the surface of contact between the igneous intrusion and the country rocks.

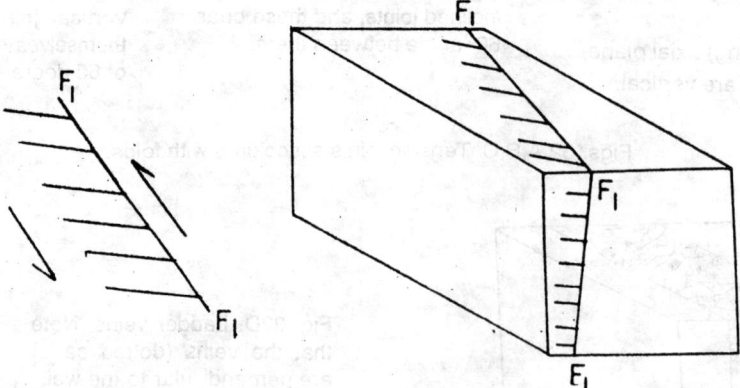

Fig. 34 A,B Shear joints associated with faults. Note that the joints are developed only on one side of the fault F_j - F_1. These joints are developed at an acute angle with the fault line.

(b) **Shear joints:** In this case, the deformative forces are acting parallel to or along the fracture planes. Joints associated with the faults belong to this category. Such joints are disposed at acute angle with fault plane. Further the joints are produced only on one side of the fault plane (Figs. 34 A,B and Photo 14). Two sets of joints are produced when coupling forces act. Such a situation is created in the case of two parallel faults. The central fault-block gets subjected to the coupling forces, and two sets of joints are produced as described by Billings (1960 Fig. 35).

Photo 14. Field photo showing development of 5-6 near parallel shear joints at an angle to the major shear-cum-fault plane trending N - S. Note that the shear joints are confined on the right hand side of the fault. These are hence classified as feather joints that are so characteristic of fault planes. A part of the narrow gorge is clearly seen to the left hand top corner of the photo. Courtesy Dr. A.H. Kouhsari.

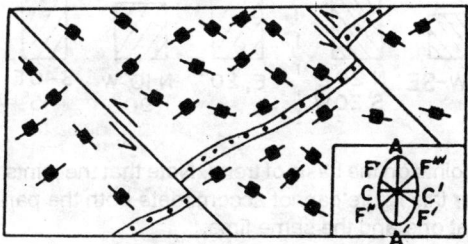

Fig. 35. (After Billings, 1960). Vertical shear joints caused by horizontal couple. Dotted part is key bed. Faults are vertical.

PORTRAYING OF JOINTS

There are several ways of portraying or representing the joints. Some are schematic (qualitative), others are of quantitative nature. Depending upon the goal, appropriate method is adopted.

(a) **Symbolic representation:** In this method, the dip direction, the strike direction and the amount of dip are reduced to certain symbols, and these are shown in Fig. 36. This method nodoubt gives an idea about the nature of the joints and their location (if shown on geological map). However this procedure cannot be adopted where a large number of readings are taken because the symbol itself occupies quite some space on the map, and it corresponds to a coverage of large area in the field, though in actuality, the symbol is for only few readings of joints recorded in the field.

(b) **Histogram:** This method can handle only one parameter at a time like the trend of the joints, dip direction, amount of dip. Height of the column is proportional to the number of readings of the joints. However, difficulty is experienced in respect of accomodating the amount of dip, because a joint dipping N 10°E may have a dip of 20°, 30° or it may even be vertical. Accordingly separate columns are required to be prepared (Figs 37 A, B). Histogram is a cumbersome mode of representation, and above all, the location of joints in the field cannot be made out from such a mode of representation.

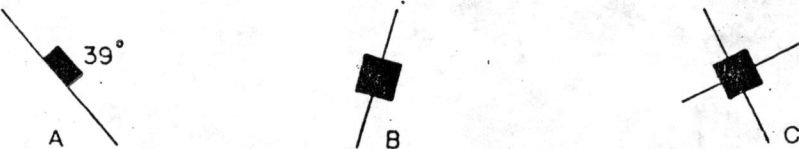

Fig. 36A. Joint plane
trending N50°W - S50°E,
dipping 39° due N40°E

Fig. 36B. Vertical Joint plane
trending N30°E - S30°W.

Fig. 36C. Horizontal joint plane.

Fig. 36A, B, C Symbolic representation of joints

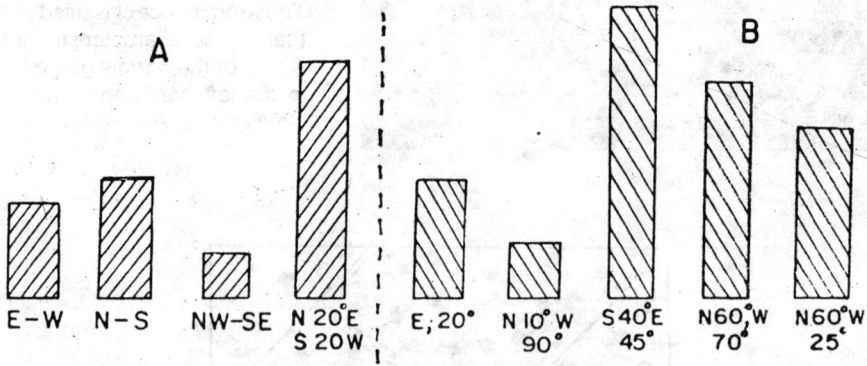

Fig. 37A. Histogram of joints on the basis of trend. Note that the joints trending N 20° E are most frequent. However this figure cannot accomodate both the parameters, namely dip direction and amount, at one and the same time.

Fig. 37B. Histogram of joints on the basis of amount of dip and direction. Note that the joints dipping 45° due S 40° E are the most frequent ones. Joints dippig due N 60° W are also frequent but those having dip amount of 70° are more frequent.

(c) **Ray or Rosette diagram:** This is a scalar mode of representation. Here two parameters could be accomodated, like the trend and the number of joints, or the dip direction and the number of joints. Actually it is an improvised version of histogram, because it effectively brings out the elements of the frequency, and the direction of joints (Figs. 38 A,B). However this mode can not handle the amount of dip of the joints. Also no knowledge about the places where the joints are produced, can be had from such diagrams.

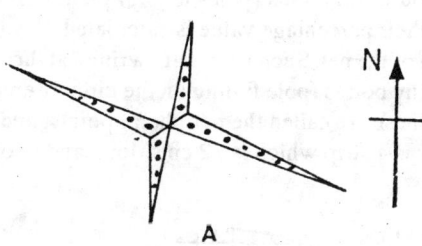

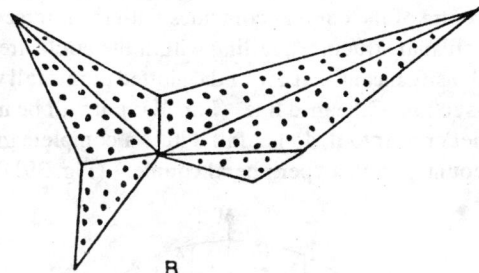

Fig. 38A. Trends of dip direction of joints. Note that the lines extend equally in opposite direction.

Fig. 38B. Dip direction and frequency of joints. Note that the lines do not extend beyond the centre of the figure, in the opposite direction, unlike that shown in 38A. This is because the direction of dip of joints is shown and a joint dipping due N 10° W (say), it need not also dip due S 10°E.

Figs. 38 A,B Rosette/Ray diagrams of joints.

(d) **Stereographic method:** This is by far the most sought after method, since it incorporates all the parameters of the joints, namely the dip direction, the amount of dip, and also the frquency of the joints. The direction of strike can be derived from that of the dip direction of the joints. In the preparation of a "point diagram, the author suggests the following changes

(i) Only one pole should be shown for all the vertical joints as against two, because a joint block is seldom symmetrical in respect of the disposition of the joint planes. If this is not true, then always a regular joint block should have been developed in the rocks, like a square block, a rectangular block, or a hexagonal block and so on. In such regular bodies, one can expect the presence of pairs of vertical joint planes. But such joints are not often developed. In fact irregular blocks of joints are commonly encountered. Further, the several joint planes together produce a "joint block". It is therefore quite obvious that a plane which is vertical due north, cannot be presumed to be so towards south direction also.

(ii) In plotting, a joint plane dipping due N 10° W is shown with a pole towards S 10° E i.e., exactly in diametrically opposite direction. This the author feels is unnecessary. The stereographic net should be considered as a mere contrivance, and hence instead of plotting the pole, the plane itself should be plotted and that too in the direction in which it is actually dipping.

Such "point diagrams" are to be later evaluated and contoured. This is done in three different ways., depending upon the number of readings of joints available. In any quantitative study, hundreds of readings are required to be considered in order to draw conclusive and an acceptable inference. But many times the readings are small in number. Therefore different methodology of evaluation is required to be followed under the different prevailing conditions. When the number of readings are between 200 and 400, *Schmidt or the Grid method* is adopted. When the readings are between 100 and 200, the *Free counter method* is adopted. When the readings are less than 100, *the Circle or the Mellis method is adopted* (Turner et. al 1963). A brief account of these methods is given below.

SCHMIDT OR THE GRID METHOD

The point diagram (Fig. 39A) is gridded, the distance between each grid being of 1 cm. The base net used is the Wulfe's stereographic net. In the case of the complete squares, the points are counted with a "central counter", which is a circle of 1 cm. radius (Fig. 39B). This counter is moved over a distance of 1 cm. each time, such that the centre of the counter coincides with the intersection of two grid lines. Points (poles to joint planes, or to any other planar structure), falling within the circle are counted and their percentage value is calculated. A value of 1% density means that 1% of the plotted points fall within 1% area of the net. Such values are written at the proper intersection of the grid lines (Fig. 39C). It will be noticed that many points (pole falling on the circumference of the net or near to it, do not fall within a complete grid of 2 cms. These are called the peripheral points, and these are counted with a "peripheral counter" (Fig. 39D). This counter is a strip which is 22 cms. long and about 2-3

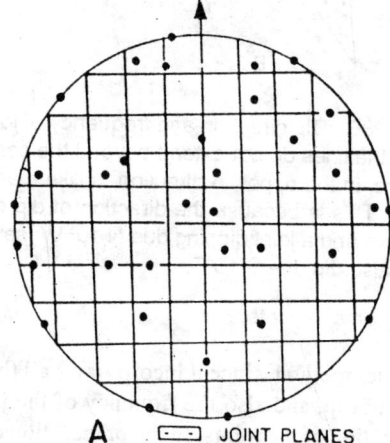

Fig. 39A Point diagram with grids drawn one centimeter apart.

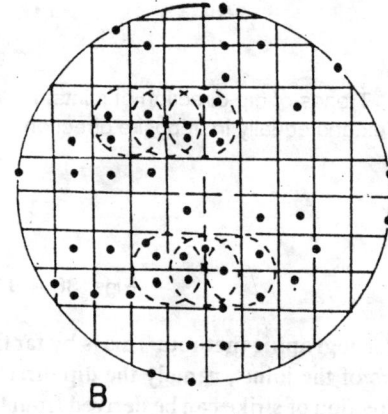

Fig. 39B Movement of central counter is shown in dashed circles. The outer frame of counter coincides with the grid lines.

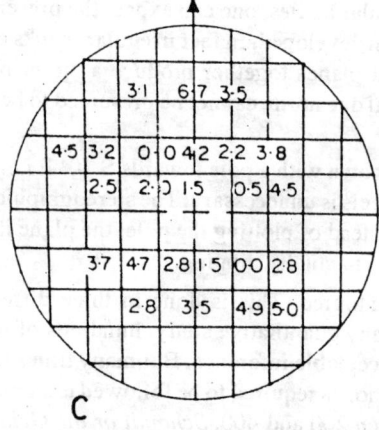

Fig. 39C Evaluated point diagram. Figures indicate density values (in percent) of joints.

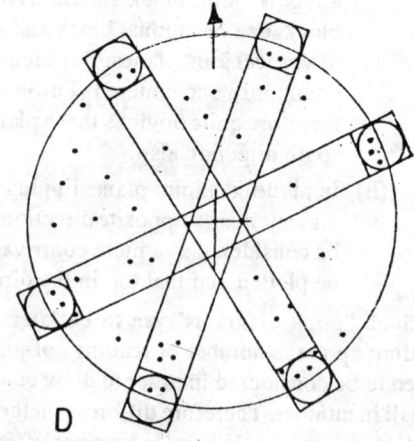

Fig. 39D. Movement of peripheral counter is indicated.

Fig. 39 A to D and 40. Stereographic method.

cms. wide. Circles are inscribed at the two ends of such a strip, such that the centres of these circles are at a distance of 10 cms. from the centre of the strip. This peripheral counter is rotated along the circumference of the circle, and the points falling in the two counters at the two ends, are counted and added. Their percentage value is calculated and such values are written at the two ends of the counter.

After evaluating the point diagram, it is contoured (Fig. 40), in the same way a surveyor prepares a contour map from the field data. The value of the contour depends upon the actual number of points plotted, and their distributions. If the points are nearly equally distributed, the density of the points become nearly similar, and no distinct contoured diagram is produced. If the density of the points be different from place to place (meaning grid to grid), then isolated, closed contours may be developed. The number of contours developed therefore depends upon the density of the plotted points, which in turn depends upon the distribution of the joints in the actual outcrop of the rocks. The author of this book further studies such sterograms of joints in an altogether different manner, which will be described at the end of this section.

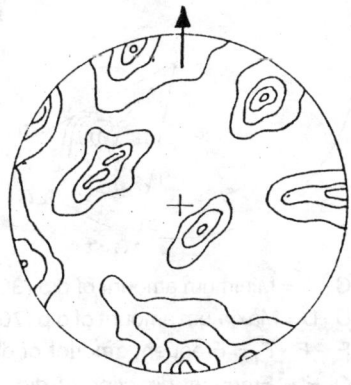

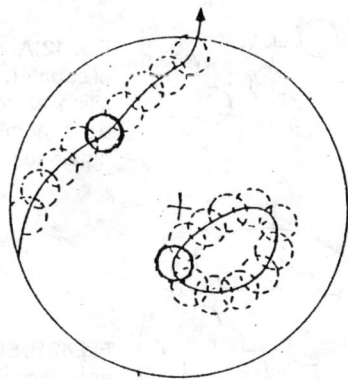

Fig. 40. A complete sterogram of joints showing disposition of maxima.

Fig. 41. Free counter method O Initial position of counter

Subsequent positions of counter where in equal number of points are obtainable within the counter.

Line (counter) joining centers of different positions of the counters.

FREE COUNTER METHOD

A point diagram is produced from the readings in the usual manner on the base of stereographic net. Such a point diagram is evaluated with a ring of 1 cm. radius, the placement of such a ring being effected in an entirely free manner. At the first stage the ring is arbitrarily placed at several places on the point diagram, and the number of maximum points falling within each placement of the ring (counter), is noted mentally. Let it be 6 points, for instance. The centre of one such position is marked with a cross. The ring is then moved from this position to another position, where again 6 points fall within the ring. The centre of this position is also marked with a cross. This procedure is repeated throughout the point diagram, and the position of the centres of the counter are marked with crosses. Such points (crosses) are then joined by a line which gives a contour of 6 points, which is further converted into percentage (Fig. 41). At the second stage, the positions of the counter where "5 or 4 or 3" points fall within the ring (counter), are noted. These positions are joined by a line which gives the next contour. This will have a value lesser than the previous one. This procedure is followed throughout the point diagram. In this manner, the point diagram is evaluated and later contoured. There are two highlights of this method namely.

(a) movement of the counter is not according to any rigid, pre-determined or biassed pattern, and

(b) counter lines get automatically defined.

MELLIS OR THE CIRCLE METHOD

A point diagram is first constructed. Then using a ring of 1 cm. radius, positions of minimum number of points falling within it, are ascertained (Fig. 42A). The outer side of the ring is traced. This line marks the position of a contour of minimum value. Others of higher values obviously will be within such a contour. This procedure is followed throughout the point diagram, and it is evaluated and contoured (Fig. 42B). In such cases, two or a maximum of three contours are obtainable, because the total number of points plotted are less, and hence these points do not produce spots or regions of higher concentraion.

There are decidedly drawbacks with every method of representation but the stereographic method has the minimum ones.

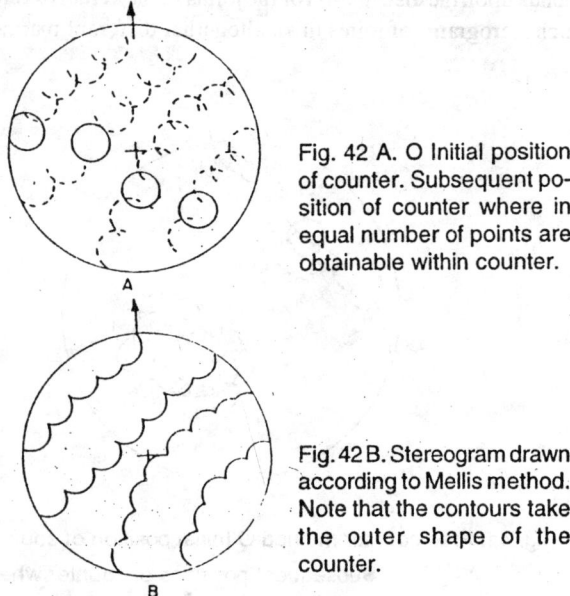

Fig. 42 A. O Initial position of counter. Subsequent position of counter where in equal number of points are obtainable within counter.

Fig. 42 B. Stereogram drawn according to Mellis method. Note that the contours take the outer shape of the counter.

Fig. 42 A,B. Mellis or Circle method.

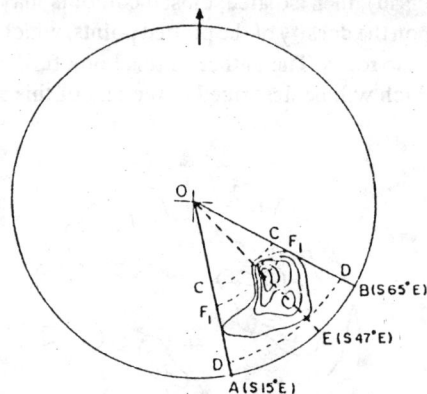

C - C = Minimum amount of dip (35°)
D - D = Maximum amount of dip (70°)
F_1 - F - F_1 = Frequent amount of dip (44°)
OFE = Frequent direction of dip (S 47° E)
Angle AOB = angular spread of the group of joints (S 15° E to S 65° E)

Fig. 43. New method of studying stereogram of joints (After Durg, 1969).

NEW APPROACH TOWARDS PROCESSING STEROGRAMS

In an earlier part of this section, it was mentioned that the stereograms could be processed further. It will be elaborated now in these paragraphs. Stereograms wherein a large number of points are plotted, contours are seen to close and produce areas of "maxima". Towards the circumference of the circle, the contours do not close, but may produce regions of maxima. Thus certain patterns of disposition of contours are observable. Besides interpreting the stereograms in respect of verticality, horizontality or inclination of the joints, it is possible to recognise *groups of joints, paired arrangement of the groups of joints* and so on. Certain other characteristics also could be read. The procedure is detailed out in the following paragraphs.

When a number of "maxima" (maxima means, isolated, closed or semi-closed contours), are developed, then such maximum may be considered as "groups" of joints. For each group, parameters like

    (a) angular spread,

    (b) range of dip amount,

    (c) frequent direction and amount of dip, and

    (d) preferred direction and amount of dip, can be determined. Recognition of these parameters has been described by Durg (1966).

A "group of joints" is said to be developed when on the stereogram two or more closed contours produce an isolated area (Fig. 43). Such a group has an angular span which is obtained by drawing lines from the centre of the stereogram to touch the outermost contour of the group (lines OA and OB in Fig. 43). In this case it is between S 15° E and S 65°E, corresponding to lines OA and OB, respectively. Each group has a "range of dip amount" for the several joints belonging to the concerned group. This is measured by a point on the outermost contour, and the contour nearest to the centre of the stereogram, and the other which is nearest to the circumference of the stereogram. These give the minimum (nearest to centre, line C-C in Fig. 43), and the maximum (nearest to the circumference, line D - D in Fig. 43), values of dip, respectively. In the present case, the values are 35° and 70° (lines C - C and D - D in Fig. 43). In many cases, a group of joints is characterised by the presence of a number of concentric contours. In such situations, the "frequent direction of dip, and the amount of dip "are determinable. The frequent direction of dip is obtained by drawing a line from the centre of the stereogram and passing it through the centralmost point of the innermost contour of the concerned group. In Fig. 43, AE is such a line and its direction is found to be S 47°E. The amount of dip is obtained from the position of the centre of the innermost contour (point F in Fig. 43), and in the present case, the value of dip is 44 degrees. It could be of any other value in the other cases.

The procedure described above, creates one more parameter. It is the "preferred direction of dip and the preferred amount of dip "for a group of joints. A stereogram may develop two or more groups of joints. Each group has a "frequent direction and amount of dip". It is therefore possible that the density of joints may vary from group to group, and one of the groups may have higher value of density. This is so because, more number of joints may be developed in that particular value of dip. The author therefore argues that such a direction should be designated as the "preferred direction of dip, and also having a preferred value of dip "in the same manner as is used in the petrofabric analysis. Thus from the foregoing account it is clear that by the new approach suggested above, it is possible to study the joints in greater details and recognise several groups within them, whereas in the field while collecting the data, so many groups might not have been apparent to the observer.

FAULTS

INTRODUCTION

Akin to the joints, the faults also belong to the class of rupture structures. These are not so numerous as are the joints. Further there is a distinct and a marked movement along the rupture plane, which feature is not seen along the joint planes. The movement may be in the vertical, the horizontal or in an oblique direction along the rupture plane. All rocks, sedimentary, metamorphic or the igneous, are faulted alike. No distinction is observed in the style or the intensity of faulting in the said three groups of rocks. However in the homogenous rocks like the igneous rocks, faults are not readily detectable, and hence it leads to the erroneous inference that these rocks are faulted to a lesser degree. But it is not so in actuality. All rocks are equally prone to faulting. Unlike the joints, the faults run over a considerable distance, but the number of faults occurring at any one place, is very much less; two, three, five or so. The distance between the faults, is considerably more. Generally, the faults do not intersect each other but this is not a pre-condition. To some extent, the pattern of joints in the three groups of rocks is dissimilar, and hence the rocks could be identified from the type of joints developed in them. However the faults do not help us in this regard, meaning that all the three verieties of rocks are faulted alike.

TECHNICAL FEATURES OF FAULTS

It is a rupture structure. Further it is a planar structure. The fault plane is generally a smooth, dipping surface. But it could be horizontal, vertical or curved also (Figs. 44 A,B,C, and Photo 15). Whenever a fault is developed, the earlier one unit (meaning the pre-existing rock) is broken into two parts, and each is called a "fault-block". Most of the fault planes are vertical in attitude, but when it has a dip, the fault-blocks (two in number) are further categorised as the "hanging block, and the foot block". The hanging block is one which rests on the other, and the block on which it rests is called as the foot block (Fig. 44A). It will be appreciated that when the dip of the fault plane is low enough, then only this distinction into foot and hanging blocks, is very apparent (Fig. 44B). Therefore when the fault plane is vertical, such a distinction is not at all possible (Fig. 44C). Besides having a dip, a fault plane has a "hade". It is the angle between the fault plane and a vertical plane in the fault plane (angle DAF in Fig. 45). Hade of a fault is measured because the faults are mostly vertical, and it is therefore the vertical plane

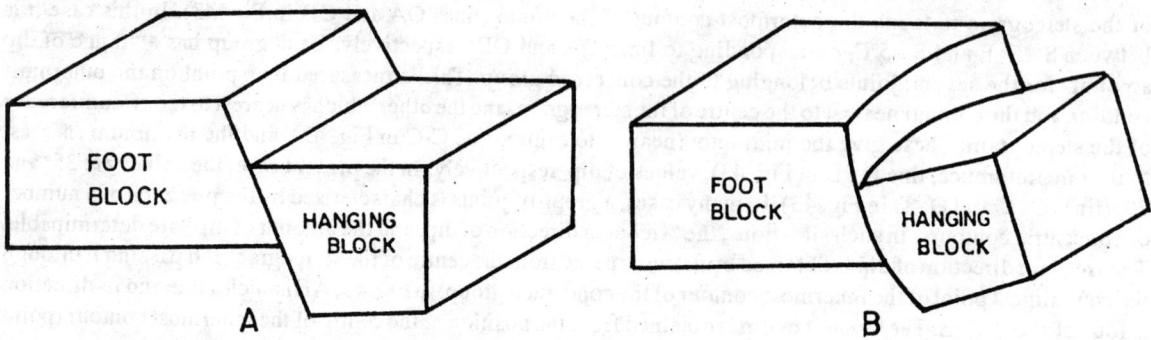

Fig. 44A. Planar fault surface. Hanging block rests on the foot block only because the fault surface is inclined one.

Fig. 44B. Curved fault surface. Hanging block rests over foot block

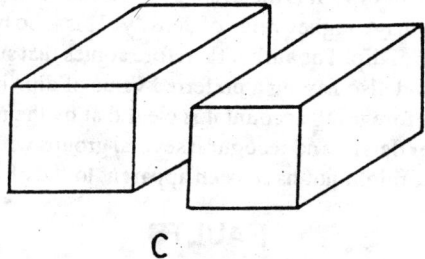

Fig. 44C. Vertical fault surface. Distinction into foot and hanging blocks is not possible because the fault surface is vertical. Therefore no block can rest over another in the strictest sense of expression

Fig. 44 A,B,C. Attitude of fault surfaces.

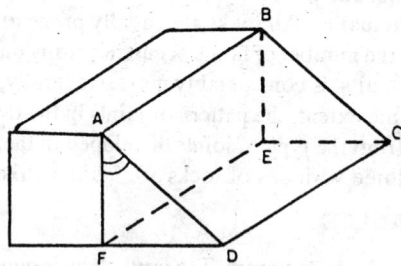

Fig. 45. Technical features of fault plane.

which is considered as the reference plane for measuring the inclination of the faults. In the case of dipping beds, the reference plane is taken to be horizontal, because when laid, the beds are horizontal in attitude. Therefore the inclination is measured with respect to a horizontal reference plane. The attitude of a fault plane whether it is the hade or the dip, it is measured in the same way as the attitude of a dipping body of sedimentary rocks. A fault plane therefore has a strike, amount of dip or the hade, and a direction of dip or the hade.

Photo 15. Field photo of a curved fault plane. Note the development of typical bluish coloured pseudotachylyte. The rocks are very coarse grained quartzarenites exposed in the vicinity of Jamkhandi town, Bijapur district, Karanataka state. Photo by author.

KINDS OF MOVEMENTS ALONG THE FAULT PLANE

These are mainly of two types, namely:

   (a) translation, and

   (b) rotation.

In the first variety, all the parts of the fault block move through equal magnitude along the fault plane. This maintains the original attitude of the rocks in the two fault blocks. In the rotational type, the magnitude of the movement is not the same along the fault plane. This causes changes in the original attitude of the rocks composing the fault-blocks. These main features have been shown in Figs. 46 A to G.

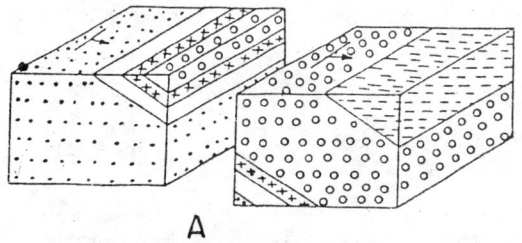

Fig. 46A. Note that there is no change in the amount of dip in the two fault blocks.

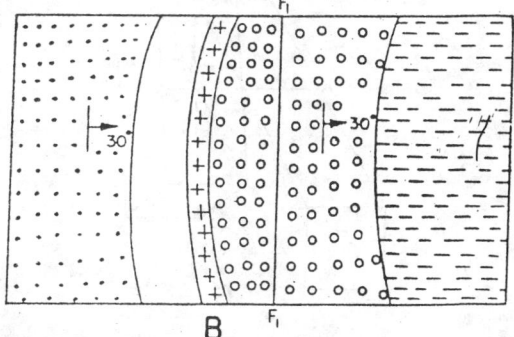

Fig. 46B. It is a map. Note that there is no change in the amount of dip in the two fault blocks.,

Figs. 46 A,B. Translatory movement along fault.

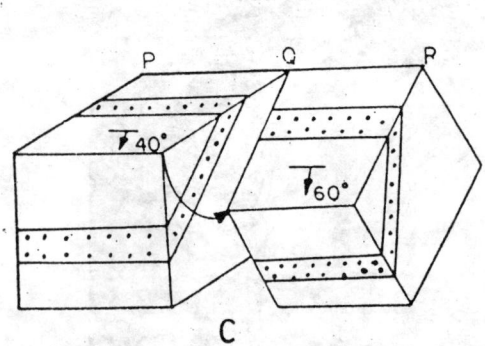

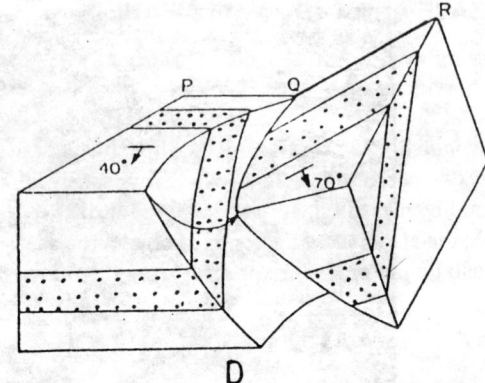

C

D

Fig. 46C. Fault surface is vertical. Note that the beds on the right hand side block have developed higher dip amount due to rotational movement. Points P,Q,R are in a straight line, and these constitute the hinge of the fault

Fig. 46D Fault surface is curved. Beds on the right hand side dip due left, owing to rotational movement on a curved fault surface. The fault has an axis at point Q. Note that points P, Q, and R, are not in a straight line. Very important feature is "unconformity like" relation created between the two fault blocks.

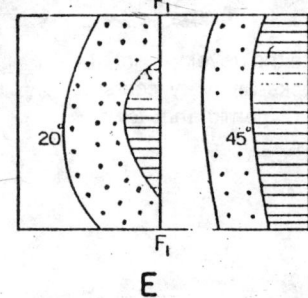

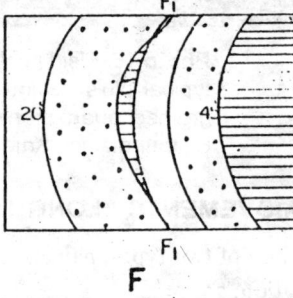

E

F

Fig. 46E Map of a rotational fault. Note the change in the amount of dip between the two fault blocks. The change in the curvature of outcrops of beds is due to different dip amounts. The fault F_1 - F_1 has a planar surface.

Fig. 46F Map of a curved rotational fault. Change in the curvature of outcrops of the beds is due to different dip amounts.

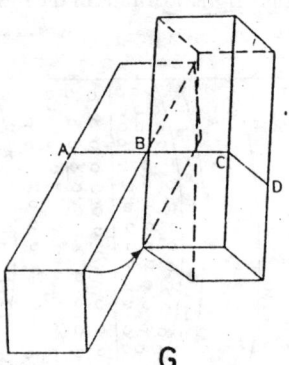

G

Fig. 46G Axial rotational fault. Movement occurs along axis located in the plane ABCD.

Figs. 46 C to G. Rotational movement along varied fault surfaces.

HEAVE AND THROW OF A FAULT

The movement during faulting is normally in the vertical direction i.e., in the direction or in the opposite direction of gravity. This is spoken off as the "throw" of a fault (Figs. 47 A, B). The movement as well could be in the horizontal direction. This is called as the "heave" of a fault (Fig. 47C, and Photo 16, 17). Some other terms are also in vogue like the separation, the slip. However in the field, the heave, the throw, the separation or the slip are not always measurable, because in many instances, "index beds or horizones" with respect to which the movement could be measured, are not existing. Further it is also to be noted that the block with edges, or the parallel sides etc., are also not available in the crustal rocks. The block diagrams drawn here are merely to simplify and not to illustrate the procedure of measuring the throw or the heave of the faults.

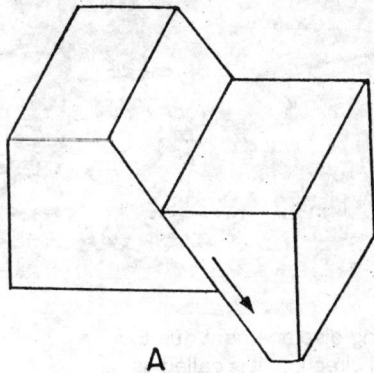

Fig. 47A Fault having a throw in the direction of gravity as indicated by the arrow.

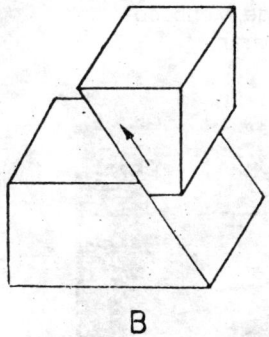

Fig. 47B Fault having a throw against the direction of gravity as indicated by the arrow.

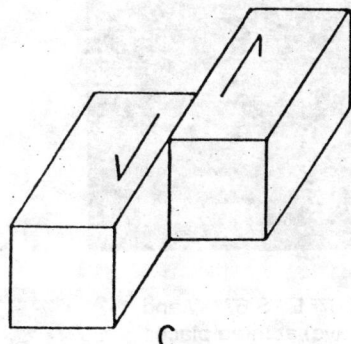

Fig. 47C Fault having a movement in horizontal direction which is called as the heave of the fault.

Figs 47 A,B,C. Throw and heave of a fault.

Photo 16. Field photo of a vertical dyke showing displacement due to faulting. As the displacement is in the horizontal direction, it is called as "heave". Also note the development of numerous parallel shears running parrallel to the trend of the dyke, and to be confined to the immediate vicinity of the dyke. This dyke is located at Akutothpalli village, Mehboobnagar district, Andhra Pradesh. Courtesy Dr. D. Muralidharan.

Photo 17. Field photo of a basaltic dyke trending N 67° E - S 67° W and traversing granitic rocks. A clear displacement (heave) at three places is observable. The exposure is located 5 km. SW of Gangavati town, Bellary district, Karnataka state. Courtesy Dr. V.N. Hegde.

CLASSIFICATION OF FAULTS

By classifying the faults, a distinction between the different varieties of the faults should be brought about. This is possible by considering bases of effecting a distinction between the several types of faults. Further, atleast in structural geology, the bases should be reasonably and readily adoptable or implimentable in the field. The main stay of the faults is the movement along the fault plane. Therefore the major basis should be the kinds of the movements. However on many occasions, the movement is not readily discernible, instead the effects produced by the faulting, could be made out more easily. The displacement of the beds or the rocks with respect to the fault plane, becomes the next important basis. The attitude or the disposition of the fault plane also needs to be considered, because all the movements, displacements etc., take place on it. Utimately the deformative forces that are responsible for the faults, those should be taken into account. These different bases are shown in the form of a schematic diagram (Fig. 48).

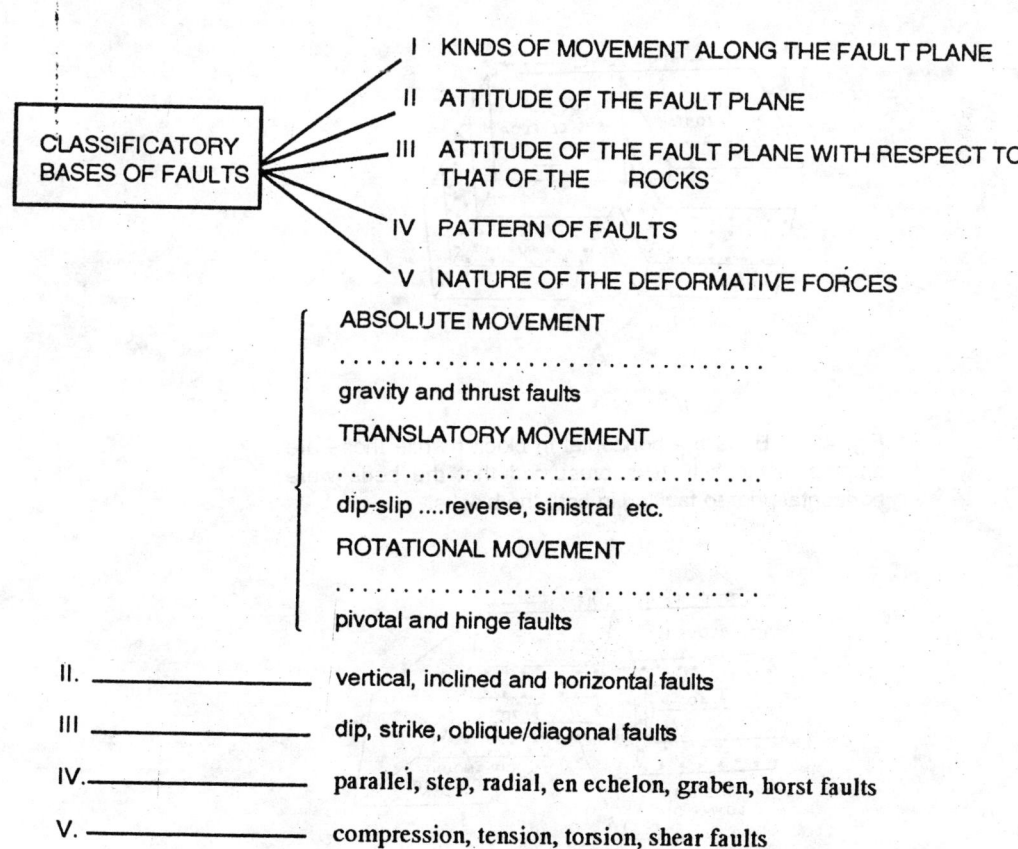

Fig. 48. Schematic diagram of classificatory bases of faults

Out of the different bases given in Fig. 48, two are not readily implimentable namely,

    (a) absolute movement, and

    (b) nature of the deformative forces.

The characteristics of the different faults will be described in the following paragraphs.

ROTATIONAL FAULTS

The rotational movement can be ascertained provided there is some key horizon. This may be a bedding plane, a sill, layered rocks and so on. In the absence of such features, change in the original attitude can not be judged. If the rocks be bedded ones, and if the bedding planes be horizontal, then due to the rotational movement, one fault block will show dip or inclination for the beds (Fig. 49A). It is also possible that the original rock may be dipping one. Then due to rotational movement, the dip amount in one fault-block may increase or decrease (Fig. 49 B). In these two cases, the strike and the dip directions have remained unchanged, but it need not be necessarily so. The attitude of the rocks in the two fault blocks may be different, especially the strike direction. This feature will be considered a little later.

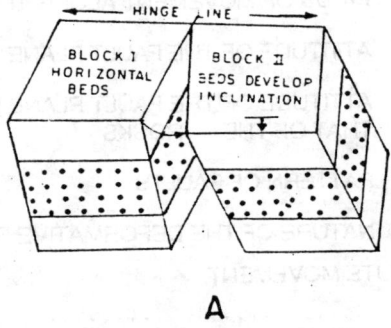

Fig. 49A. Beds are horizontal in block I while those are dipping in block II. It is presumed that the beds were horizontal prior to faulting in both the blocks.

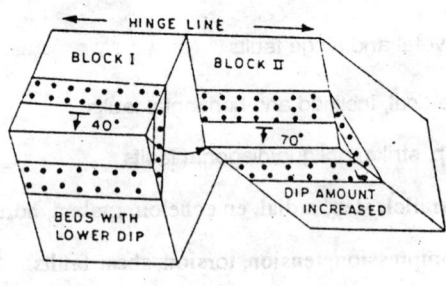

Fig. 49B. Beds had low dip in both the blocks prior to faulting. Due to rotational movement, beds in block II have developed higher amount of dip.

Figs. 49 A,B. Rotational faults with hinge.

Rupture Structures **49**

The rotational faults are further splitable into:

(a) pivotal or hinge faults, and

(b) axial faults.

In the former type, there is a hinge beyond which displacement is not continued (Figs. 49 A,B, 51 and 52). The other feature of such faults is that the downthrow and the upthrow sides are not changed along the fault plane. Thus in Figs. 49 A,B, block II is the downthrow side all along and upto the hinge line. In the case of the axial fault, the downthrow and the upthrow sides change on the two sides of the hinge-cum-axis of the fault (line ABCD in Fig. 50). In this figure, block I is the downthrow side upto the hinge line ABCD, but beyond it and on the other side of the hinge line, the same block becomes the upthrow side. This is the most salient feature of the axial faults.

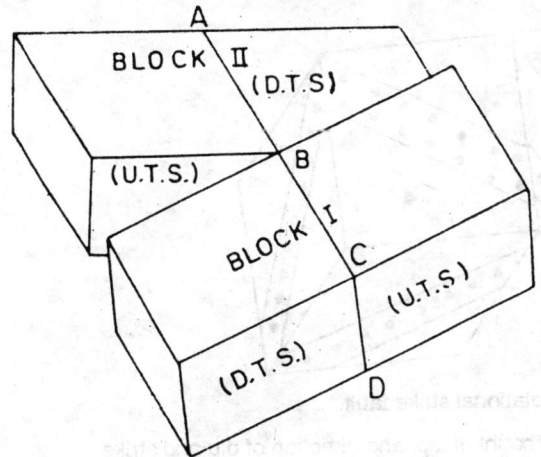

ABCD = plane containing axis of fault.

U.T.S. = up throw side.

D.T.S. = down throw side

Fig. 50. Axial fault

Note that D.T.S. and U.T.S. change across the line ABC. Same block becomes U.T.S. on one side and D.T.S. on the other side of the line.

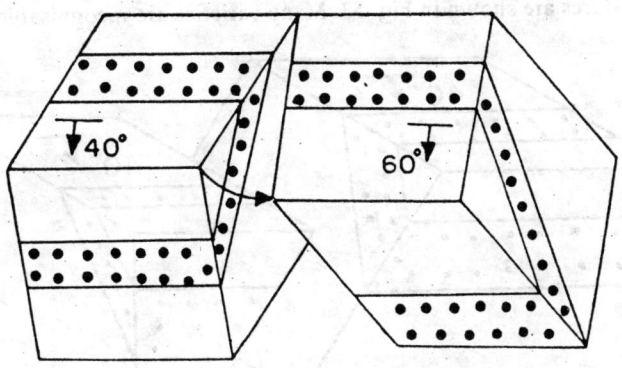

Fig. 51. Rotational dip fault.

Note that the amount of dip is more in the right hand side block which is due to rotational movement along the fault surface. Trend of the fault is parallel to the dip direction of the beds. There is no change in the direction of dip and strike.

The rotational faults are classifiable further into:

(a) rotational dip fault, and

(b) rotational strike fault.

In the case of "rotational dip fault", the trend of the fault is parallel to the dip direction of the beds affected by the fault (Fig. 51.). In the "rotational strike fault", the trend of the fault is parallel to the strike of the beds affected by the fault (Fig. 52).

In every variety of the rotational faults described above, there is one very important feature to be mentioned, and that is the amount of throw. In this kind of fault, the throw goes on increasing as one proceeds away from the "hinge or the pivot" of the fault. In fact, the rotational movements of any fault are judged only by a change in the amount of throw, and also of the amount of dip of the beds. This feature is needed to be observed in the field.

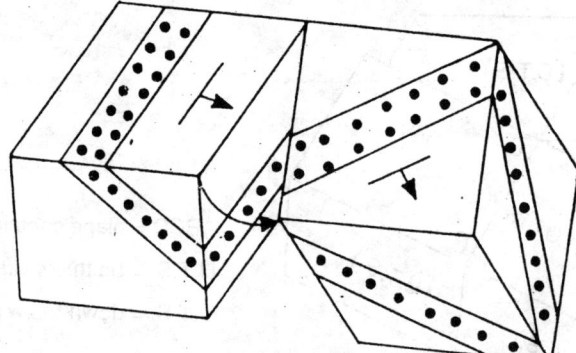

Fig. 52. Rotational strike fault.

Note that there is change in the amount of dip, and direction of dip and strike between the two fault blocks.

FAULTS OF TRANSLATORY MOVEMENT

Majority of the faults belong to this category. As already noted, the attitude of the rocks in the two or more fault-blocks, remains unchanged. Amount of throw and the downthrow or the upthrow sides are also not changed along the fault plane. These features are shown in Fig. 53. Many varieties are recognisable under the faults of

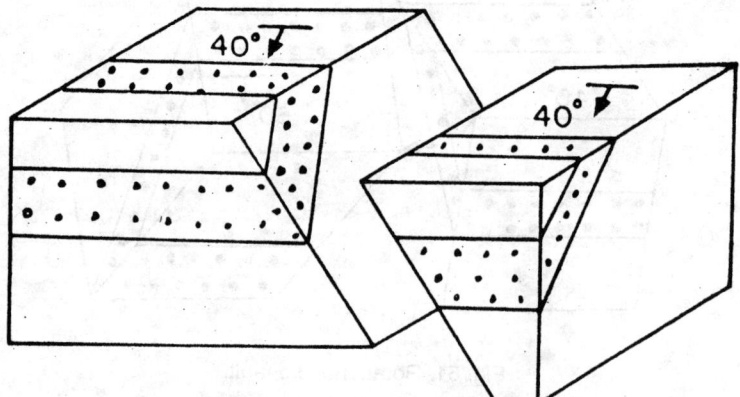

Fig. 53. Translotary movement along fault surface.

Note that parallelism exists between the lines and the edges of the two fault blocks. Further there is no change in the attitude of the beds occurring in the two fault blocks.

"translation". But these are considered under different headings like the "attitude of the fault plane, attitude of the rocks with respect to that of the fault plane" and so on. Therefore these are likewise described under the concerned headings.

BASIS OF ATTITUDE OF FAULT PLANE

In this case, the hade or the dip of the faults is considered, and accordingly three types of faults are recognised namely,

(a) vertical or high angle faults,

(b) medium angle faults, and

(c) low angle faults (Figs. 54 A,B,C).

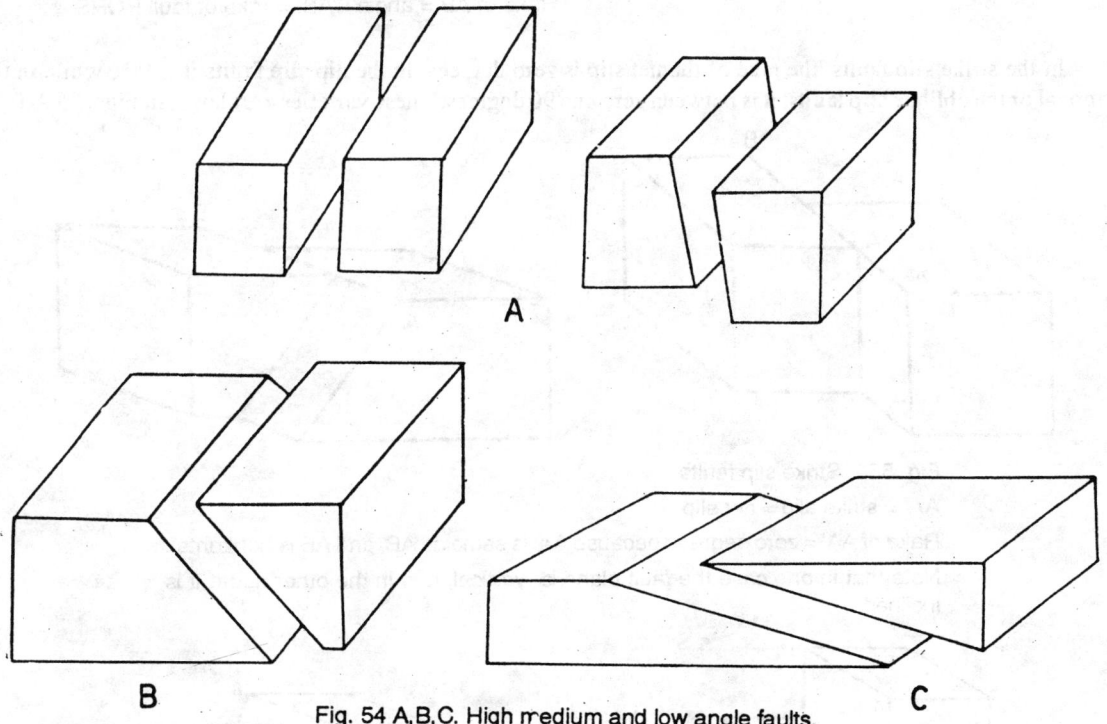

Fig. 54 A,B,C. High medium and low angle faults.

BASIS OF MOVEMENT ALONG THE FAULT PLANE

This classification is effected on the "rake of the net-slip" of a fault. Rake of a line (in this case it corresponds to the actual direction of movement on the fault plane), is the angle made by it with a horizontal line in that plane. In Fig. 55, the rake of line AB (seen on the inclined surface PQRS) is the angle CAB, because AC is the horizontal line in the fault plane PQRS. It is argued that due to the movement along the fault plane, scratches, striations or lines will be produced due to the movement of one block over the other. Such lines therefore indicate the actual directions of the movement produced during faulting. Hence the "rake of the net-slip" is a very useful element in the classification of the faults. Three types are recognised according to the value of the "rake of net-slip" namely,

(a) strike slip,

(b) dip slip, and

(c) oblique or diagonal slip faults.

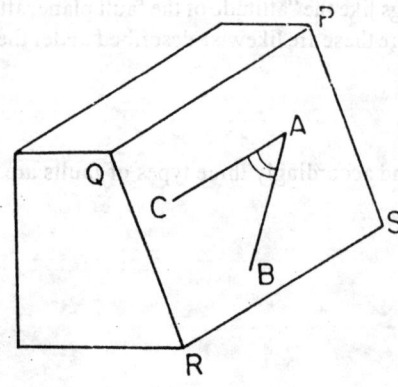

Fig. 55. Rake of a fault.

PQRS = .Inclined fault

AB = any line in PQRS

AC = horizontal line through A.

Rake of AB = angle CAB. = rake of fault PQRS

In the strike slip faults, the rake of the net slip is zero degrees, in the dip slip faults it is 90°, while in the diagonal or the oblique slip faults, it is between zero and 90 degrees. These varieties are shown in Figs. 56 A,B,C.

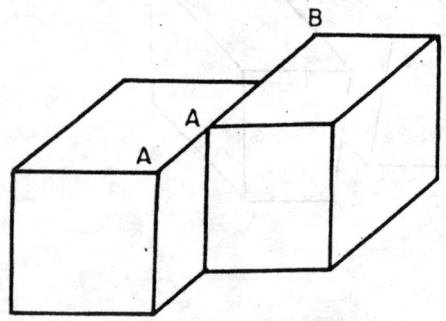

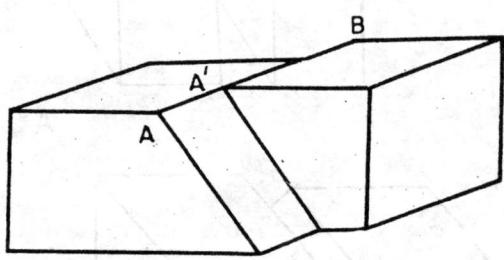

Fig. 56A. Strike slip faults

AA' = strike slip = net slip

Rake of AA' = zero degrees because AA' is same as AB, and AB is horizontal.

Note that in one case the fault plane is vertical, and in the other figure it is inclined.

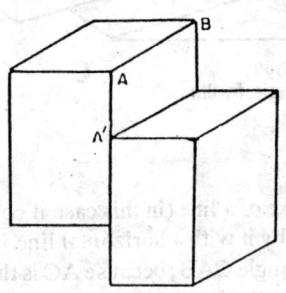

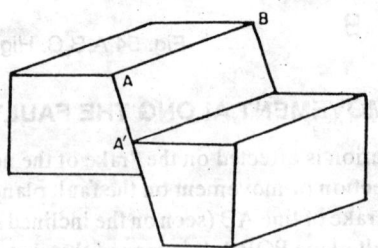

Fig. 56B. Dip slip faults.

AA' = dip slip = net slip

Rake of AA' = 90°, because AA' is perpendicular to AB, and AB is a horizontal line.

In one figure the fault plane is vertical while in another it is inclined.

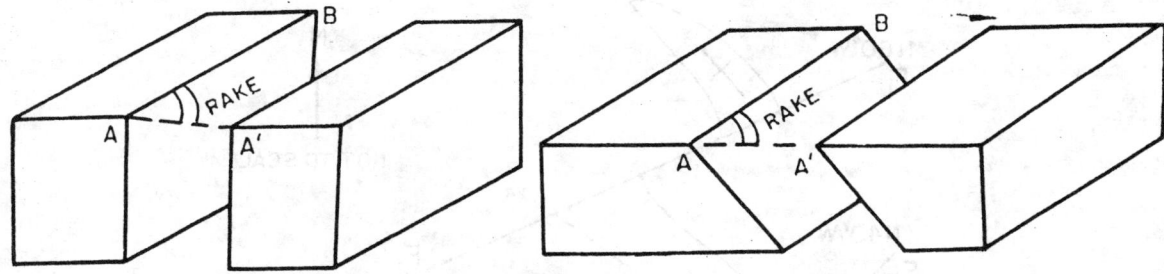

Fig. 56C. Oblique/diagonal slip faults.

In one figure, the fault plane is vertical, in the other it is inclined.

AB = horizontal line in the fault plane

AA' = direction of movement due to faulting (slip)

Rake of AA' = angle BAA'. It is less than 90°, but is more than zero degrees.

Note that AA' is at angle to the strike as well as the dip direction of the fault plane. It is thus oblique or diagonal slip.

DEXTRAL AND SINISTRAL FAULTS

Movement along the fault plane can be used in yet another way to recognise different types of faults. It is argued that by a careful study of the fault plane, the actual direction of movement during the faulting could be ascertained. Looking along the extension of the fault plane, if the block on the right hand side of the observer appears to have moved away from him (moved in the direction of the extension of the fault plane), then such faults are called as "sinistral or right handed "faults (Fig. 57 A,B). Deendar (1982) has described 3 near-parallel sinistral faults wherein the vertical quartzite band has been affected by it (Fig. 57C). Gokhale (1977) has described sinistral fault from the quartzites exposed near saundatti, Belgaum district, Karnataka state. If however the block on the right hand side of the observer appears to have moved towards him, then such faults are called as "dextral or left handed "faults (Figs. 58 A,B and Photo 18).

It will be noticed from the foregoing description that the dextral and the sinistral faults are basically strike slip faults. But since the rake of the net slip of the fault is not taken into account, and only the direction of the movement of the block with respect to the position of the observer, these are called as the dextral and the sinistral faults. To some extent, the oblique slip faults also could be categorised likewise, provided the rake is around 20 to 40 degrees, i.e., it is less than 45 degrees. The dip slip faults can not be categorised into the dextral or the sinistral

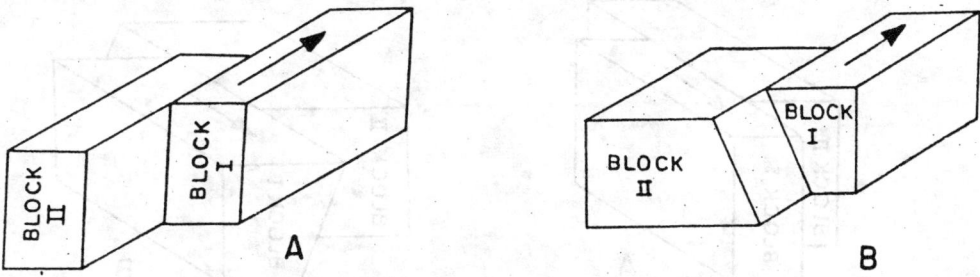

Fig. 57 A,B. Vertical and inclined sinistral faults, respectively.

Figs. 57 A,B,C. Sinistral faults.

Note that in figures A and B, block I appears to have moved away from the observer, on looking along the trend of the fault. In C, the quartzite band (outcrop) is clearly seen to have moved away from the observer, when observed along the trend of the faults.

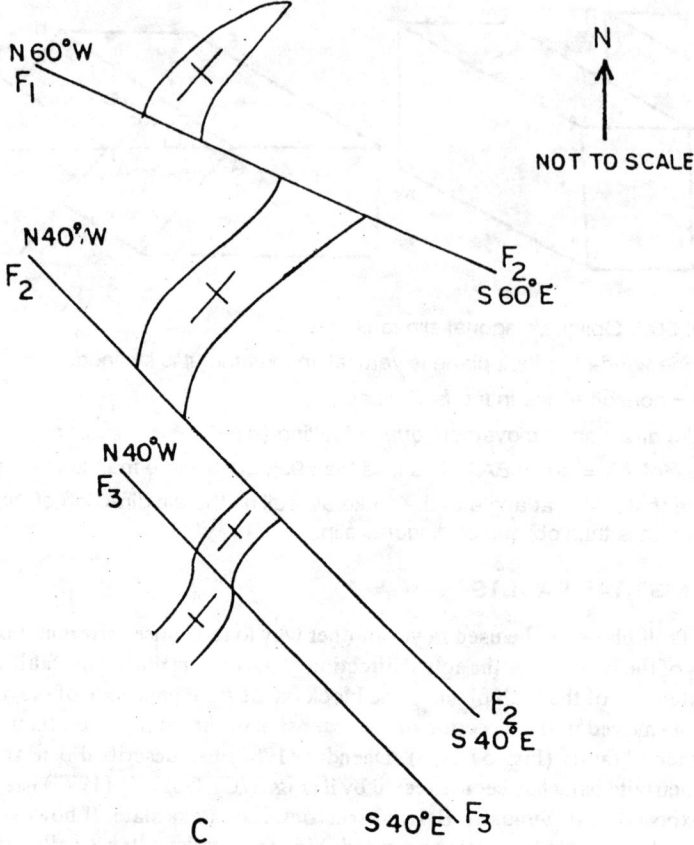

Fig. 57 C. after Deendar, 1982.

The vertical quartzite is faulted thrice and in each case the outcrop on the right hand side of the observer appears to have moved away from the observer. These faults also can be classified as wrench or tear faults because the rocks are torn perpendicular to the trend of the rocks.

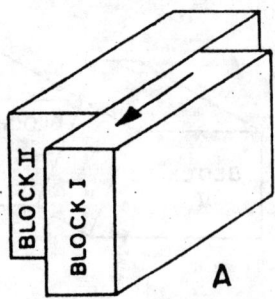

Fig. 58A. Vertical dextral fault

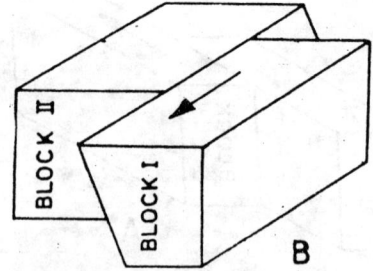

Fig. 58B Inclined dextral fault

Figs. 58 A,B Dextral faults.

Note that in the figures, block I appears to have moved towards the observer, on looking along the trend of the fault.

Photo 18. Field photo showing a band of banded hematite quartzite trending N 45° W - S 45° E offset by a fault trending N 45° E - S 45°W. It is a dextral fault. View looking along the trend of the rocks. Pile of rocks on the left hand side of the photo, marks the boundary between two mines, and it is not an outcrop of rocks. The dextral fault is located 4 km. S35° W of Vadrahalli, N.E.B. range, Bellary district, karnataka state. Courtesy Dr. H.D. Desai.

varieties, since the movement is not along the trend of the fault, but it is up or down the fault plane. Such movements give rise to the gravity and the thrust faults, which will be described at the proper place.

WRENCH OR TEAR FAULTS

The dextral and the sinistral faults can be further categorised into the wrench or the tear varieties, if the fault affects bedded formations. In such cases, if the trend of the dextral or the sinistral fault be perpendicular to the strike direction of the formations, then the rocks will appear as though they have been torn along the direction of dip of the formations. Such faults are then described as the wrench or the tear faults. In Fig. 67 A,B, dip faults are shown. But these also get categorised as the "wrench or tear" faults, because these faults also have a strike slip component. Sinistral faults of Saundatti area described by Gokhale (1977) are also wrench faults.

NORMAL AND REVERSE FAULTS

This distinction is made on the basis of the movement of the hanging block or the foot block. As previously noted, the fault plane must possess inclination which should not be of 90 degrees value. A normal fault results when the hanging blcok has moved down with respect to the foot block. In a reverse fault, the hanging block has moved up with respect to the foot block. The author finds it more appropriate to define these faults with respect to the dip direction of the fault plane alone. Thus a normal fault results when the hanging block has moved down and in the direction of the dip of the fault plane. In a reverse fault, the hanging block has moved up and against the direction of the dip of the fault plane (Figs. 59 A,B)

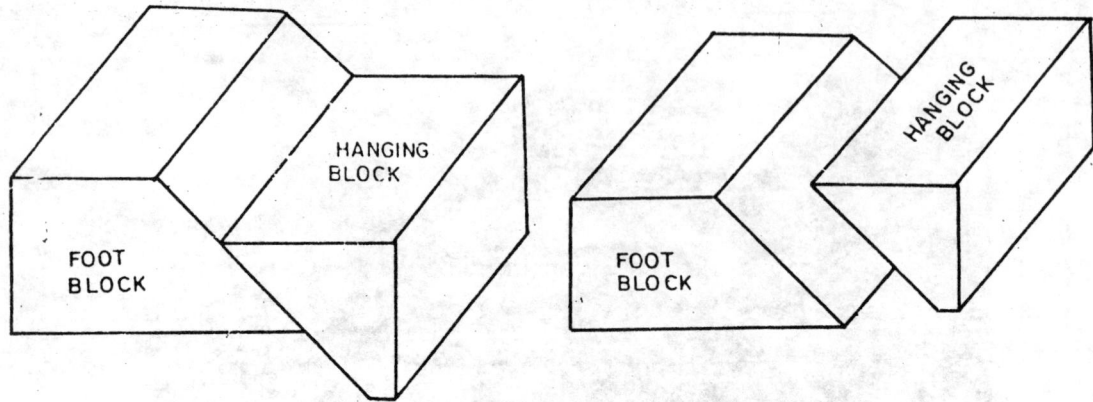

Fig. 59A Normal faults

Note that in the two figures, the hanging block appears to have moved down the direction of dip of the fault plane. Note that arrows indicating direction of movement of the blocks are not shown because which block has actually moved is not known.

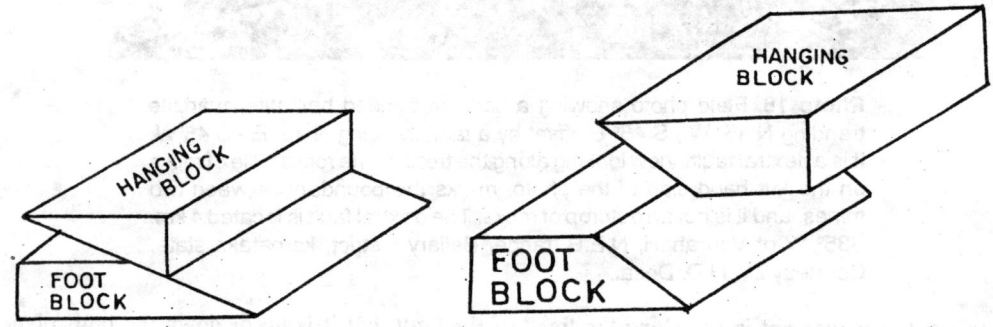

Fig. 59B Reverse faults.

In the two figures, hanging block appears to have apparently moved up and against the direction of dip of the fault plane. Note that arrows indicating direction of movement of the blocks are not shown because which block has actually moved is not known.

THRUST AND GRAVITY FAUTLS

The actual direction of movement along the fault plane is many times not determinable. Therefore though the hanging block at present appears to have moved down, it may be that the block did not move down, but the foot block might have moved up, producing similar ultimate appearance. However, at times, it is possible to establish the direction of movement, and then the normal and the reverse fautls get designated as the gravity and the thrust faults, respectively. Accordingly their characteristics are described below.

GRAVITY FAULTS

These are faults where in the hangingblock has moved in the direction of the dip of the fault plane. If the fault plane is vertical, then it is called as a vertical gravity fault; if it is inclined, then it is called as an inclined gravity fault (Figs. 60 A,B,C). In these cases it is presumed that the hanging block alone had moved down. However it is possible that both the blocks might have moved, but in different directions and through different amounts. Accordingly, FIVE kinds of gravity faults are recognised, considering the actual, absolute movements, and these are described below.

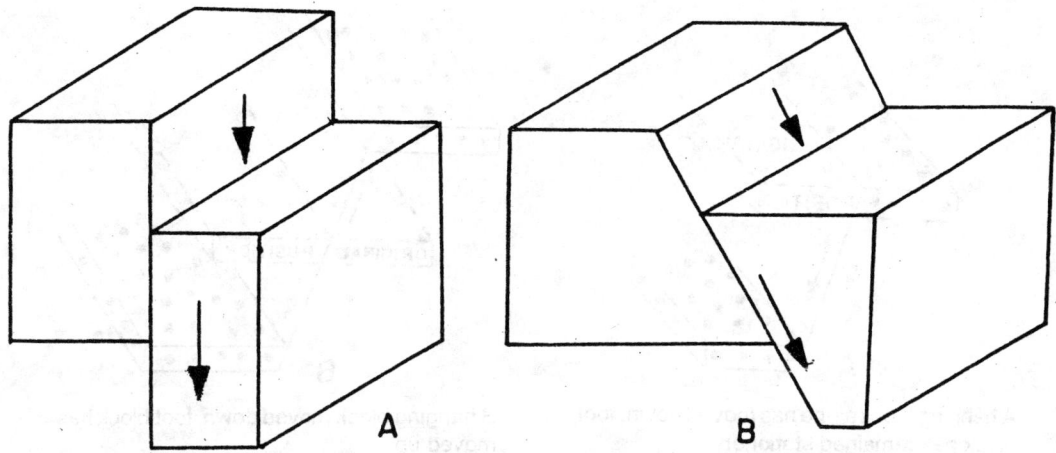

Fig. 60 A vertical gravity fault, with dip slip movement Fig. 60 B inclined gravity fault with dip slip movement

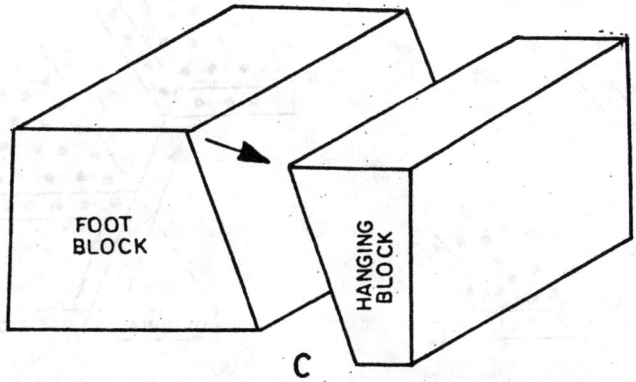

Fig. 60 C inclined gravity fault with diagonal slip movement.

Note that in figures 60 A,B, the movement is in the direction of dip of the fault plane as indicated by the arrows. In figure Fig 60C, though it is diagonal, the block all the same has moved down the dip of the fault plane.

Fig. 60 A,B,C. Gravity faults

(a) hanging block alone has moved down, the foot block remaining stationary (Fig. 61A),

(b) hanging block moved down and foot block moved up (Fig. 61B),

(c) hanging block remained stationary, but the foot block moved up (Fig. 61C).

(d) both the blocks moved up but the foot block moved more than the hanging block (Fig. 61D), and

(e) both the blocks moved down, but the hanging block moved more than the foot block (Fig. 61E).

THRUST FAULTS

These are characterised by the active displacements of the hanging block against the direction the dip of the fault plane. There are more varieties within the thrust fault than those encountered in the gravity faults. Ordinarily when the actual direction of the movement is not establishable, if the effective movement is against the direction of dip the fault plane, then such faults are called as the "reverse faults" to which a mention has been already made.

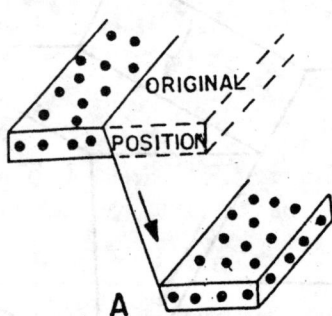

A hanging block alone has moved down, foot block has remained stationery

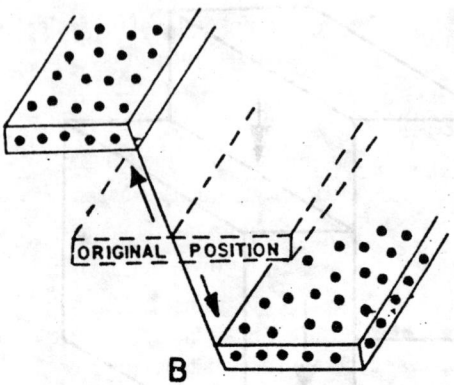

B hanging block moved down, foot block has moved up

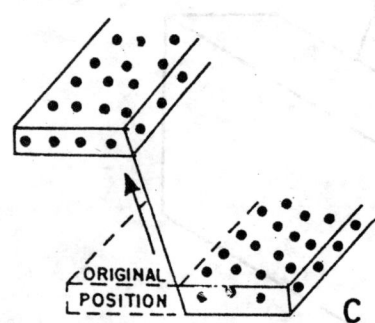

C foot block alone has moved up, hanging block has remained stationery

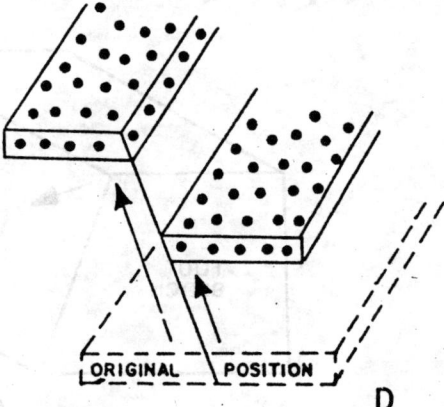

D both blocks have moved up, but foot block has moved more

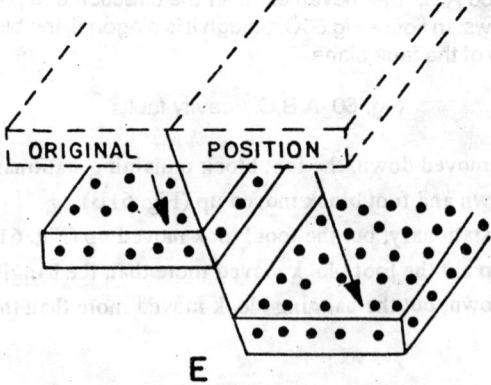

E both blocks have moved down, but the hanging block has moved more

Figs. 61 A,B,C,D,E Varieties of gravity faults

Further, considering the amount of dip of the fault plane coupled with the actual direction of movement, varieties like upthrust, the downthrust, the underthrust, the overthrust and the nappe are distinguished. Akin to the gravity faults, there are five sub-varieties within the general variety of thrust fault, and these are described below.

(a) foot block alone moved down, the hanging block remaining stationary (Fig. 62A),

(b) foot block has moved down, the hanging block has moved up (Fig. 62B)

(c) foot block remained stationary, but the hanging block has moved up (Fig. 62C)

(d) both the blocks moved up but the hanging block has moved more than the foot block (Fig. 62D), and

(e) both the blocks moved down, but the foot block has moved more than the hanging block (Fig. 62E)

Yet another type of overthrust is distinguishable. It is called as "nappe", which is a low angle overthrust. In this variety, overthrusting is measurable in a few tens of kilometers. This produces complexities in the stratigraphical succession. Rocks belonging to a totally different series may be found to be in contact with a series either considerably older than or younger to it. Through overthrusting, rocks will be elevated to higher topographical heights. Such places are called by the term "klippe". The place from where the rocks are removed and taken away, that place exposes "older"rocks. These latter places are called as "fensters". These are fantasies in structural geology, and these abound in the Himalayan terrane. Gokhale et. al. (1971) have reported such structures in the rocks of the Gadag schist belt. The two boulder beds are construed by them as the effect of thrust faulting.

The development of the nappe structure gives rise to two more terms namely, "autochthonous" and "allochthonous" nappes. This is so because due to the the thrust faulting, rocks formed somewhere else are brought to a locality where similar or totally dissimilar rocks might be present. Thus if the thrusting movement does not take the original rocks, to rocks of different age, then such "thrusts" are described as the "autochthonous thrusts". If the rocks differ in age, such "thrusts" are designated as the "allochthonous thrusts". For example, if the rocks of Cambrian age are thrust over rocks of same age, then this situation gives rise to "autochthonous thrust." If on the other hand, Cambrian rocks are thrust over rocks of Permian, Carboniferous or even younger ages, then this situation produces "allochthonous thrusts". The basic concept here is that the place of formation of the rock is different than where it is found presently. Thus from one basin of formation, they may be thrust over another basin of formation of rocks.

As noted earlier, nappe is a low angle overthrust. It is therefore possible that such a thrust plane may coincide with the bedding plane of the rock series. Such thrusts are given the name of "bedding thrust", which are difficult to recognise, because discontinuity of structures is not created in such faults. Imbricate structure or the shingle-block structure is developed when a number of thrust faults dip in the same direction and are parallel to each other. Schematic diagram of the different varieties of thrusts, is given in Fig. 66 for a quick reference and appraisal.

BASIS OF ATTITUDE OF FAULT PLANE WITH RESPECT TO THAT OF THE ROCKS

This basis is applicable to the sedimentary rocks in the main, and to the layered or banded metamorphic and igneous rocks. If the rocks be homogeneous in composition and devoid of any index horizons, then this basis can not be utilised. Here three types of faults are recognised namely, the dip, the strike, and the oblique or the diagonal faults. In a *dip fault,* the trend of the fault plane is perpendicular to the strike of the beds, or parallel to the dip direction of the rocks. This variety is readily recognisable in the field, since a distinct off setting of the beds is produced (Fig. 67 A,B). In a *strike fault,* the trend of the fault is parallel to the strike of the beds (Fig. 68 A,B). If the fault plane were to coincide with the bedding plane, then such a strike fault is further called as a *bedding fault* (Fig. 69 A,B,C, and Photo 21). Gokhale et. al. (1989) have described such a fault from Basidoni area, Saundatti taluka, Belgaum district, Karnataka state. In an *oblique or a diagonal fault,* the trend of the fault is diagonal or oblique to that of the beds (Figs. 70 A,B).

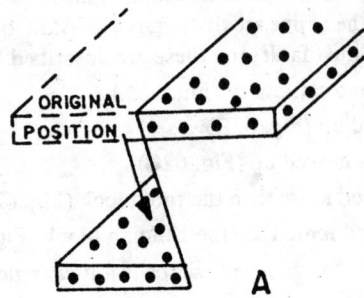

A. foot block alone has moved down, hanging block has remained stationery

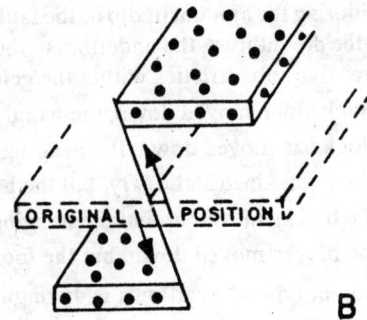

B. hanging block has moved up, foot block has moved down

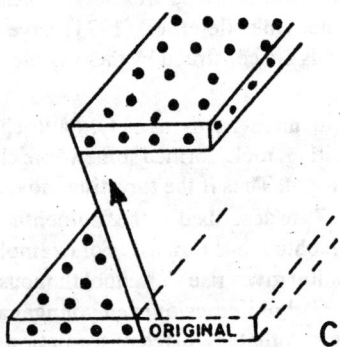

C. hanging block has moved up, foot block has remained stationery

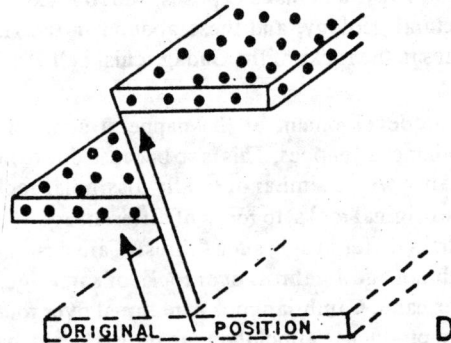

D. both blocks have moved up, but the hanging block has moved more

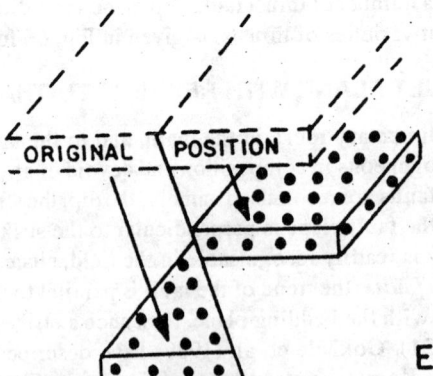

E. fault plane is nearly vertical. The hanging block has moved up and against gravity, and therefore against the direction of dip of the fault plane.

Figs. 62. A,B,C,D,E. Varieties of thrust faults.

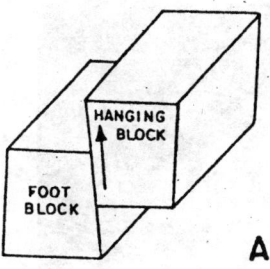

Fig. 63.A Upthrust.

Note that the fault plane is nearly vertical. The hanging block has moved up and against gravity, and therefore against the direction of dip of the fault plane.

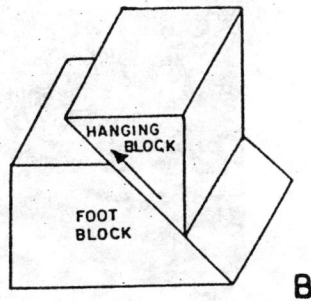

Fig. 63B Upthrust

Note that the fault plane has moderate inclination and the hanging block has moved up and against the gravity. It is therefore against the direction of dip of the fault plane.

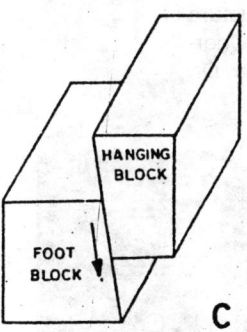

Fig. 63C. Downthrust

Note that the fault plane is nearly vertical and the foot block has moved down and in the direction of gravity. It is therefore down the dip direction of the fault plane.

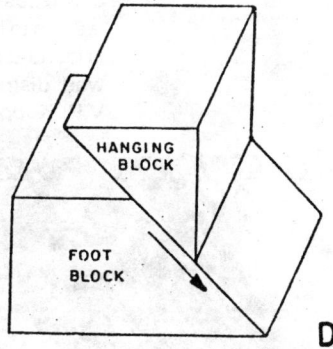

Fig. 63D. Downthrust

Note that the fault plane has a moderate inclination. The foot block has moved down and in the direction of gravity. It is therefore down the dip direction of the fault plane.

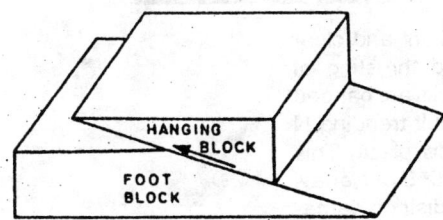

Fig. 64. Overthrust

Note that the fault plane has a low dip, and the hanging block has moved up and against the direction of dip of the fault plane. On comparing it with Figures 63 A, B, it is clearly seen that the hanging block "rides over" the foot block. It is hence called an "overthrust fault".

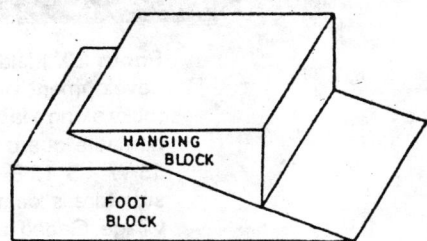

Fig. 65 Underthrust.

Note that the fault plane has a low dip, and the foot block has moved down and in the direction of dip of the fault plane. On comparing Figures 63 C, D with the present one, it is observed that the foot block is "thrust under" the hanging block. It is hence called an "underthrust" fault.

Photo 19. Field photo documenting excellent thrusting and overriding of banded hematite quartzites along a fault plane. This latter plane is seen to be filled by quartz vein. The structure is located in the Tarikoppa hills, Belhatti, Dharwad district, Karnataka state. Courtesy Dr. V.B. Koppad.

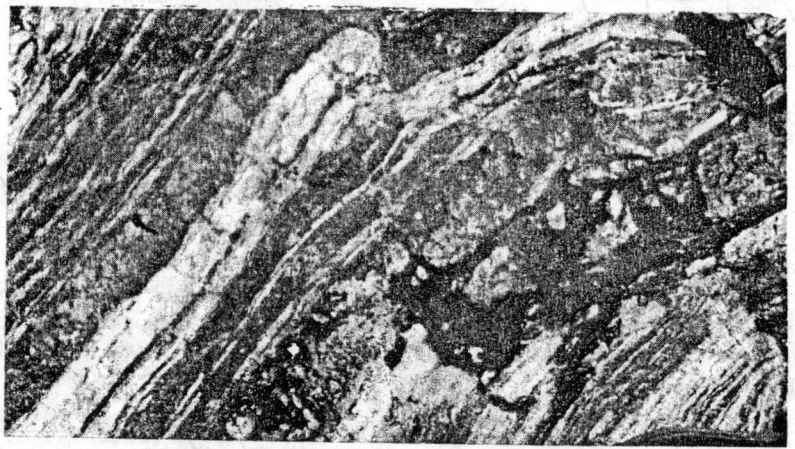

Photo 20. Field photo of excellent and clear development of overriding and thrusting of rocks along a fault plane. The rocks are banded hematite quartzites. A thrust fault trending N 15°W - S 15°E is located at this place. This structure is located 1.5 Km. S 25° E of Nagavi village, Gadag taluka, Dharwad district, Karnataka state. Courtesy Dr. S.C. Puranik.

Upthrust and the downthrust are formed when the fault plane is vertical, or has a very high angle of incination. In the case of upthrust, the hanging block has moved up, while in the downthrust variety, the foot block has moved down (Figs. 63 A to D). When the inclination of the fault is around 45°, then the "thrusting action" is more clearly discernible. In the case of an "overthrust", the hanging block has moved up and thrust over the foot block (Fig, 64, Photos 19, 20). In the case of an "under thrust" the foot block has moved in the direction of the dip of the fault plane, but in so doing, the block gets thrust below the hanging block (Fig. 65).

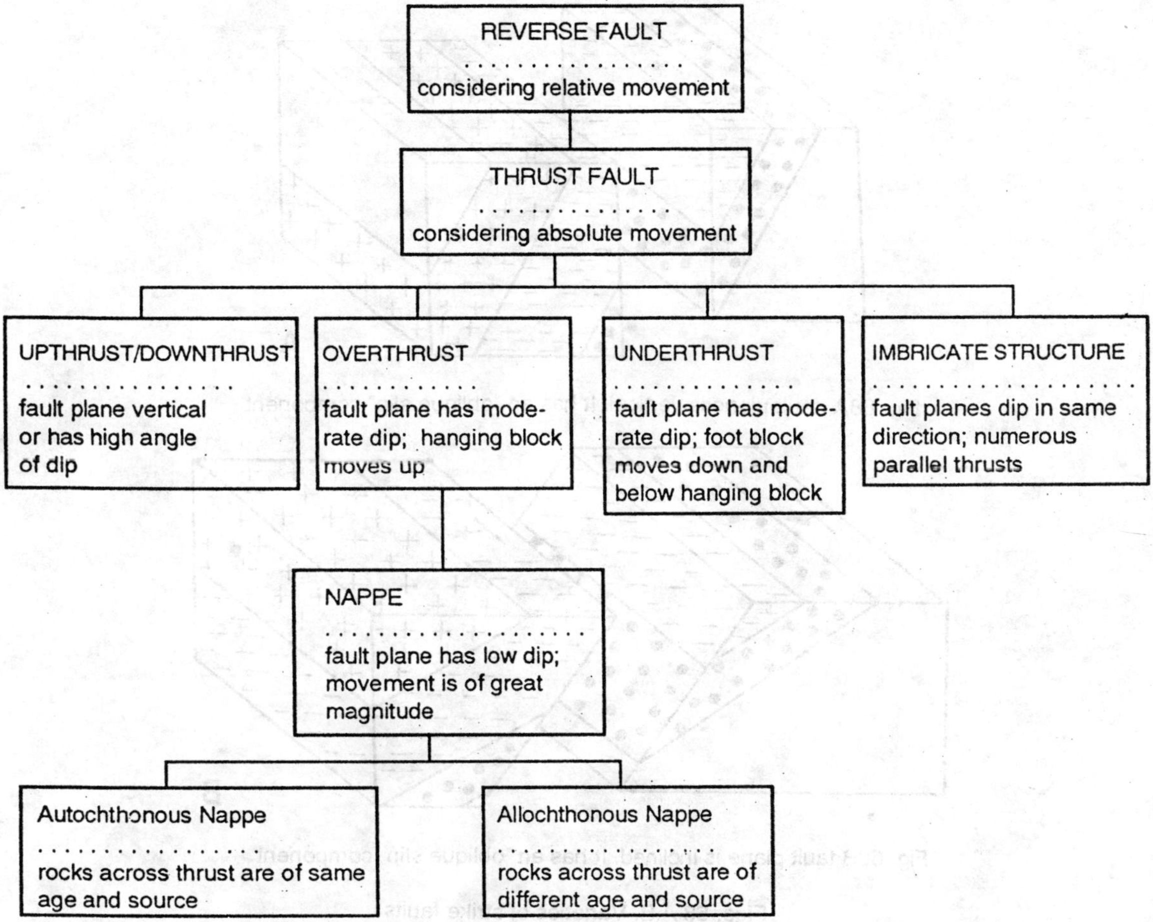

```
┌─────────────────────────────────┐
│         REVERSE FAULT           │
│        · · · · · · · · · ·      │
│  considering relative movement  │
└─────────────────────────────────┘
                 │
┌─────────────────────────────────┐
│         THRUST FAULT            │
│        · · · · · · · · · ·      │
│  considering absolute movement  │
└─────────────────────────────────┘
```

UPTHRUST/DOWNTHRUST	OVERTHRUST	UNDERTHRUST	IMBRICATE STRUCTURE
· · · · · · · · · · · · ·	· · · · · · · · · · ·	· · · · · · · · · · ·	· · · · · · · · · · ·
fault plane vertical or has high angle of dip	fault plane has moderate dip; hanging block moves up	fault plane has moderate dip; foot block moves down and below hanging block	fault planes dip in same direction; numerous parallel thrusts

```
┌─────────────────────────────────┐
│            NAPPE                │
│     · · · · · · · · · · · · ·   │
│  fault plane has low dip;       │
│  movement is of great           │
│  magnitude                      │
└─────────────────────────────────┘
```

Autochthonous Nappe	Allochthonous Nappe
· ·	· · · · · · · · · · · · · ·
rocks across thrust are of same age and source	rocks across thrust are of different age and source

Fig. 66 Schematic Classification of Thrusts

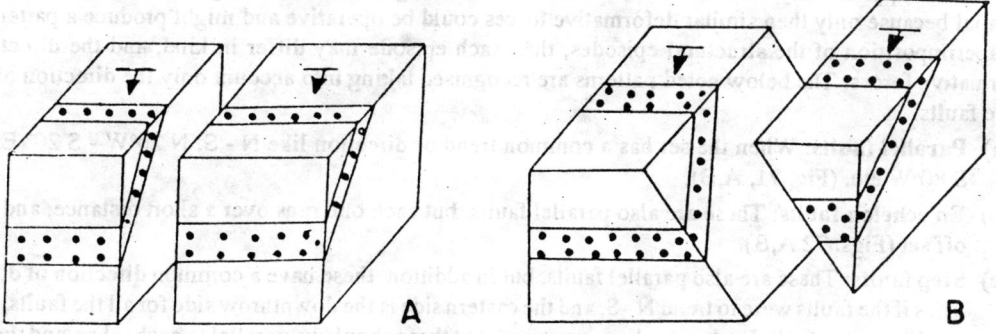

Fig. 67A. fault plane is vertical. It has an "oblique slip" component.

Fig. 67B. fault plane is inclined. It has an "oblique slip" component.

Figs. 67 A,B. Varieties of dip faults.

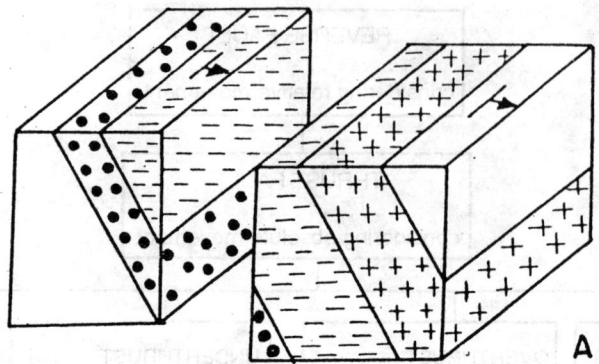

Fig. 68A. fault plane is vertical. It has an "oblique slip" component

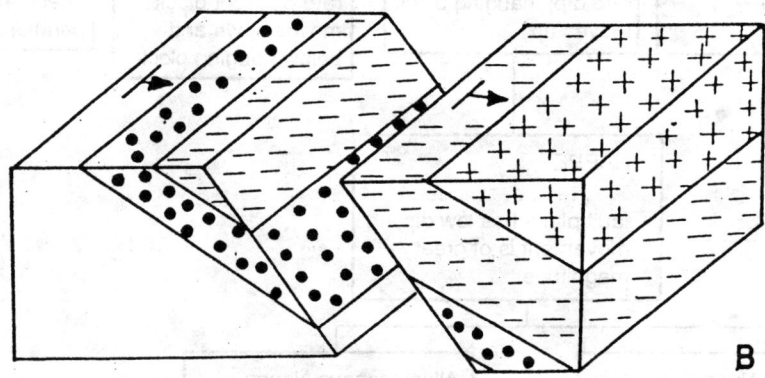

Fig. 68B fault plane is inclined. It has an "oblique slip "component

Figs. 68 A,B. Varieties of strike faults

BASIS OF PATTERN OF FAULTS

This basis is applied only when numerous faults are occurring and these belong to the same age. Same age is essential because only then similar deformative forces could be operative and might produce a pattern. If there be superimposition of the structural episodes, then each episode may differ in kind, and the direction of the deformative forces. The below noted patterns are recognised taking into account only the direction of the trend of the faults.

    (a) **Parallel faults:** When the set has a common trend or direction like N - S, N 20°W - S 20° E, N 80°E - S 80°W etc. (Fig. 71, A,B).

    (b) **En echelon faults:** These are also parallel faults, but each one runs over a short distance, and itself gets off set (Figs. 72 A,B).

    (c) **Step faults:** These are also parallel faults, but in addition, these have a common direction of downthrow. Thus if the faults were to trend N - S, and the eastern side is the downthrow side for all the faults, then these produce a step fault. Faults may have any trend, but these should be parallel to each other, and the direction of downthrow side must be same (Figs. 73 A,B and Photo 22).

    (d) **Radial faults:** If the faults radiate out from a common point, in several directions, then such a pattern is produced (Fig. 74). The individual faults may be vertical or have high angle of dip.

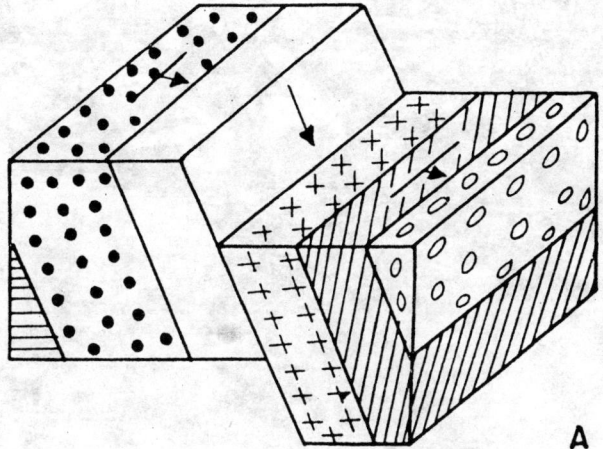

Fig. 69A. it has dip slip component

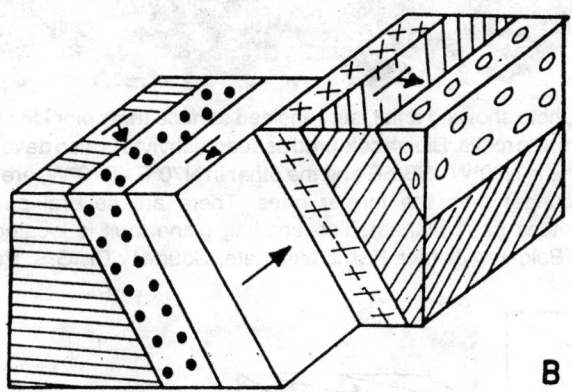

Fig. 69B. it has strike slip component

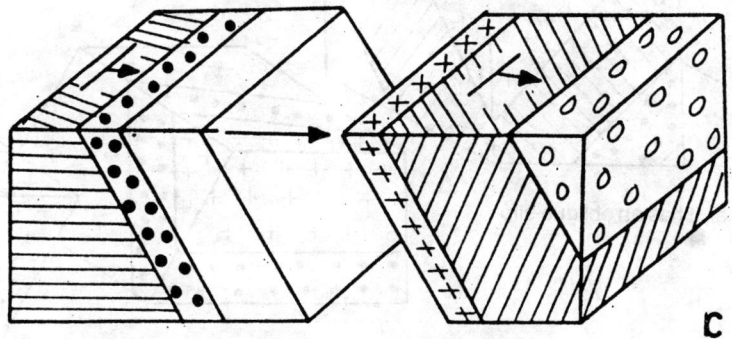

Fig. 69C. it has oblique slip component

Figs. 69 A,B,C. Varieties of strike-cum-bedding plane faults

Photo 21. Field photo showing a flat, slickensided surface that coincides with the bedding plane of quartzarenitic rocks. Bluish coloured pseudotachylyte is also developed. Striations in 2 directions, one in N30°W - S30°E and the other in N70°E - S70°W, are seen. The latter striations are younger than the former ones. There are several parallely disposed slickensided surfaces (fault planes). The bedding plane fault is located 2 km. west of Basidoni village, Belgaum district, Karnataka state. Courtesy Dr. G.S. Pujar.

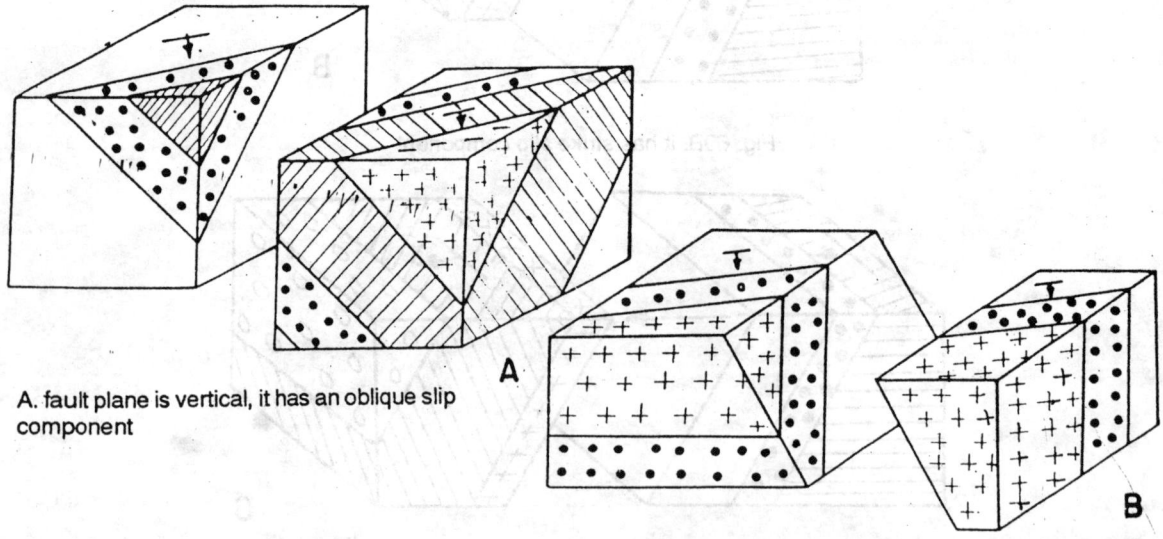

A. fault plane is vertical, it has an oblique slip component

B. fault plane is inclined, it has an oblique slip component

Figs. 70 A,B. Varieties of oblique/diagonal faults.

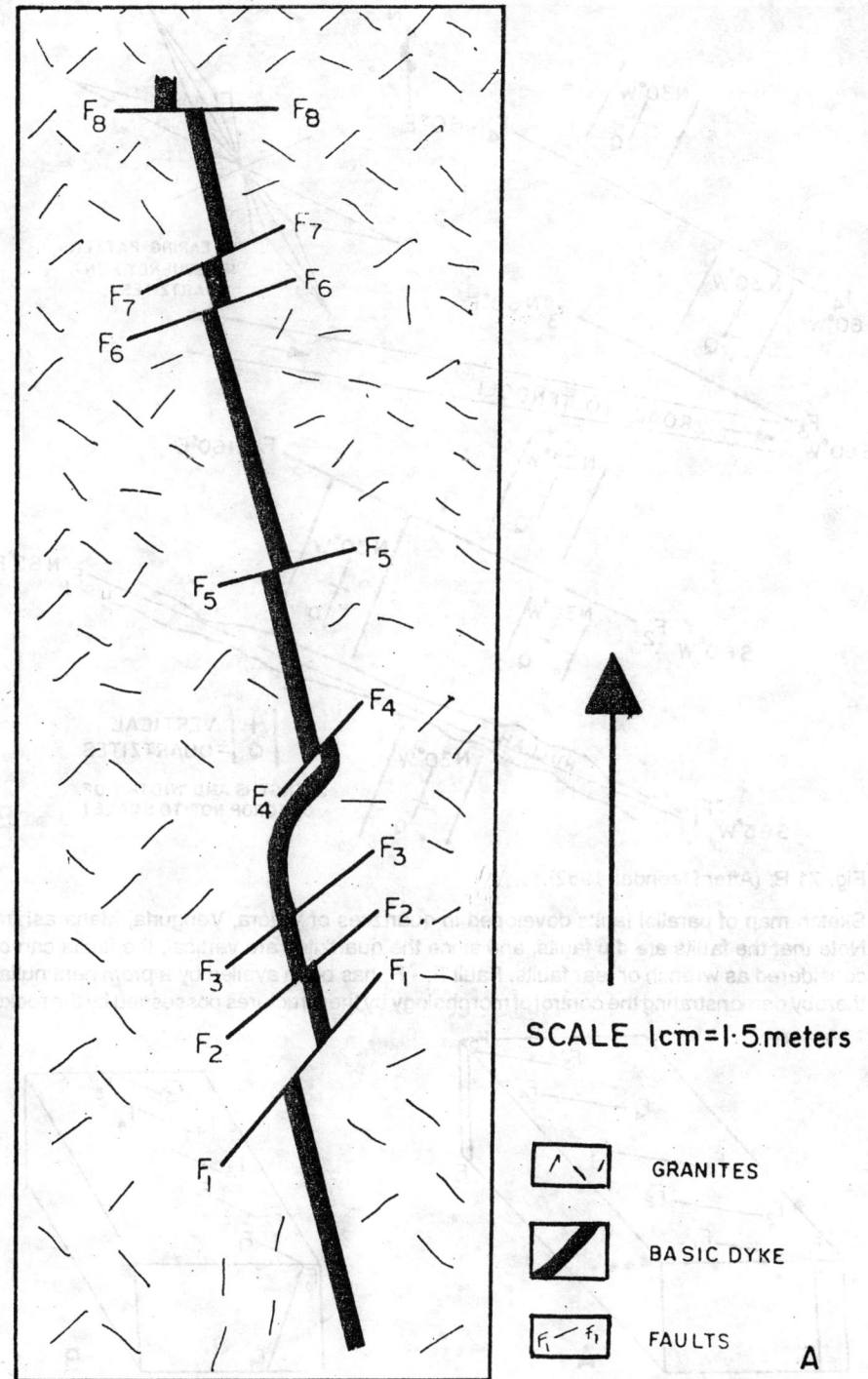

SCALE 1cm = 1·5 meters

∕ ⟍ ∕	GRANITES
∕	BASIC DYKE
F₁ ⟋ F₁	FAULTS

A

Fig. 71 A. (After Muralidharan, 1991).

A pencil thick dyke exposed 4 km. west of Akutothpalli, Mehboobnagar district, A.P. Note that the dyke is faulted at 8 places, and the trends of the several faults are nearly parallel to one another

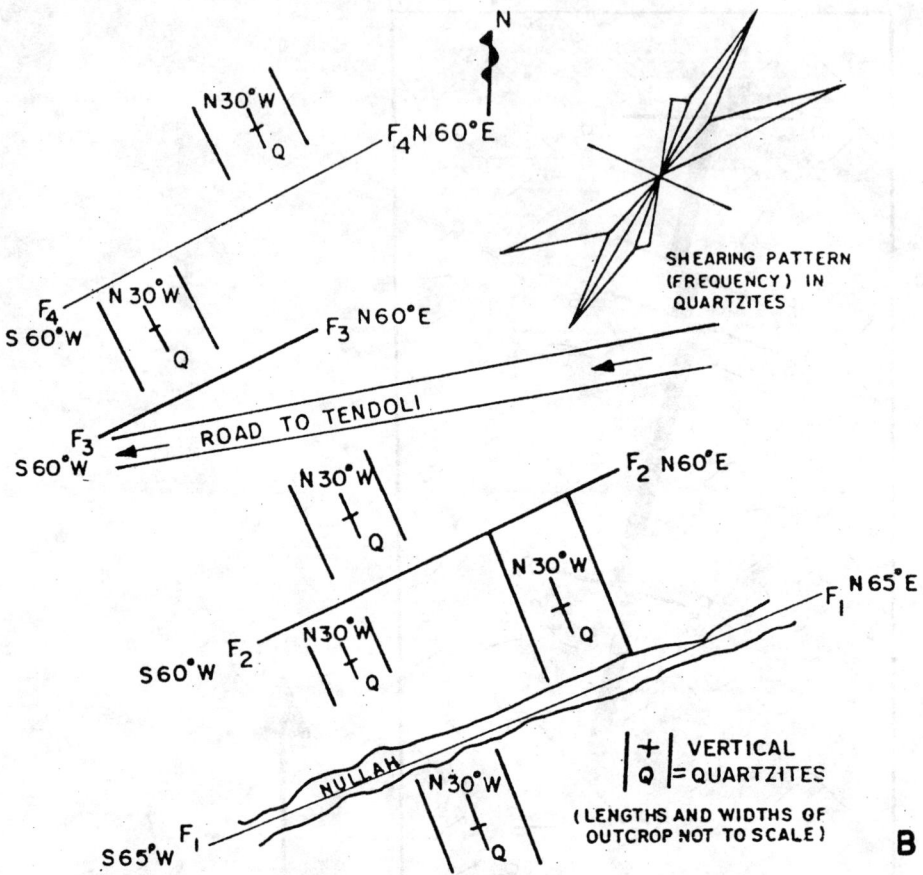

Fig. 71 B. (After Deendar 1982).

Sketch map of parallel faults developed in quartzites of Vetora, Vengurla, Maharashtra. Note that the faults are dip faults, and since the quartzites are vertical, the faults can be considered as wrench or tear faults. Fault F_1 - F_1 has been availed by a prominent nullah thereby demonstrating the control of morphology by the structures possessed by the rocks.

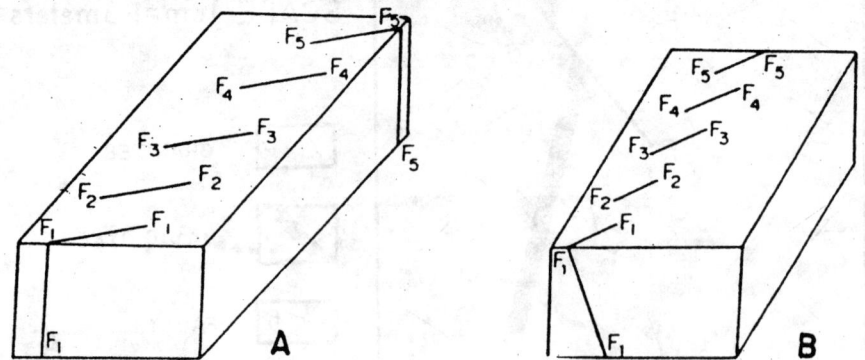

Figs. 72 A,B. En Echelon faults.

Note that the faults are parallel to one another and run for a short distance, and appear to be off set. However it is not a case of the effect of faulting, but it is a pattern. Faults are vertical in figure A, while these are inclined in figure B.

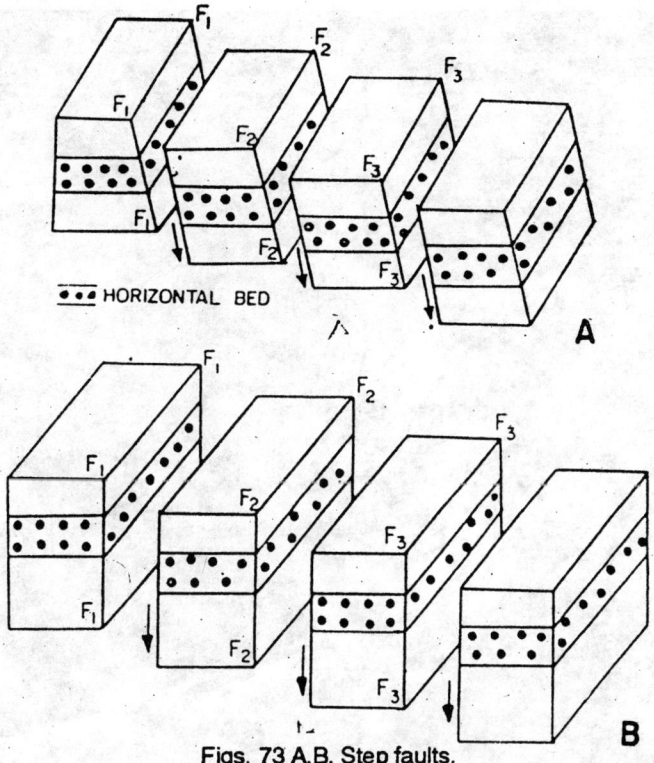

Figs. 73 A,B. Step faults.

Note that the block on the right hand side of each fault plane is thrown down. These faults also could be further classified as oblique/diagonal slip faults. Faults in figure A, are inclined, while those in figure B, are vertical ones.

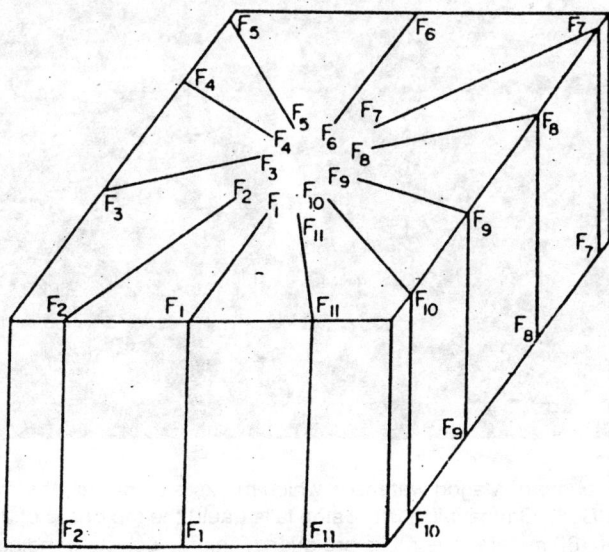

Fig. 74. Radial faults

Note that numerous faults have originated from a common centre. All are vertical faults. Just a few faults can not produce a convincing radial pattern

Photo 22. Field photo of Magod waterfalls which marks a step fault, the river causing it being Shalmala-Bedti-Gangavali. It is a 2 step falls (fault) the top one is of 20 meters, the bottom one is of 180 meters. The rocks are quartz, chlorite schists trending NNW - SSE with an esterly dip of 35 degrees. Courtesy; Technical data by the Karnataka Power Corporation Ltd., photo by Dr. V.C. Chavadi and Shri. A.R. Nalavadi. Location : 17 km. east of Yellapur, Uttara Kannada district, Karnataka state.

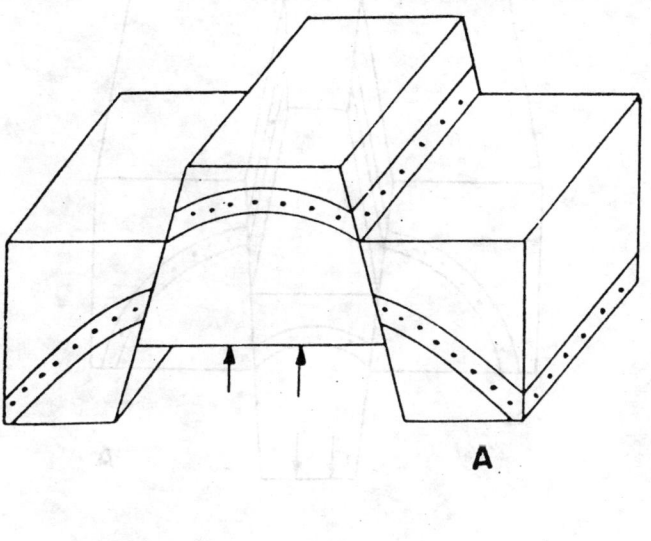

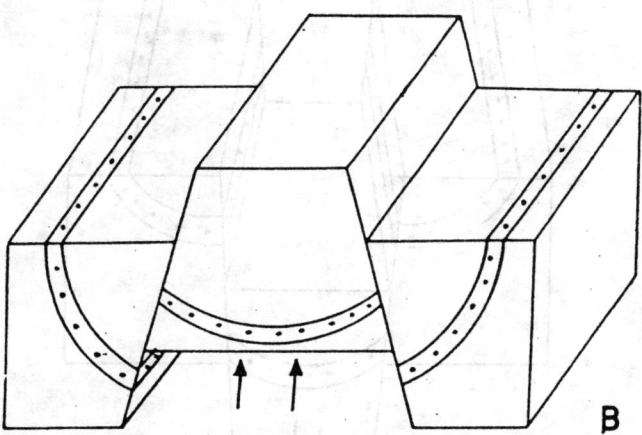

Figs. 76 A,B. Horst/Ridge faults

Displacement upwards inbetween two parallel fault planes produces this structure. Faulted anticline and syncline are shown in figures A and B, respectively.

(e) **Peripheral faults:** This pattern is produced in rising igneous bodies and in the sedimentary basins. The faults are some what confined to the peripheral parts of the body, possibly due to the creation of maximum shearing stresses between the rising body and the resisting country rocks. Good example of such a **fault** is afforded by the southern part of the Kaladgi basin between Halgatti and Lakhmapur villages, a distance of 20 km. wherein there are as many as 36 faults (Fig. 75) which has been reported by Hedge (1984). These are nearly perpendicular to the bedding planes and are therefore dip faults, in general.

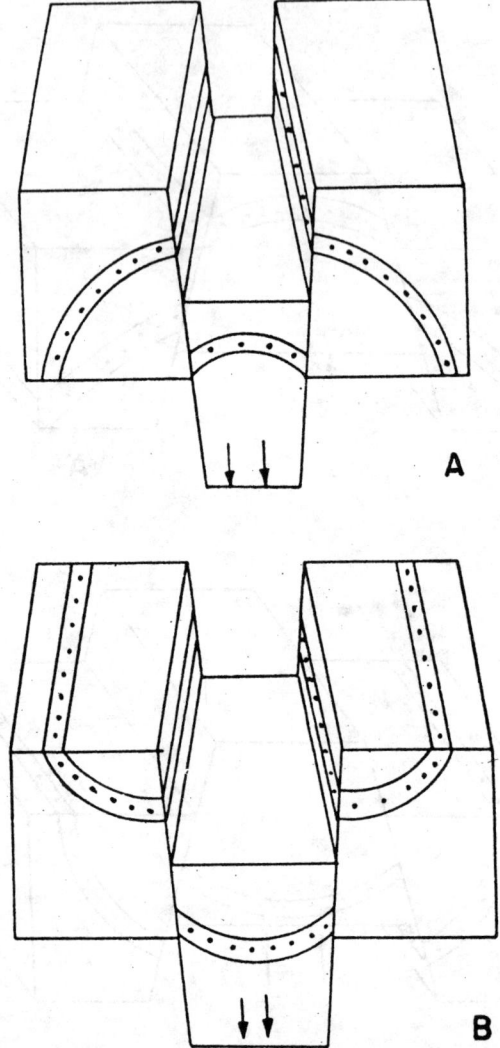

Figs. 77 A,B. Graben/Trough/Trench/ faults.

Displacement downwards inbetween two parallel fault planes produces this structure. Faulted anticline and syncline are shown in figures A and B, respectively.

(f) **Horst/ridge faults:** Generally, two, near parallel faults having a central upthrown block, produce such a fault (Figs. 76 A,B). Morpologically, these give rise to hills or ridges, and are therefore called as ridge faults. If there be more than two parallel faults, then more number of "horsts or ridges" will be produced.

(g) **Graben/trough/trench faults:** Such a fault is produced when two, near parallel faults have a central block thrown down (Figs. 77 A,B). If there be more than two parallel faults, then more grabens will be formed. Morphologically such structures give rise to valleys which are very long. Narmada river is a good example of a valley controlled by a graben structure.

BASIS OF NATURE OF DEFORMATIVE FORCES

This is a truely genetic classification, because it takes into account the causative factors which have produced the faults. However it is not always possible to establish whether compression, tension or shear had produced the faults. Therefore such a distinction is not possible in most of the cases.

RECOGNITION OF FAULTS

Recognition or identification of a structure or structures, is the main pursuit of structural studies. Unless exposures are plentiful, recognition of the structure in the field, becomes difficult. In the case of joints, there is hardly any problem about their establishment, because these could be seen. However recognition of faults faces many difficulties. The main problem is that the displacement due to faulting is required to be proved, if not diectly observable. Displacements become undeterminable, if there are no index beds, horizons or key beds. Thus in homogeneous rocks like the massive intrusives or the extrusive lava flows, faults are not detectable owing to the absence of the index horizon. However there are other effects of the faults, and these could be made use of. The criteria utilised for the three units of rocks namely, the igneous, the metamorphic and the sedimentary, differ from one another. Two major criteria are used, namely.

1. displacement produced by the faulting, and or,

2. other effects of faults.

Whereas in the sedimentary rocks, the first criterion as applied, the second criterion can be applied to any rock especially to the igneous and the metamorphic rocks, as these latter rocks are devoid of planar structures or index horizons. The methods adopted in recognising faults, are shown in the from of a schematic diagram (Fig. 78).

DISPLACEMENT

This being the most characteristic feature of the faults, it forms the first criterion. When the rocks are bedded or layered ones, these feaatures naturally show continuity as long as the rocks extend. Likewise in a sedimentary terraine, rocks occur in a definite order of sedimentation. The normal order is conglomerate, grit, coarse grained sandstone, moderately grained sandstone, fine grained sandstone, shale, limestone and so on. Each rock therefore, has a lateral or a strike extension without any break. By faulting, this may get up set. Metamorphic rocks too show an orderly gradation of occurrence. Low grade, intermediate grade and then high grade. These features are spoken off as the facies of sedimentary and metamorphic rocks, respectively. In the case of igneous rocks such distinct and regular facies are not noticed. Instead, there are intrusive dykes like dolerite, basalt, aplite, pegmatite, quartz veins and so on. These may show break or discontinuity of their extension abruptly. The displacement or the

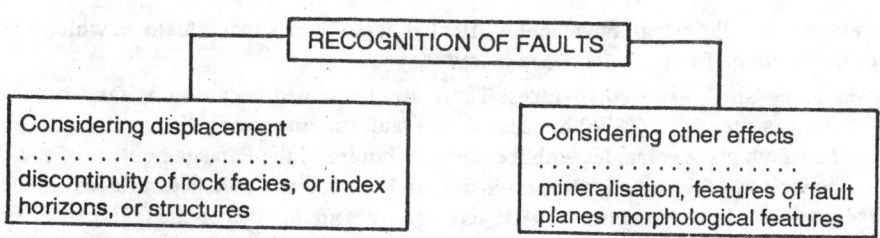

Fig. 78. Schematic representation of bases used in recognising faults in the field.

discontinuity thus noticed is therefore utilised as an evidence of faulting. Greater the displacement or discontinuity, stronger or intense is the faulting. The different types of discontinuities of rocks facie are shown in Figs. 79 A,B, 80 A,B and Photos 16, 17, 18, 19 and 20.

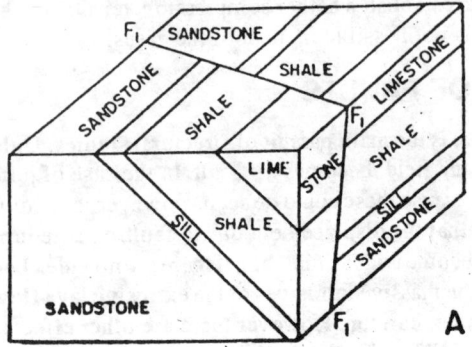

A. Note the discontinuity of rock types across the fault, either on the top surface or on the vertical face. face.

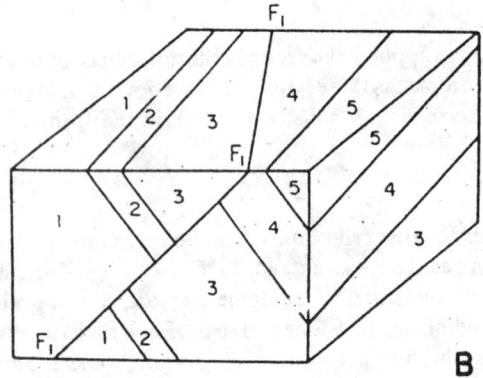

B. Note omission of beds 1 and 2 on the top surface of the block after crossing the fault F_1 - F_1, and proceeding due right. Also discontinuity of beds can be seen across the fault line on the front vertical surface of the block.

Figs. 79 A,B. Effects of faulting on outcrops

EVIDENCES FROM THE FAULT PLANE

The fault plane along which the actual movement is effected, it developes many features which are regarded as the outccome of the faulting. These are described below.

(a) **Slickensided, polished, grooved surfaces:** These are developed because one fault block moves over another, with pressure. Foote (1876) has described a fault running over a distance of 11 Km. in an W - E direction. This fault plane coincides with the northern border of the Parasgad outlier of rocks of Kaladgi group. In the vicinity of Saundatti town, extensive slickensided and polished surfaces can be seen on the sandstones exposed there. Many times striations or grooves are noticed on the fault plane. These indicate the trend of movement along the fault plane. Thus if the striations or grooves be horizontal, the movement likewise was in that direction. The fault thus can be classified as a "strike slip fault". If the striations be trending in NW - SE direction, then the movement also is taken to have occurred in that direction. It has been experienced by the field geologists that on running the fingers on the fault plane, a smooth feeling is obtained only in one direction, which marks the actual direction of movement of the fault block. Thus

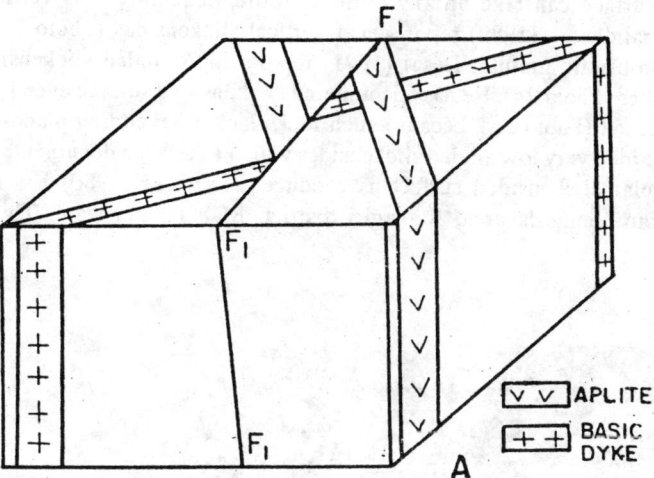

Fig. 80A Faults recognised through intrusions.

Note that the fault F_1 - F_1 shows discontinuity of aplite (VVV) and the basic dyke (+++). However the country rock (blank) which is homogeneous in composition and texture, does not apparently show any discontinuity, though it has also undergone faulting.

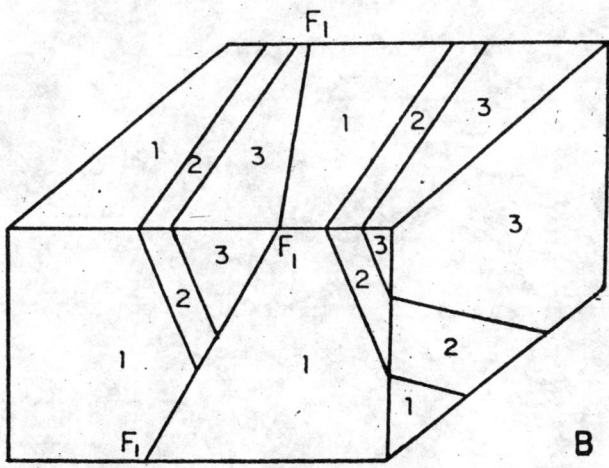

Fig. 80B Fault recognised throgh repetition of beds.

Note that the fault plane F_1 - F_1 has brought about repetition of beds 1, 2 and 3 after crossing it and proceeding to the right hand side of it, on the top surface of the block. On the front vertical face discontinuity of 1, 2 and 3 is observed across the fault line.

on a fault plane a movement from west to east be smooth, then the movement due to faulting also is from west to east. In such an instance, it will be experienced that a movement from east to west is rough to touch or feel. Therefore it will be necessary to find out the direction in which maximum "smooth" feeling is realised. Such a study requires a large exposure of the fault plane, and a considerable patience and experience on the part of the field geologist. Only by this process, faults such as dextral, sinistral, gravity and the thrust, can be detected.

Slickensided surface can take up any attitude. More frequently it is vertical or has high angle of inclination. Muralidharan (1989) reports near-vertical slickensides (Photo 23) in an extremely coarse grained porphyroblastic granite, Desai (1991) reports high angled slickensides produced in banded hematite quartzites (Photo 24). He has encountered slickensided surface even in a conglomerate which is very unusual indeed (Photo 25), because such rocks lack any bedding planes. Desai (1991) and Pujar (1988) have recorded very low angled slickensides which have been documented in Photos 21, 26 and 27. Usually a singular slickensided surface is produced. Deendar (1980) has reported several parallel slickensides from Vengurla area, Ratnagiri district, Maharashtra state. This has been presented in Photo 28.

Photo 23. Field photo of very clearly developed slickensided surface in an extremely coarse grained, pink, porphyroblastic granite. Such a smooth surface is possible only through faulting movements. The slickensided surface has a very high inclination of about 70 degrees. A fault has been located along the surface. The structure is located at about 2.5 km. SW of Akutothpalli village, Mehboobnagar district, Andhra Pradesh. Courtesy Dr. D. Muralidharan.

Photo 24. Field photo of pseudotachylyte developed in banded heamtite quartzites exposed in the Kardikolla region. Lenticular, elongated white portions (quartzite) of the rock have produced a lineation plunging down the dip of the surface of pseudotachylyte. Development of several parallel shears on the surface of the pseudotachylyte can be seen. The handle of the hammer indicates a dip fo 55° due S 35°W for the pseudotachylyte. Locality, Jambunath Temple, Bellary district, Karnataka state. Courtesy Dr. H.D. Desai.

Photo 25. Field photo of development of slickensides and pseudotachylyte in such a rock like a conglomerate which is very unusual indeed, because of the pebbly nature of the rock. The slickensided surface is dipping 70° due S 60°W. A fault located at this place is trending N30°W - S 30°E. This structure is exposed 3.5 km. S 40°W of Vadrahalli village, Sandur taluka, Bellary district, Karnataka state. Courtesy Dr. H.D. Desai.

Photo 26. Field photo of a medium angled (40°) slickensided surface with development of bluish coloured pseudotachylyte, exposed 3.5 km. S 30°E of Vadrahalli village, Sandur taluka, Bellary district, Karnataka state. The rocks are thinnly laminated brittle schists dipping 75° due S 35°W. The slickensided surface dips 40° S 50°E, and marks a medium angled fault. Corutesy Dr. H.D. Desai.

Photo 27. Field photo of extensively developed low angled (10° dip due N 45°W) slickensided surface possessing striations (indicated by ball pen). The striations also plunge in the direction of dip. The slickensided surface marks a dip-slip fault possessing very low dip of 10 degrees. The rocks are schists trending N 55°W - S 55°E and having a dip of 75° due S 35°W. The fault further turns out to be a dip fault. The structure is located in the Dalmia Iron Ore mines, Hospet, Bellary district, Karnataka state. Courtesy Dr. H.D. Desai.

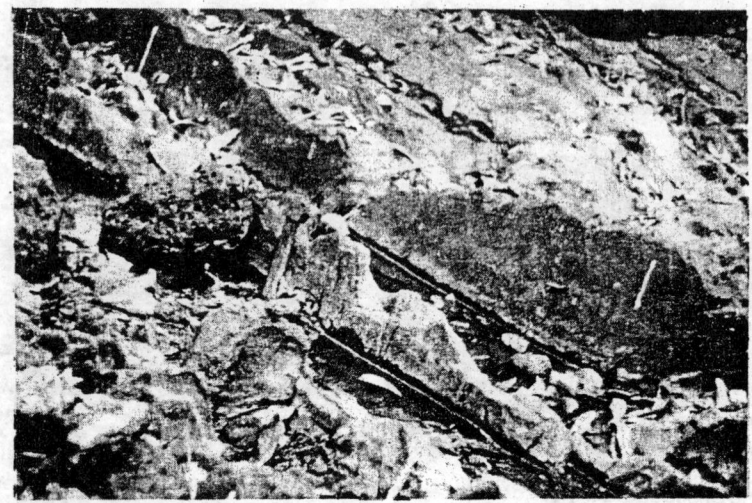

Photo 28. Field photo of slickensided surfaces produced in the hornblende biotite schists of Dhakurwadi, Vengurla, Mahrashtra state. Commonly only one surface is produced, but at this place atleast 3-4 surfaces running parallel to each other are developed, which is unusual indeed. Trend of slickensided surfaces is same as the strike of the rocks. The slickensided surfaces thus mark parallel strike faults. Courtesy Dr. D.I. Deendar.

(b) **Breccia, horses, slices, pseudotachylyte, mylonite, etc.:** Faults belong to the category of "rupture" structures. The deformative forces many times do not culminate into the development of a rupture plane, but into many intersecting fracture planes. This therefore gives rise to a "jumbled up" mass of angular pieces of rocks cemented in still smaller pieces of the same rock. Such a rock is hence regarded as the effect of faulting movement, and the rock is called a "fault breccia" (Photos 29, 30, 31). It is to be noted that in such cases, if the movement along the fault plane is considerable, then it produces a powdery rock which is called a "mylonite". Heat may be generated due to friction and this heat may melt portions of the rock which is later quenched to a glassy rock. Such rocks are called "pseudotachylyte" (Photo 32). Sometimes large blocks of rocks get caught up along faults, which may be or may not be accompanied by breccia. Such blocks are called as horses or slices. Blocks associated with gravity faults are called as "horses", while those associated with the thrust faults are called as "slices". Some of the pulverised fine grained rock resembles a clay, and it is called as "gouge" (Photo 33). This has a similarity to a mylonite, but the mylonite has coherence and is not like clay. Thus breccia, mylonite, gouge, pseudotachylyte etc., indicate the presence of fault.

(c) **Drag:** Many a times, sedimentary beds are pulled in the direction of movement along the fault plane. This is called as dragging of the rocks along the fault plane. This development is more in the cases of rocks which are a bit plastic in nature. However this is not a pre-requisite for the development of a drag.

(d) **Steps:** On the Surface of some fault planes, parallel, small ridges are noticed. These are called as steps. Excellent example is seen in the quartzarenitic rocks exposed 2 km. NE of Gorvankolla village, Saundatti taluka, Belgaum district, Karnataka state, as observed by Pujar (1989). This has been documented in Photo 34.

(e) **Feather joints:** With some of the fault planes, joints are developed at acute angle, and only on one side of the fault plane. This has been noticed in the granitic exposures of Kamlapuram area (Hegde, 1984) and in the granitic rocks of Vengurla area (Deendar, 1982).

Photo 29. Field photo of a cataclasite (breccia) which has given rise to a conspicuous and an isolated hillock. Even from a distance, the presence of very large and quite small sized blocks of rocks (Quartzarenites) can be made out. The exposure is located 500 meters east of Nilagund village, Bijapur district, Karnataka state. The structure is unusual because of the enormous areal extent. Courtesy Dr. G.V. Hegde.

Photo 30. Field photo of an abnormal cataclasite developed 500 meters east of Nilagund village, Bijapur district, Karnataka state. Note a giant sized rock fragment measuring 125 cm in length (indicated by 2 hammers) and 70 cm. in width (indicated by stick). Also note the angularity of rock pieces surrounding the giant sized rock fragment. The rocks are quartzarenites belonging to Kaladgi Group. Courtesy Dr. G.V. Hegde.

Photo 31. Field photo showing development of very wide zone of fractured and ruptured graintes located near Rannanwadi, Kelus, Vengurla, Maharashtra state. The zone is one meter in width. Unless the rocks (granites) were subjected to strong deformative forces (shearing), these cannot get ruptured to the extent as seen in the photo. This is possible only along fault planes. It is pertient to note that not a plane but a fault affected zone is developed. Courtesy Dr. D.I. Deendar.

Photo 32. Field photo of slickensides developed in basaltic flows of Kudgaon, Shrivardhan, Kolaba district, Maharashtra state. Note the developement of bluish coloured pseudotachylyte even in dark coloured basalts. Also note that several parallel slickensided surfaces are developed. A fault trending N 55oE - S 55oW is located at this place. A spring is typically present at the same place. Courtesy Dr. A.H. Kouhsari.

Photo 33. Field photo of mylonite produced in a conglomerate. Hammer is along the zone of mylonite, which is 4-5 meters long and trends N 50°W - S50°E. This marks the site of a fault located 4 km. S 50°W of Vadrahalli village, Sandur taluka, Bellary district, Karnataka state. Courtesy Dr. H.D. Desai.

Photo 34. Field photo documenting excellently developed steps and slickensided surface in pink coloured quartzarenites belonging to Kaladgi Group of rocks. Slickensided surface is situated to the back side of the hammer. Note that distinct difference in level exists inbetween the consecutive steps. The structure is located 2 km. NE of Gorvankolla village, Belgaum district, Karnataka state. Courtesy Dr. G.S. Pujar.

Photo 35. Field photo showing a zone of fault breccia produced due to shearing and fracturing in the granites exposed in the vicinity of Light House hills of Vengurla, Ratnagiri district, Maharashtra state. Note "fracture cleavage like" appearance which is restricted to the central portion of the photo. Trend of brecciation is N 70°W - S 70°E as indicated by the ball pen. Also note felspar laths paralleling the trend of shearing. Courtesy Dr. D.I. Deendar.

(f) **Development of fracture cleavage:** This is especially developed where the deformative forces do not confine to a plane, but effect a wider zone. In rocks like granites, it is very difficult and almost impossible to expect the development of cleavage. But many workers have noticed such a feature. Thus in Vengurla area, this feature has been clearly observed by Deendar (1982) in the granitic rocks (Photo 35).

MORPHOLOICAL FEATURES

Geology being a field science, many aspects are first observable in the field. The morphological or the topographical features are controlled by the structural characters possessed by the rocks. Faults belong to the category of major structures, and as such these could be expected to influence the development of valleys, hills and other geomorphic forms. Some of the morphological features are described below.

(a) **Offset ridges:** Ridges are hillocks that are very long but are remarkably narrow. Naturally such ridges are expected to run for longer distances without any change in their trend. But if these be observed to get shifted and if the shift is not due to the refolding of the axis of the fold (in case the rocks are folded ones), then it may be construed as an evidence of faulting. Care also should be taken to isolate "en echelon" pattern of hillocks which are themselves produced by "en echelon" folds. The rocks comprising ridges are often basic dykes like dolerite or basalt. However rocks with high angle of dip, also can produce "ridges". Thus the offsetting of ridges gains conviction only if the rocks are almost vertical in attitude. Rocks with moderate to low dip have their outcrops shifted, but not necessarily due to faulting.

(b) **Development of windgap:** Ridges are often found to be broken into isolated hillocks like the beads on a string of thread. Erosion naturally is initiated along the fault planes cutting across the ridge and soon it gives rise to a narrow pass called a "windgap". In the Chandvan area of Vengurla, quartzitic sandstone ridges have produced such a wind gap along the fault plane as is described by Deendar (1982). This has been presented in Photo 36. Excellent windgap is noticed to the north of Saundatti town where quartzarenites of Kaladgi group are exposed. The rocks there are very strongly sheared thereby indicating the presence of a fault. There may be other associated features too, but notwithstanding any other cause for the narrow pass, the windgap may stand for a fault (Photo 37).

(c) **Gorge:** This is a narrow, deep valley with very steep sides. These are generally ascribed to fast erosion along some weak zones such as faults. Erosion can not proceed so rapidly and that too vertically down, unless weaker and a favourable plane is present. River Malaprabha which earlier has a meandering course of flow over tens of Km. of distance over the Archaean granites and gneisses, and the Dharwarian schists, it is suddenly found to cut a narrow gorge, that too in hard, resistant quartzarenites. The place is known as Navilutirth, and is situated about 6 km. NW of Saundatti town, Belgaum district, Karnataka state. Foote (1876) has made a mention to the presence of the gorge. It in fact stands for a fault trending nearly N - S. It is not necessry that gorge alone should be developed. At many places streams and valleys may be initiated at such places, and other evidences like breccia, slickensides may also be noticed. In the southeastern part of the Kaladgi basin, Hegde (1984) has reported many such valley-cum-fault planes. The steep valley sides cut by the Narmada river, near Jabalpur is an excellent example of gorge. In south India, the river Kaveri cuts a very deep gorge called as Mekedatu (goat's leap). This gorge is atleast 30 meters deep and is cut in charnockitic rocks. It is about 100 km south of Bangalore, the nearest place being Kanakapura. Narihalla stream has cut a deep gorge across the trend of the hard, competent banded hemitite quartzites and the other rocks (Photo 38). These rocks form the Sandur schist belt. Kouhsari (1986) reports a deep gorge in the vicinity of Harihareshwar temple, 6 km. south of Shrivardhan, Kolaba district, Maharashtra state. It is about 50 meters deep (Photo 39). All these gorges are good examples of the faults.

Photo 36. Field photo of an excellently developed windgap in a ridge of vertical quartzites exposed in the vicinity of Chandvan, Vengurla, Ratnagiri district, Maharashtra state. Based on this and other evidences 2 faults, one trending N 30°E - S 30°W and another trending N 45°E - S 45°W have been located in the windgap. Courtesy Dr. D.I. Deendar.

Photo 37. Field photo of windgap produced in resistant banded hematite quartzites occurring in Kardikolla region. In the windgap, are present other features like slickensides and pseudotachylyte. The rocks are nearly vertical. The gap is more than 500 meters wide. A fault trending N 65°E - S 65°W is therefore located in the windgap. The structure is situated 4 km. S 40°W of Vadrahalli, Sandur taluka, Bellary district, Karnatak state. Courtesy Dr. H.D. Desai.

Photo 38. Field photo showing narrow gorge cut across resistant banded hematite quartzites forming an important part of the Sandur schist belt. A stream called Narihalla has carved the gorge. Slickensided surface can be clearly seen on the left hand wall of the gorge, and also in a block fallen in the foreground of the photo. The gorge thus marks a dip fault, and as the rocks are nearly vertical, it is further classifiable as a wrench or a tear fault. The structure is located near Sandur town, Bellary district, Karnataka state Courtesy Dr. B.C. Prabhakar.

Photo 39. Field photo of an excellently developed gorge in hard, resistant basaltic flows of Harihareshwar Temple area, Shrivardhan, Kolaba district Maharashtra state. The gorge is 1.5 m. wide and 50 m. deep. The exposure is on the Arabian sea coast, and the sea waves have carved a deep gorge along a fault trending N - S.A. prominent shear shown in Photo 14, is associated with the gorge shown in the present Photo 39 Courtesy Dr. A.H. Kouhsari

(d) **Scarp surfaces:** These are steep slopes that are associated with some hillocks. The geomorphologists hold the view that the erosional processes develop gentle to moderate slopes, both for the valleys as well as for the hills. Gorges are therefore regarded to be the outcome of erosion along the fault planes. Likewise the hill slopes normally are low to moderate in inclination. As such extensively developed steep slopes are suspected to represent fault planes. Fault scarps are noticeable at numerous places. The western ghat of India is regarded to represent a fault scarp facing the Arabian Sea. The World famous Gersoppa waterfalls of the Sharavati river, the Gokak water falls of the Ghataprabha river have very steep and deep rock surfaces over which the water leaps down. In the case of the Gersoppa waterfalls, the leap is more than 300 meters. These therefore may stand for fault scraps. Horizontal beds and those possessing low dips also produce scarp surfaces. Care therefore should be taken to rule out such possibilities and then attribute the scarp surfaces to the faulting movements.

(e) **Meandering streams and rivers:** Streams and rivers take to meandering pattern only during the "oldage stage" of the river development. Hence if in a hilly tract, and in the youth stage, a tendency for mendering be noticed for the streams or the rivers, the bends may be controlled by the weak structural elements like faults. River Hiranyakeshi in its initial upper part of flow shows as many as 10 bends over a length of 56 km. One of the bend is atleast due to faulting, because the other features like sudden change in the strike direction of the rocks, are also noticed. This fault is nearly two km long and the meandering takes a N - S trend, whereas the river is flowing in a general west - east direction. This feature has been described by Gokhale et. al. (1987), and it is presented in Fig. 81.

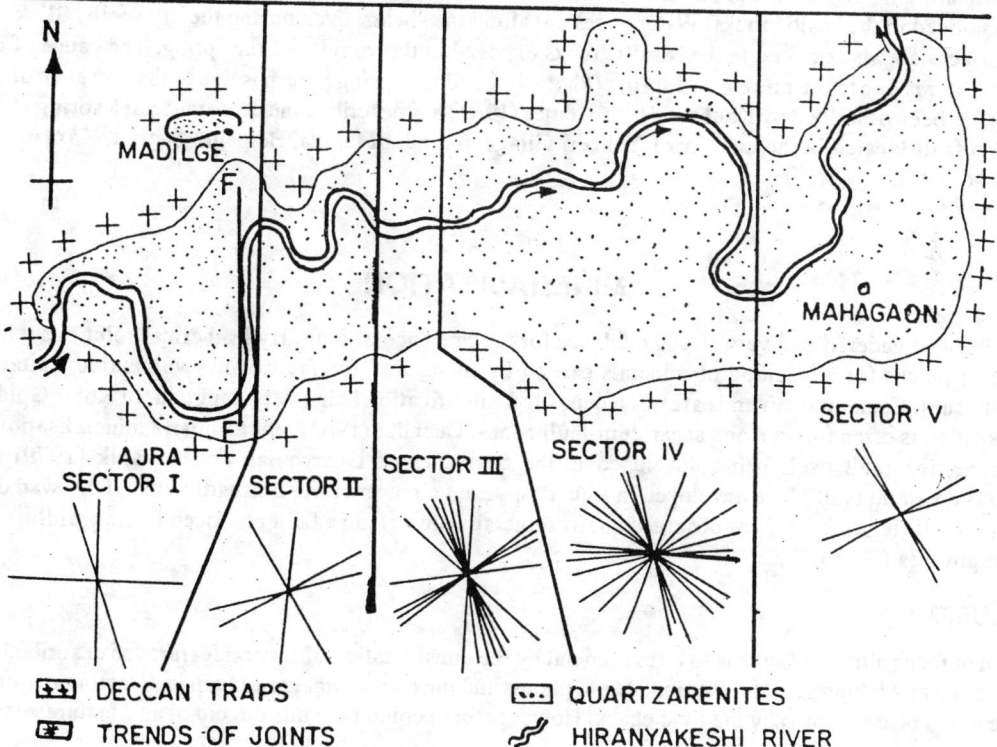

Fig. 81. Morphology (meanders) controlled by structures (faults/shears). Hiranyakeshi river develops several meanders while flowing over sedimentaries. Along one such meander, faulting has been suspected (F - F). The pattern of shear joints shows conspicuous change in the vicinity of the meander (patterns in sector II and III). Thus morphology (meander) may be utilised to detect fault in the rocks alongside with other significant features (shears, joints).

(f) **Straight courses of streams and rivers:** Like the meandering streams and the rivers, even unusually straight and long courses of streams and the rivers, as well as the sea coasts, are deemed to be due to faults, because faults can run over longer distances in a near-straight line. River Narmada has a near E - W course of flow right from its origin at Jabalpur, till it joins the Arabian Sea at Surat. It is further argued that it probably represents a trough or a trench fault, which has been followed by the Narmada river. West coast of India is remarkably straight and this straightness is attributed to a fault along which a part of the Deccan Basalts have been down faulted (Guha et. al. 1970). It certainly cannot be said that the fault extends all along the course of any stream, river or sea coast. A greater part could have been a fault and the initial stream or river took to that course, and that controlled the further course of flow of the river in a near straight direction, even beyond the actual extent of the fault line.

(g) **Spring:** Presence of a spring invariably indicates a fault, because by the displacement of the crustal rocks, permeable and impervious rocks may be brought into juxtaposition of each other. Along such a junction, water is found to ooze out in the form of a spring. In the Parasgad (Saundatti taluka, Belgaum district, Karnataka state) outlier of Kaladgi group of rocks, a narrow valley is carved, in the lower reaches of which, water is observed to flow throughout the year, though in the top portion, there is no flowage of water at all. This valley is thus situated along a fault and has given rise to a spring. The entire Parasgad outlier forms a large isolated hillock, and the region receives scanty rainfall, inspite of which the spring is formed. Foote (1876) reports the presence of a spring in the Parasgad fort area, where quartzarenitic sandstones of Kaladgi group are found to be sheared and structurally much disturbed, suggesting faulting. Along the Shrivardhan - Dighi road at Kudgaon, 3 km. south of Dighi (Ratnagiri district, Maharashtra) shear zone is noticed in the basltic rocks. Water oozes out along the shears, even during the dry spells. Slickensided surfaces are also noticed in the basaltic rocks exposed in the vicinity of the spring. The cause of oozing of water is due to the presence of spring (Photo 40), and the spring in turn is due to the presence of a fault, as has been described by Kouhsari (1986). Pujar (1989) has described the occurrence of a spring at the site of a fault located in the vicinity of Yekkeri village, Saundatti taluka, Belgaum district, Karnataka state (Photo 41).

MINERALISATION

This has been considered as a very strong evidence for the presence of fault. Hydrothermal solutions which are in search of places for deposition of minerals carried by them, find the fault planes vulnerable, suitable and favourable ones. Economic minerals are often deposited, silicification being very common. In Kolar Gold Field, gold deposition is often found along shear-cum-fault zones. Deendar (1982) reports quartz mineralisation along a fault zone, the fault itself being developed in the quartzites of Dharwarian age. Puranik (1979) reports mineralisation along fault plane developed in a quartz-pyrite vein near Harti, Shirhatti taluka, Dharwad district, Karnataka state. Hegde (1984) has observed quartz mineralisation along a fault produced in the granitic rocks of Kamlapuram area (Photo 42).

CONCLUSION

Detection or recognition of faults is to be carried out by the consideration of several features as described above, in conjunction and taking great precaution. Aerial photos and the landsat imageries help in detecting major faults, but these are to be confirmed by the field check. However there could be faults devoid of any features described above.

Photo 40. Field photo of an extremely well developed shear plane-cum-fault in the basaltic flows of Dighi area, Shrivardhan, Kolaba district, Maharashtra state. Trend of shear is N 65°E - S 65°W. A perenial spring is associated with this shear, and in July 1984, a landslide was caused probably by the presence of the said fault. In the lower reaches (foreground) typical slickensides and pseudotachylyte (Photo 32) are developed confirming the existing of fault noted above. Courtesy Dr. A.H. Kouhsari.

Photo 41. Field photo of a well developed near vertical slickensided surface which is 3 meters in height. Rocks are also shattered at this spot. Water is found to ooze out even in the dry months which is evidenced by the collection of water in the form of a pond. The spring and the slickensided surface together confirm the existence of a fault at this place. The rocks are quartzarenites. Locality 1.5 km. S 45° E of Magnur village, Belgaum district, Karnataka state. Courtesy Dr. G.S. Pujar.

Photo 42. Field photo of an excellently developed slickensided surface and mineralisation along it in medium grained gray granitic rocks. The structure is exposed 2.5 km. NE of Dharam sagar village, Bellary district, Karnataka state. A fault trending N 45°E - S 45°W is located along the mineralised zone. Courtesy Dr. V.N. Hegde.

Photo 43. Field photo of a complex pattern of faulting produced by aplitic and pegmatitic veins traversing gray coloured migmatitic granites composing the Mundargi hillock, Dharwad district, Karnataka state. Two pegmatites trending in different directions have been faulted twice. One fault has affected the thinner pegmatite (indicated by the ball pens), while the other fault having a different trend has diplaced the thicker pegmatite. This latter fault is however concealed by the gray coloured thick aplitic vein. Photo by the author.

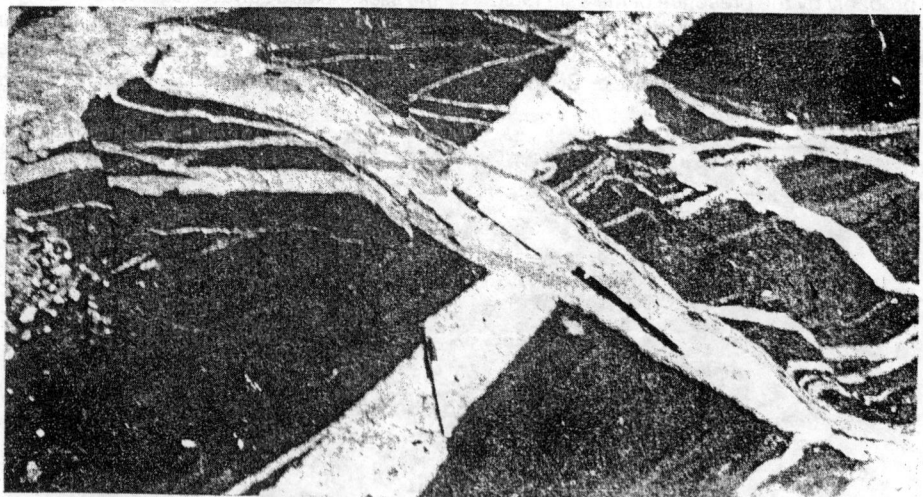

Photo 44. Field photo showing 2 pegmatites traversing migmatitic granites (Mundargi hillock, Dharwad distirict, Karnataka state), affected by complicated pattern of faulting. Pegmatite vein (following red pencil) has concealed a fault that has displaced another pegmatite. The 3 ball pens are along 3 parallel faults. The faults in the foreground and in the background, are not concealed. There is yet another fault located on the left hand side of the photo which has displaced yet another pegmatite vein. The trends of the several faults produce a criss-cross pattern. Photo by the author.

COMPLICATED FAULTING AND ESTABLISHMENT OF RELATIVE AGES OF DISPLACEMENT

When the faults are not parallel, these might either intersect each other or might get themselves displaced. It is also likely that some of them get concealed by the acidic or the basic intrusions. Even curved faults displaced by a straight one might be encountered. Such instances many times are helpful in establishing the relative age of faulting movements. A few actual instances are described below.

FAULTED APLITIC AND PEGMATITIC INTRUSIONS OF MUNDARGI AREA

Massive migmatitic granites composing the massive, solitary hillock located in the immediate western vicinity of Mundargi town, Gadag taluka, Dharwad district, Karnataka state (Survery of India topopsheet No 48 M/12) are observed to be traversed by numerous intrusions of aplites and pegmatits. In Photo 43 two pegmatites trending in different directions have been faulted twice. One fault has affected the thinner pegmatite (indicated by the ball pens), while the other fault having a different trend, has displaced the gray coloured thick aplitic vein. In Photo 44, two pegmatites traversing the migmatite, display a complicated pattern of displacement. Trends of the faults have been indicated by the ball pens, and a red pencil. Pegmatite vein (following red pencil) has concealed a fault that has displaced another pegmatite. The three blue ball pens are along 3 parallel faults. The faults in the foreground and in the background, are not concealed. There is yet another fault located in the left hand side of the photo which has displaced yet another thin pegmatite vein. The trends of the several faults produce a criss-cross pettern.

In Photo 45 a sort of curved-cum-intersecting-cum-concealed fault is noticeable. Aplite vein shows a curved shape, but it displaces another thin aplitic vein (more top part of the photo). In the central part of the photo, the bigger aplite vein displaces a pegmatite. Thus the main, bigger aplite has concealed a curved fault. Towards the more left hand side of the photo, several smaller faults displacing thin veins of pegmatite, are noticeable. The pattern of the several faults is decidedly complicated one. Gothe (1973) has reported an instance of a concealed

Photo 45. Field photo documenting a sort of a curved-cum-intersecting-cum concealed fault in the migmatitic granites composing Mundargi hillock, Dharwad district. Karnataka state. Curved aplite vein displaces a thin aplite vein (more top portion of the photo). In the central part, bigger aplite vein displaces a pegmatite. Thus the main bigger vein of aplite has concealed a curved fault. Towards the more left hand side of the photo, several smaller faults displace thin veins of pegmatite. The pattern of faulting in totality is complicated one. Photo by the author.

fault. A big aplitic vein has concealed the fault that had displaced veins of pegmatite (Photo 46). Towards the right hand corner of the photo, another concealed fault is observable that has displaced a thin vein of aplite.

In Fig. 82 yet another instance of a complicated fault is documented. Two quartz veins have been faulted by 3 faults. F_1 - F_1 and F_2 - F_2 are parallel faults, but these intersect fault F_3 - F_3 and displace it. The disposition of the displaced veins clearly helps to determine the relative ages of the faults. F_3 - F_3 is older than the faults F_1 - F_1 and F_2 - F_2. Fault F_3 - F_3 has displaced a small vein towards its eastern end. Gokhale et. al. (1989) have reported 3 intersecting basic dykes, one of which is faulted and dragged at the intersection of the 2 dykes. The fault is again concealed by the dyke, existence of which is proved by the dragging and displacement of the dyke.

Examples described above clearly demonstrate the existence of complicated types of faults. It is also pertinent to note that not only the acidic and basic intrusions indicate the existence of faults, but it also means that the country rocks are also faulted and displaced.

Photo 46. Field photo showing a thick vein of pegmatite concealing a fault that has displaced a gray, smaller pegmatite and a thinner aplite vein. Towards the right hand corner of the photo, another concealed fault is noticeable that has displaced a thin vein of aplite. This structure is from the migmatitic granites of Mundargi hillock, Dharwad district, Karnataka state. Courtesy Dr. N.N. Gothe.

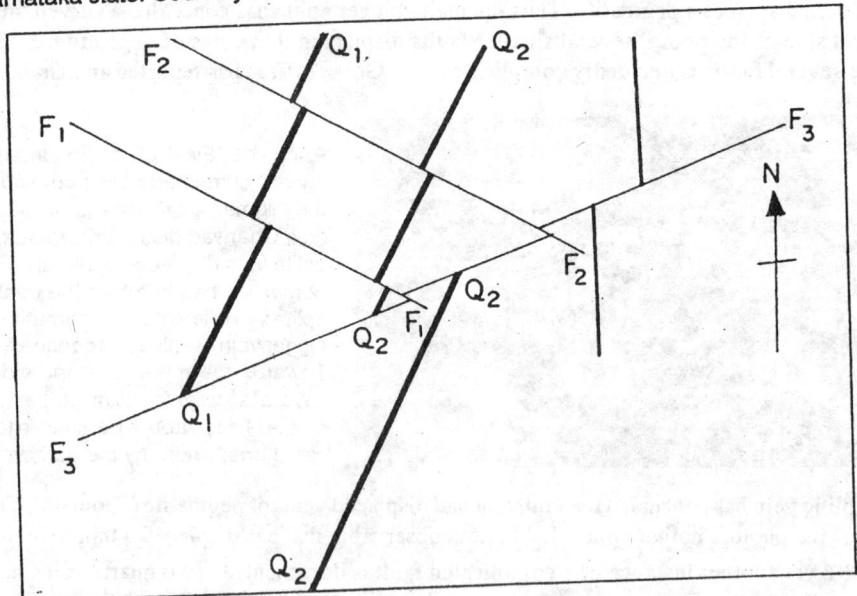

Fig. 82. Relative age of faults (after Muralidharan 1991).

Quartz veins exposed in the vicinity of Aurepalle, Mehboobnagar district, Hyderabad, A.P. register displacements clearly indicating difference in age. Three faults F_1 - F_1, F_2 - F_2 and F_3 - F_3 differ in age. Faults F_1 - F_1 and F_2 - F_2 are of same age and are parallel to each other. Fault F_3 - F_3 itself is off set by the other two faults. Therefore F_3 - F_3 is the oldest fault. In the absence of displaced quartz veins Q_1, Q_2 and very thin vein towards the eastern part of fault F_3 - F_3, relative ages of faults could not have been determinable.

EXAMPLES OF DIFFERENT KINDS OF FAULTS FROM INDIA

So far a detailed technical account of the faults has been given. It is desired in this section to describe some faults actually encountered in the different parts of India.

1. The Dauki fault is a dextral transcurrent fault along which the Assam Plateau has been moved to the east over a distance of 250 km. (Krishnan, 1968, p. 52).

2. Along the western margin of Vindhyan basin is a large reversed fault (the Great Boundary Fault of Rajasthan) which brings the Arvallis on the western side, against the Upper Vindhyan Bhander sandstones on the eastern side, and which has been traced over a length of 800 km. (Krishnan 1968 p. 54).

3. The chief coal fields of India owe their preservation to the block faulting (Krishnan, 1968, p. 56)

4. In the Extra Peninsular region of India, several thrust faults have been recognised. *Murree thrust* is located in Kashmir and it is an *autochthonous thrust.* In the Simla-Garhwal region, *Krol, Jutogh, Giri, and Chail thrusts* are recognised. According to Heim and Gansser (1939), there are atleast 4 superimposed thrust sheets in the Lesser Himalayas of Garhwal region. The Main Boundary Fault (MBF) separates Siwaliks from the earlier Tertiary and older rocks. There are imbricate thrusts along the Himalayan border. Mt. Everest and Kanchanganga peaks are located in the *Central Himalayan thrust.* There are thrusts in the *Tethys Himalayan zone, Flysch and exotic zone and Counter thrust of Darchen zone* (Krishnan, 1968, p. 59).

5. The edge of the continental shelf of the western coast is remarkably straight as it as a fault line formed in the Late Pliocene (Krishnan, 1968, p. 64).

6. The Cambay area lies in a trough fault running N - S in which the Deccan Traps have been dropped down to a depth of about 2000 meters (Krishnan, 1968 p. 64).

7. Major E - W trending faults have been found bounding Kathiawar on the north and south, the latter one by geophysical methods. There is also a prominent fault seen for a distance of 40 miles or more along the northern margin of Rann (Krishnan, 1968 p. 66).

8. E - W trending strike fault runs over a distance of more than 10 km. affecting the rocks of Kaladgi formations (Foote 1876). This is located to the immediate north of Saundatti town, Belgaum district, Karnataka state.

9. Three major faults have been reported by Pujar (1989), from quartzarenites exposed between Yekkeri, Hulikatti and Manikatti villages, Saundatti taluka, Belgaum district, Karnataka state. One fault is 4.5 km long and trends in a N 45°W - S 45°E direction, the second one has a length of 2.5 km. and a N 15°E S 15°W trend. The third one is the longest, it being 6 km. in length. It has a W - E trend and is further classified as a sinistral fault. The first two faults are oblique faults, while the third one is a near strike fault.

10. A N 50°W - S 50°E trending thrust fault occurs on the eastern border of the Bagewadi conglomerate, this rock constituting a member of the Gadag schist belt. This fault is about 10 to 15 km. long (Gokhale et. al., 1971).

11. A N 65°E - S 65°W trending dip rotational fault is developed in the Kaladgi formations composing the Nargund hill, Dharwad district, Karnatak state. The central part of the hill has developed a sag which can be seen even from a far off distance (Puranik et. al. 1982).

12. The Navilutirth gorge carved by the Malaprabha river is a dip fault trending nearly N - S which has been availed by the said river. The gorge is located in the northwestern vicinity of the Saundatti town, Belgaum district, Karnataka state.

13. Mekedatu (Goat's leap) which is 1.6 km. long and 40 meters deep, is cut by the Kaveri river, the rocks being Closepet granites. Trend of the gorge is nearly N - S. Notwithstanding any other reason, the gorge marks a fault plane. The gorge-cum fault is located 100 km. south of Bangalore, Karnataka state.

14. Maged waterfalls of Bedti river is a two step falls, the upper one of 20 meters, and the lower one of 180 meters. The country rocks are metagraywackes. The two step falls represents a step fault and is located about 17 km. east of Yellapur, North Kanara district, Karnataka state (Photo 22).

15. Kunchikallu Abbi waterfalls of Varahi river has 9 cascades over a drop of 400 meters. This in all probabilities stands for a "multiple step fault", developed in the biotite granite gneiss. The cascading-cum-multiple step fault is located 5 km. from Hulikal, Hosnagar taluka, Shimoga district, Karnataka state.

16. An unusual strike rotational fault is exposed in the coarse grained quartzarenitic rocks occurring in the southern vicinity of Jamkhandi town, Bijapur district, Karnataka state. Rocks in between 2 parallel, E-W trending faults are rotated as a result of which the original dip of 21° is increased to 51° (Patil, 1989). This has been documented in Photos 47, 48, 49 and Fig. 83.

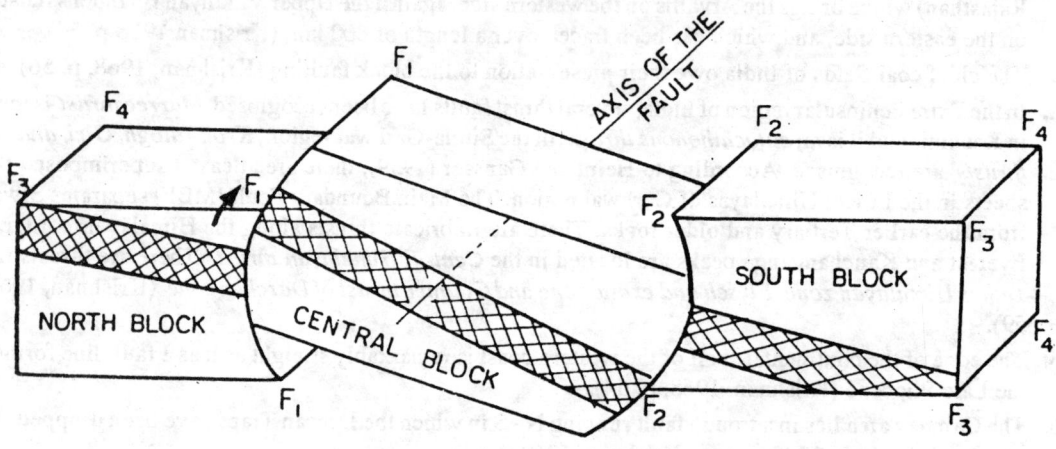

Fig. 83. Axial rotational fault.
Mechanism of rotational movement.

17. The Narihalla stream having a near E-W flow, it has cut a deep gorge (Photo 38) in the hard and resistant banded hematite quartzites and the associated rocks trending in an NW-SE direction. The valley sides display unmistakable slickensided surfaces. These features clearly demonstrate that the Narihalla stream clearly flows along an E-W trending fault (Sandur taluka, Bellary distict, Karnataka state).

18. The Tungabhadra river cuts across NW-SE trending rocks of the Gadag schist belt, near Shingatalur, Shirhatti taluka, Dharwad district, Karnataka state. The rocks are strong and resistant banded hematite quartzites and the associated schistose rocks possessing high angle dip. The river in the southern vicinity of Shingatalur has a near perpendicular direction of flow with respect to that of the trend of the rocks. The river must be flowing along a NE-SW trending dip fault. This incidentally is a wrench or a tear fault.

19. Krishnan (1968) has described several rivers flowing along faults. The course of *Jhelum* along the eastern side of the plateau is apparently controlled by a fault (p. 53). The *Brahmaputra* flows along a N-S fault bordering the western foot of the Garo hills (p. 58). The *Narmada* and the *Tapi* valleys are in the graben formed due to the E-W trending faults bounding Kathiwar on the north and south, and a prominent fault seen for a distance of 40 miles or more along the northern margin of Rann (p. 66). South of *Damodar* river, mass of anorthosite extends in an east-west direction for about 32 km. with a maximum width of 10 km. In the eastern part its northern boundary is faulted against the Gondwanas, and is more or less marked by Damodar river (p. 133).

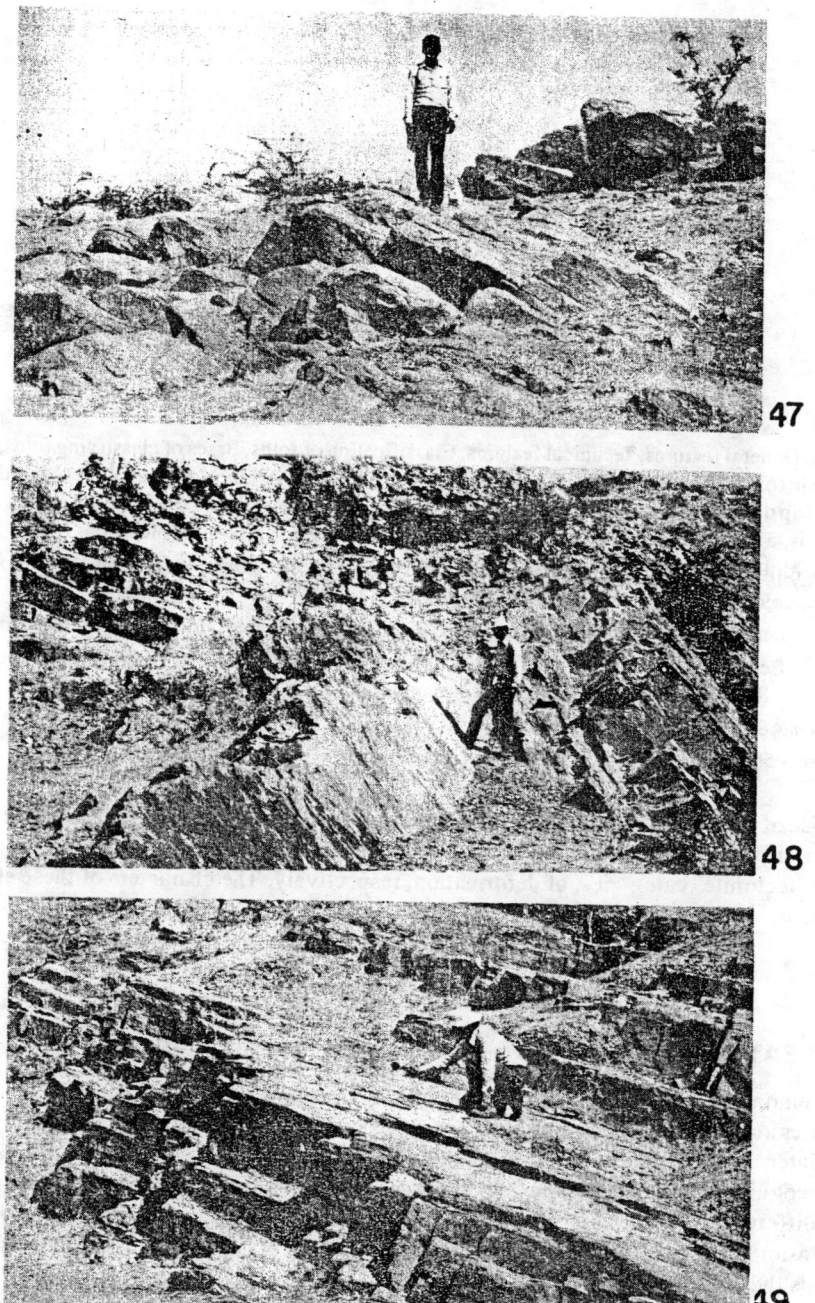

Photos 47, 48, 49. Field photos clearly documenting rotational fault developed in the coarse grained quartzarenites of Jamkhandi town, Bijapur district, Karnataka state. In Photos 47 and 49, the rocks possess a low dip of 22° due south, while in Photo 48, the dip is as high as 51 degrees. At the place of change of dip, shearing and slickensides are clearly produced. Thus the central block (Photo 48) is rotated in between the northern and the southern parts resulting in rotational fault. Curved fault planes are located in the near vicinity of the outcrop (see Photo 15). Photos by the author.

3

PLASTIC STRUCTURES

Folds, General features, Technical features, Classification of folds, Bases of classifying folds, Definition of basic forms, Forms according to non-genetic consideration, Attitude of axial plane, Attitude of axis of fold, Dip direction of limbs, Dip amount of limbs. Shape of folds, Thickness of limbs, Trend of axis, Depth as basis, Size as basis, Combination of several parameters, Examples of folds from India.

Recognition of folds, Physiographic studies, Structural studies, Outcrop patterns, Direct observation, Photogeological and landsat imageries, Drilling and mining.

Mechanism of folding, Flexure folds, Flow folds, Shear folds, Folds due to vertical movement.

Kinds of cleavage, schistocity and foliation, Schematic classification of cleavage, Lineation, Mineral oriented, Rock oriented, Lineation not due to minerals.

Figures - 84 to 139

Photos - 50 to 63

These comprise of folds in the main, and the development of foliation, schistocity, lineation in the schistose rocks, and in the gneissic rocks. The folds belong to the "non-tectonite", while the foliation, lineation, and the schistocity, belong to the "tectonite" categories, of deformation, respectively. The characters of these structures have been described below.

FOLDS

GENERAL FEATURES

Folds are common in the bedded formations like the sedimentary and the metamorphosed sedimentary rocks. These structures are noted to a lesser degree in the igneous rocks. Fold is defined as an undulation in the rock. The style and the intensity of folding differs from place to place, and from rock to rock. Accordingly different varieties of folds are recognised. One and the same rock like a sandstone or a shale may show one kind of fold at one place, and a totally different style of folding at another place. Unlike the faults and the joints, a fold needs to be studied in three dimensions and therefore the behaviour of this structure at depth assumes significance and importance, both as regards the style and its continuation at depth.

TECHNICAL FEATURES OF FOLDS

A fold being an undulation or a bend, it is divided into two parts called as the "limbs". Due to folding, the original flat or horizontal attitude of the rock or a bed, is turned into a convex or a concave form. The radius of concavity or the convexity is called the height of the fold. The point of maximum convexity or the concavity is called the axis of the fold, while the point of maximum elevation is called the "crest", and the point of minimum level is called the "trough" of the fold. These features are shown in Fig. 84.

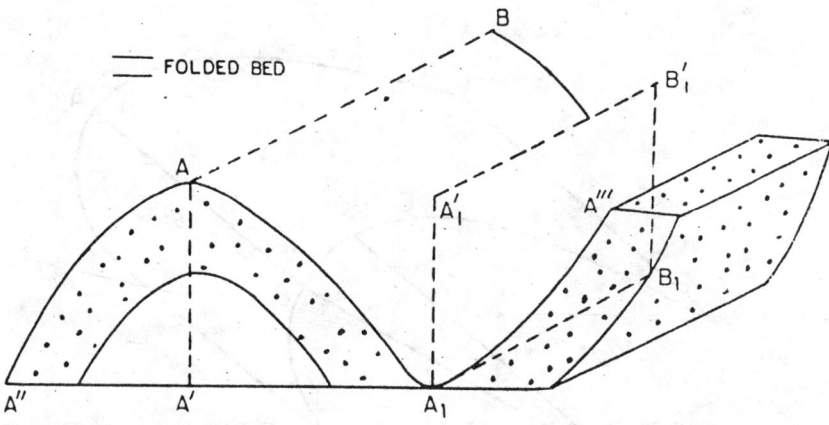

Fig. 84 Parts of a fold

$A''AA_1$ = anticline

AA_1A''' = syncline

AB = trace of axial plane of anticline

A_1B_1 = trace of axial plane of syncline

$A'AB$ = axial plane of anticline

$A_1A'_1B'_1B_1$ = axial plane of syncline

A = axis and crest of anticline

A_1 = axis of trough of syncline

Usually the axis and the crest are one and the same (as in a vertical fold), but the axis could be at a level lower than the crest (as in an inclined fold or a recumbent fold). This aspect has been shown in Figs. 85 A,B. A syncline has a trough, it being the lowest part of the fold. In this case too the trough and the axis coincide, but if the fold be overturned, the axis is at a level higher than the trough (Figs. 86 A,B)

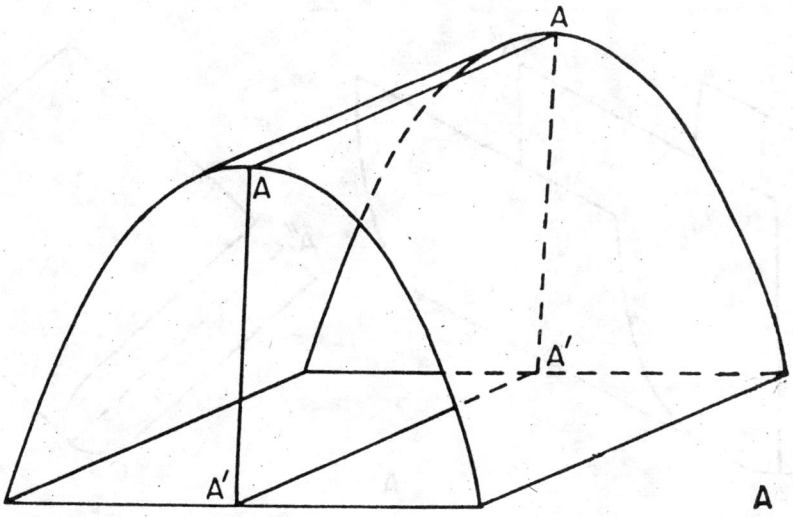

Fig. 85. A Vertical anticline

$AAA'A'$ = axial plane

AA' = height of fold

A = axis and crest of fold. This is realised only in vertical fold

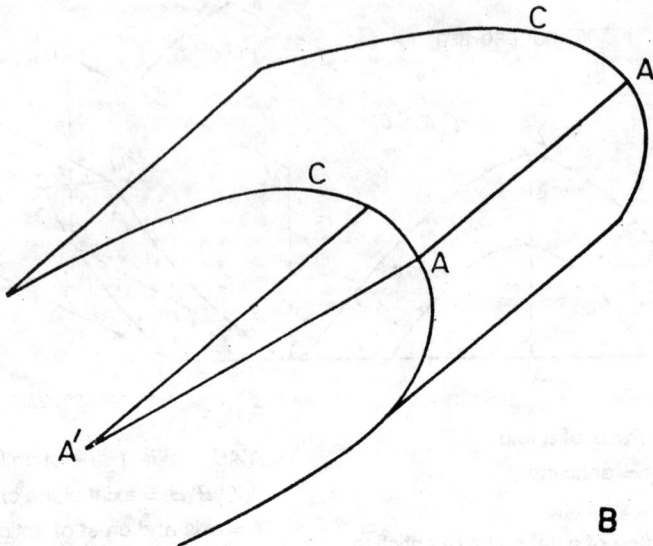

Fig. 85 B Inclined anticline

AAA' = axial plane,

A = axis of fold,

C = crest of fold

Note that "A" is at a lower level than "C". This is realised in case of inclined axial plane, and when height of the fold is low.

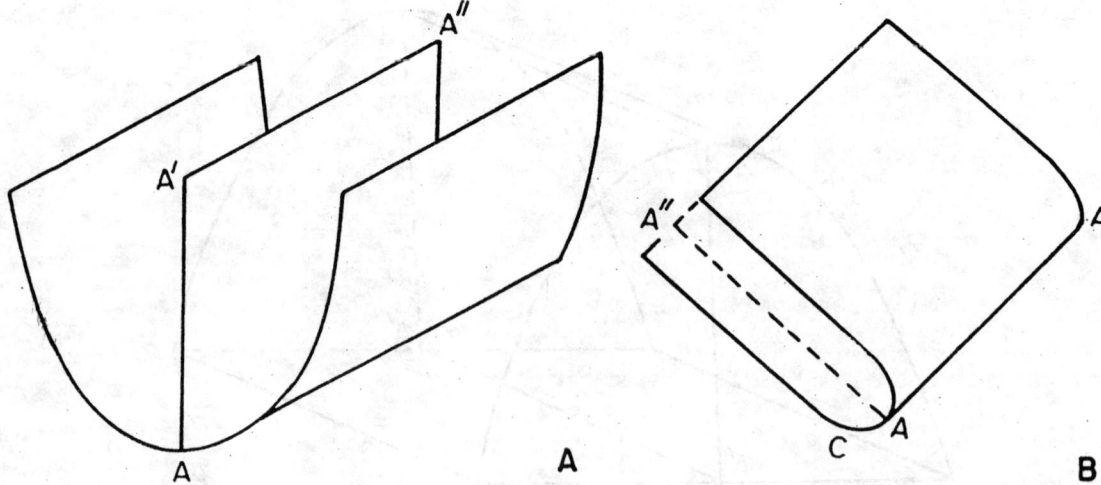

Fig. 86. A Vertical syncline

AA'A" = axial plane,

AA' = height of fold

A = axis and trough of fold. This is realised only in vertical fold.

Fig. 86 B Inclined syncline

A'AA" = axial plane,

A = axis of fold

C = trough which is at a lower level than A. This situation arises only when fold is inclined.

The line joining the axes of the fold is called the "axial line" while the plane containing the axial line is called the "axial plane". These features are documented in Figs. 87 A,B

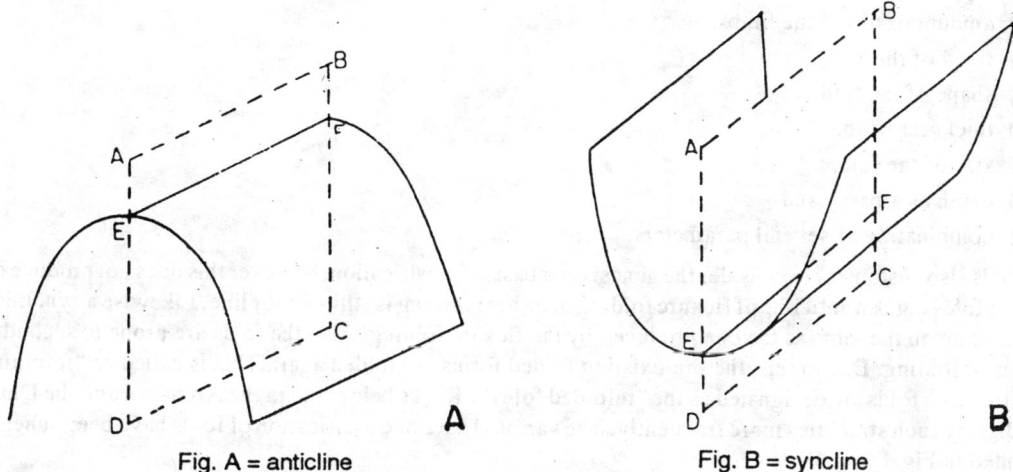

Fig. A = anticline   Fig. B = syncline

Fig. 87 A,B. ABCD = axial plane, EF = axial line

The size of the fold depends upon the height and the distance between the limbs of the fold. Smaller the parameters, smaller is the size of the fold, bigger the parameters, bigger is the size of the fold (Fig. 88). The number of synclines and the anticlines produced will depend upon the size, besides the other factors like the intensity of folding movements, nature of the rocks and so on.

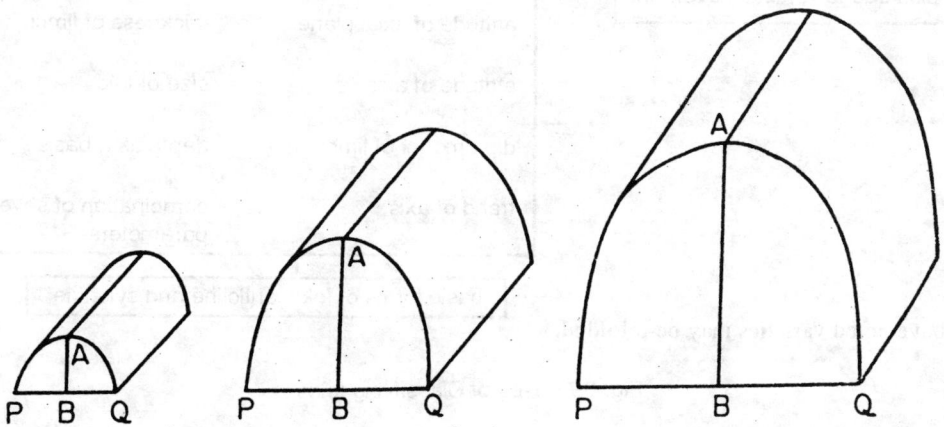

Fig. 88. Controls of Size of fold

Note that size of fold depends upon the height (AB) and the distance between the limbs of the fold (PQ). As the height and the distance increases, the size of the fold also increases.

CLASSIFICATION OR KINDS OF FOLDS

Strictly speaking, there are only two basic forms of folds namely, the anticline and the syncline. The other varieties recognised are only the modifications of these two basic forms. The bases considered in recognising varieties of folds are the following ones.

(a) attitude of the axial plane,

(b) Attitude of the axis,

(c) dip direction of the limbs,

(d) amount of dip of the limbs,

(e) trend of the axis,

(f) shape of the fold,

(g) thickness of the limbs,

(h) size of the fold,

(i) depth as a basis, and

(j) combination of several parameters.

It is also customary to consider the genesis as a basis of clasification, however this does not produce different kinds of folds e.g., an anticline of flexure folding or a shear folding is still an anticline. Likewise a syncline of flow folding is much the same as the one produced by the flexure folding. Also the folds are prone to second or more periods of folding. Due to this, the pre-existing folded forms get folded again. This is called as "refolding", and therefore such folds are designated as the "refolded folds". Rocks belonging to the Archaean and the Dharwarian ages display such structures more frequently. The various bases of classification of folds have been schematically presented in Fig. 89.

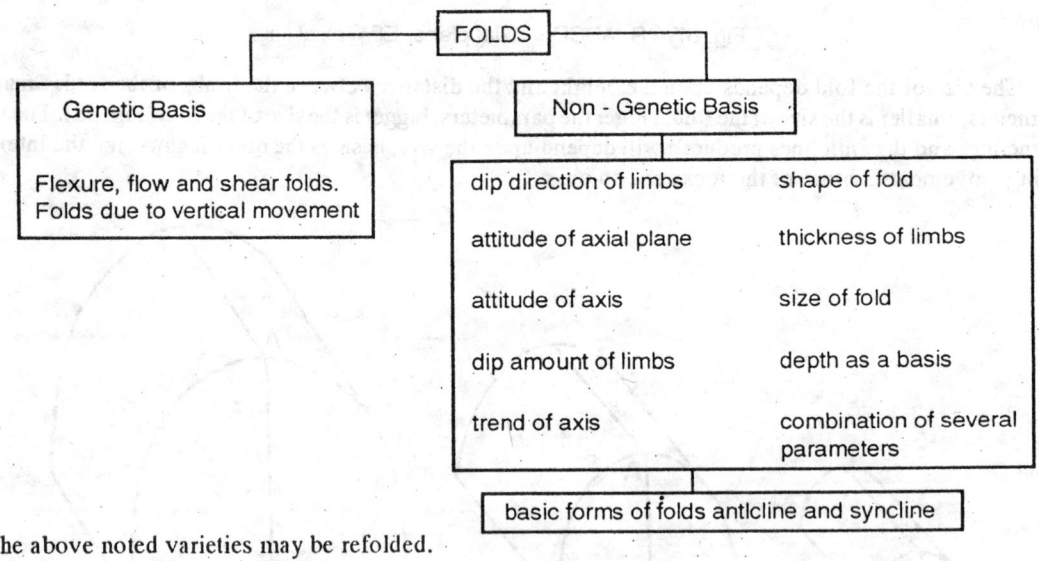

The above noted varieties may be refolded.

Fig. 89 Bases of Classifying folds

DEFINITION OF BASIC FORMS OF FOLDS

Anticline is a fold which is convex upwards, or is one where in the limbs generally dip away from the axial plane, or is one where in the oldest bed is found to occur in the centre of the form. In case the relative ages of the rocks composing the fold be not determinable, then a non-commital term "antiform" is used (Figs. 90 A,B).

Syncline is fold which is concave upwards, or is one wherein the youngest bed or rock is found in the centre of the form, or is one wherein the limbs generally dip towards the axial plane. In case the relative ages of the rocks composing the fold be not determinable, then the non-commital term "synform" is used (Figs. 90 C,D)

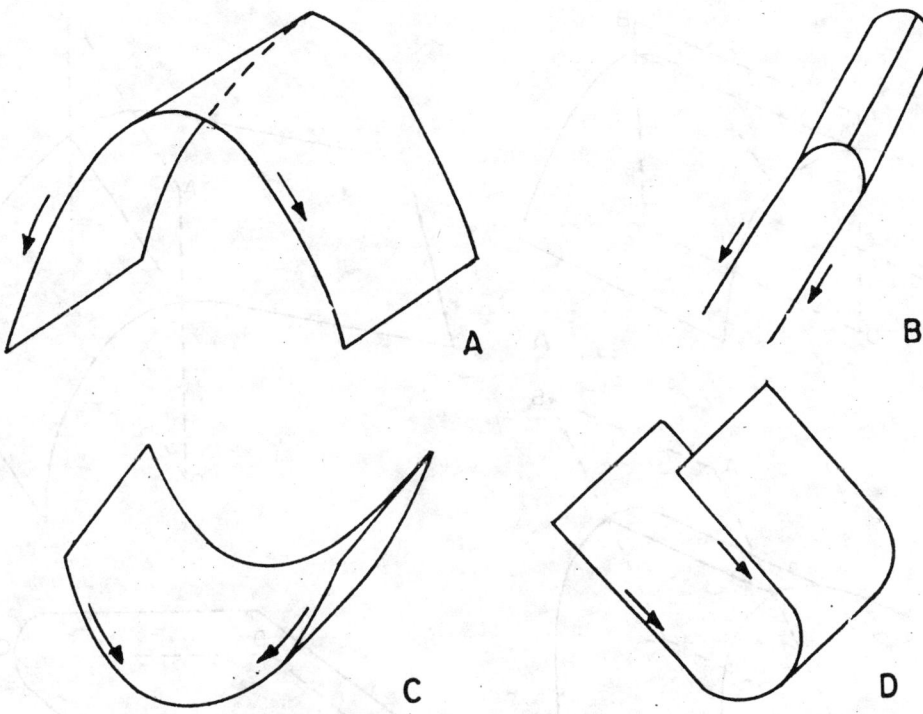

Fig. 90. A,B,C,D. Basic forms of folds.

A = anticline. note that both the limbs dip away from each other.

B = antiform. Note that both the limbs dip in the same direction, but the fold is convex upwards.

C = syncline. Note that both the limbs dip towards each other.

D = synform. Note that both the limbs dip in the same direction, but the fold is concave upwards.

FORMS ACCORDING TO NON-GENETIC (GEOMETRIC) CONSIDERATION

(a) **Attitude of axial plane:** Axial plane is an imaginary plane which divides the fold into two nearly symmetrical parts. The axial plane could be vertical, inclined or horizontal. Accordingly three varieties of folds are recognisable. The axial plane can divide the fold into two symmetrical parts, provided the amount of dip for the two limbs is same. When the amount of dip is quite different, then the symmetrical disposition is not possible. Considering these various controls, four types of folds are realised which are shown in Figs. 91 A to D. In the symmetrical and the asymmetrical types, vertical, inclined and recumbent subtypes are recognisable. Thus there will be

   (i) vertical symmetrical fold (Photo 50),

  (ii) inclined symmetrical fold,

 (iii) vertical asymmetrical fold,

 (iv) inclined asymmetrical fold,

  (v) recumbent symmetrical fold, and

 (vi) recumbent asymmetrical fold.

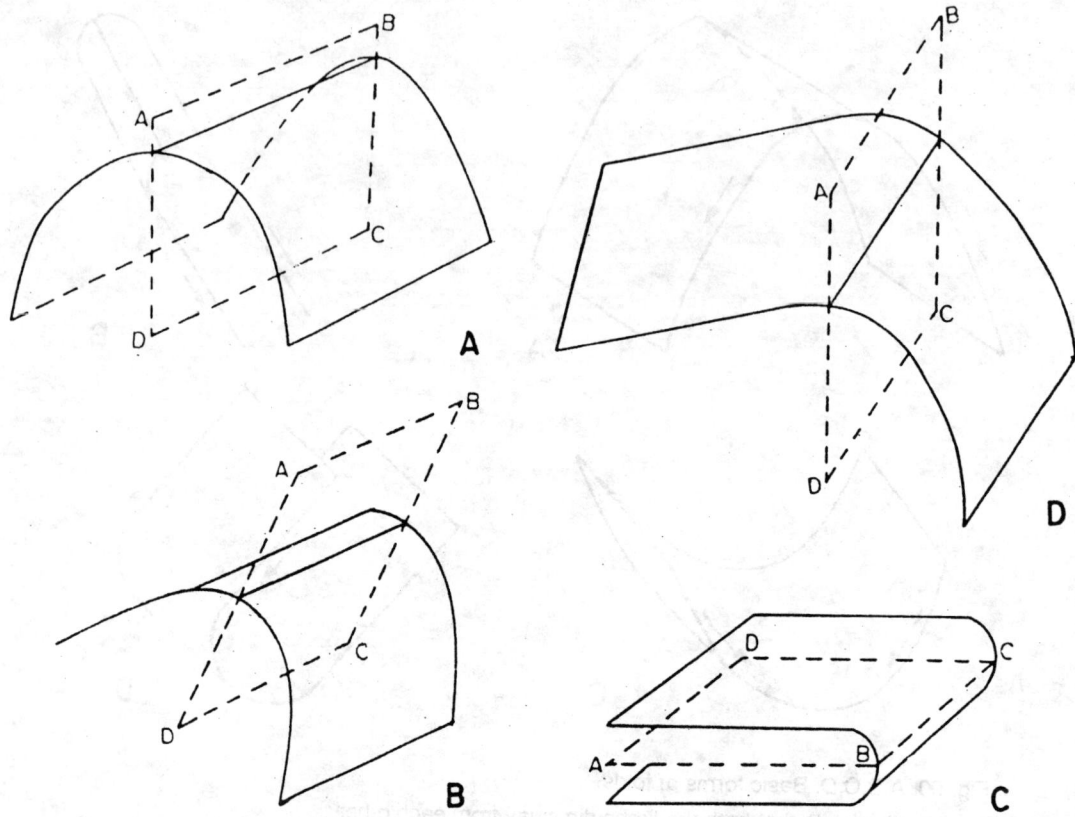

Fig. 91 A,B,C,D. Different attitudes of axial planes resulting in different types of folds.

A = symmetrical anticline with a vertical axial plane (ABCD).

B = symmetrical anticline with an inclined axial plane (ABCD)..

C = recumbent fold which could be anticline or a syncline depending upon whether oldest or youngest rock is in the core of the fold, respectively. Note that axial plane ABCD is horizontal in attitude.

D = asymmetrical anticline. Note that the two limbs dip with different amounts of dip. Axial plane ABCD is inclined.

(b) **Attitude of the axis of the fold:** The axis may be horizontal or inclined. When it is inclined, it may be inclined in one direction, or in both the directions. Based on these features the folds are divided into

   (i) plunging, and the

 (ii) non-plunging varieties. The plunging variety is further split into two sub-types namely,

    (a) plunging in one direction, and

    (b) plunging in both the directions, the latter one is called as a "doubly plunging fold". These different varieties are shown in Figs. 92 A to D and Photo 51. Also see frontis piece photo of this chapter. These figures and the photo are giving the appearance clearly of plunging or non-plunging folds, because these are drawn in three dimensions. However the geoscientists often observe these and the other structural features as outcrops, that means in two dimensions. There fore the outcrop patterns of these are more useful in the field, and the same are shown in Figs. 92 A_1, B_1, C_1 and D_1.

Photo 50. Field photo documenting unmistakable folded structure developed in the granitic rocks of Kamalapuram township, Bellary district, Karnataka state. Typical near symmetrical syncline and anticline can be seen. As the granitic roocks cannot produce folds, such structures are called as "relict or ghost" feature by the geologists. The ball pen is along the axis of the anticline. Note that the axial plane is highly inclined, it being nearly vertical. Courtesy Dr. V.N. Hegde.

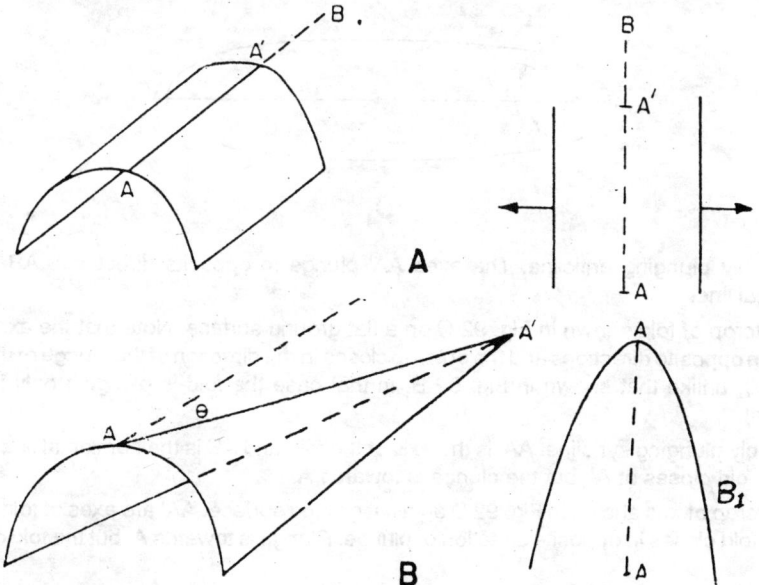

Figs. 92 A, A₁ to D, D₁.

A = non-plunging anticline. Axis AA' coincides with horizontal line AA'B.

A₁ = outcrop of anticline shown in Fig. 92A, on a flat ground surface. Note that the two limbs crop out as 2 lines parallel to axis AA'.

B = singularly plunging anticline. Axis AA' of the fold is at an angle to the horizontal line AB. Angle A'AB is the plunge of the fold axis.

B₁ = outcrop of fold shown in Fig. 92 B, (on a flat ground surface. Note that the two limbs of the fold meet at point A'. AA' is the axis of the fold.

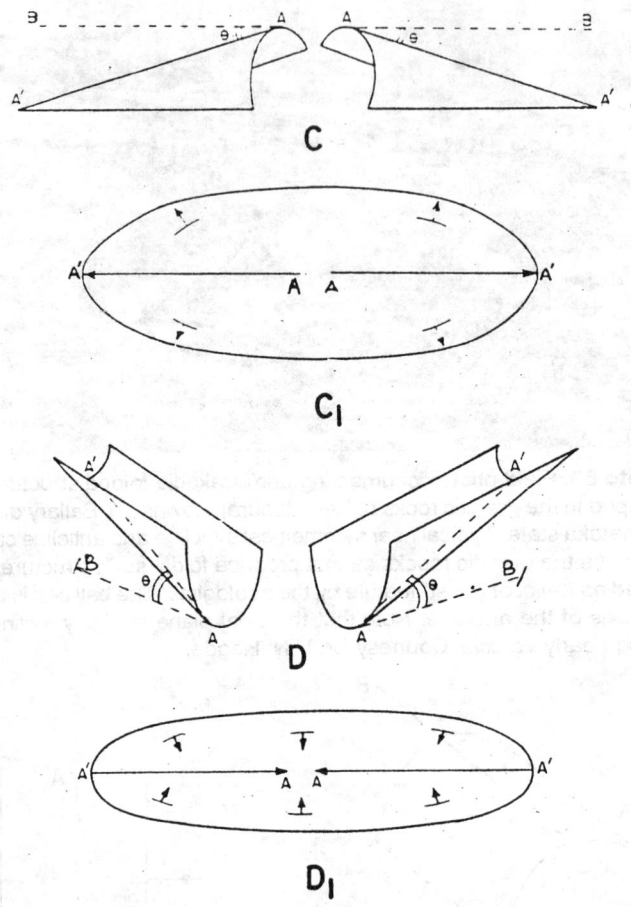

C = doubly plunging anticline. The axes AA' plunge in opposite directions. AB is the horizontal line.

C_1 = outcrop of fold shown in Fig. 92 C on a flat ground surface. Note that the axes AA' plunge in opposite directions and the outcrop closes in the direction of the plunge of the fold (point A'), unlike that shown in Fig. 92 B_1 in that case the fold is plunging only in one direction).

D = doubly plunging syncline. AA' is the axis of the fold and AB is the horizontal line. Note that the fold closes at A', but the plunge is towards A.

D_1 = outcrop of fold shown in Fig. 92 D on a flat ground surface. AA' are axes of fold. Note that the fold closes in opposite direction of plunge. Plunge is towards A, but the fold closes at A'.

(c) **Dip direction of limbs of fold:** This basis nodoubt gives rise to two basic forms namely the anticline and the syncline. A further variety is developed when both the limbs dip in the same direction, unlike that noticed in the basic forms. The new variety is called the "isoclinal fold" - iso meaning same, and clinal meaning the direction in this particular case. The isoclinal fold is further divisible into

   (i) inclined, and

   (ii) recumbent or horizontal folds.

These varieties are shown in Figs. 93 A,B,C. In Photo 52 inclined isoclinal fold has been presented.

Photo 51. Field photo of a plunging anticline developed in banded hematite quartzites exposed in the Tarikoppa hills, Shirhatti taluka, Dharwad district, Karnataka state. The fold plunges (indicated by the hammer) 45° due S 10° E, while the limbs are almost vertical. The quartzite layer has developed boudinage structure as can be seen in the vicinity of the head of the hammer. Courtesy Dr. V.B. Koppad.

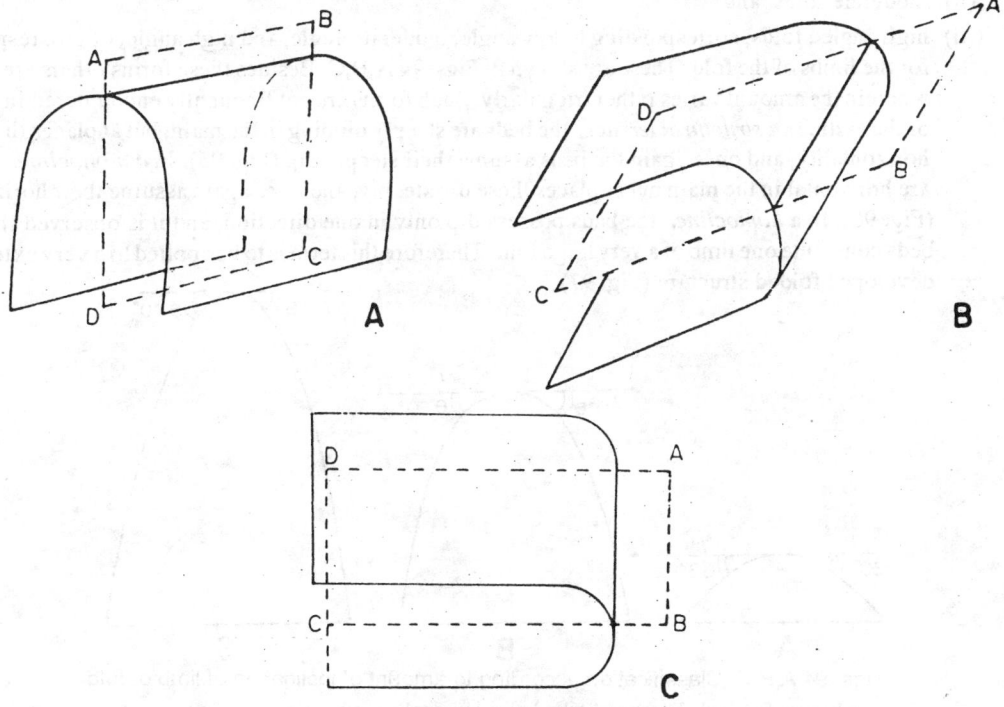

Fig. 93 A,B,C. Varieties of isoclinal folds

Photo 52. Field photo of an isoclinally folded and plunging (hammer indicates direction of plunge) banded hematite quartizite exposed in a hillock located 3 km. N 75°W of Hirevadavatti village, Dharwad district, Karnataka state. Note the development of boundins of quartzite from the banded hematite quartzite. Courtesy Dr. N.N. Gothe.

(d) Dip amount of limbs of fold: Here three varieties are recognised namely,

    (i) low or flat folds

    (ii) moderate folds, and

    (iii) high angled folds, corresponding to low angle, moderate angle, and high angle, of dip, respectively for the limbs of the fold. These are shown in Figs. 94 A,B,C. Besides these forms others are possible wherein the amount varies rather irregularly. Such forms are not frequently encountered in the crust of the earth. In a *structural terrace,* the beds are steeply dipping in the main, but at places these show horizontality, and once again the beds assume their steeper dip (Fig. 95). In *a monocline,* the beds are horizontal in the main but at places these dip steeply, and once again assume their horizontality (Fig. 96). In a *homocline,* the beds possess dip only in one direction, and it is observed that these beds constitute one limb of a very large fold. Therefore this term is to be applied to a very extensively developed folded structure (Fig. 97).

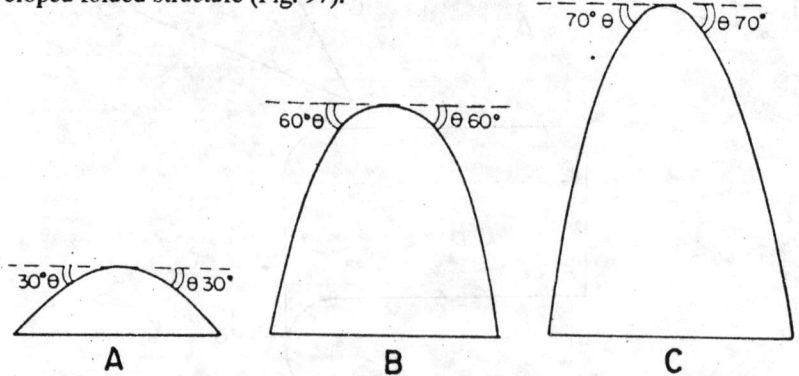

Figs. 94 A,B,C. Classification according to amount of inclination of limb of fold.

A = low or flat fold. The two limbs have low inclination.

B = moderate or medium fold. The two limbs have moderate inclination

C = High fold. The two limbs have high inclination.

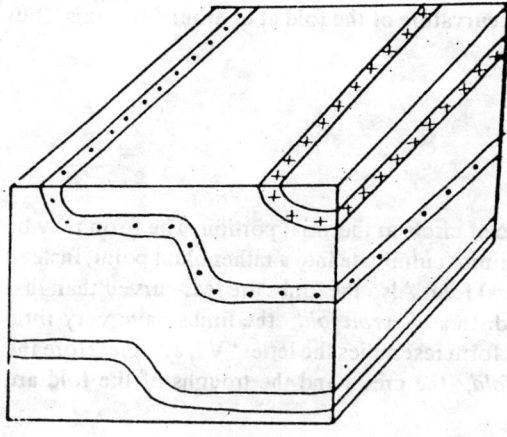

Fig. 95. Structural terrace.
Beds are mainly steeply dipping ones, but suddenly at some spots these become flat. This results in the development of "terrace like" form.

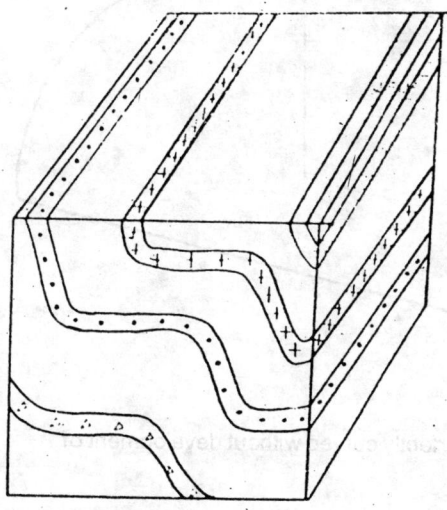

Fig. 96. Monocline.
Beds are mainly horizontal, but suddenly become steep at some spots and then once again maintain horizontal form.

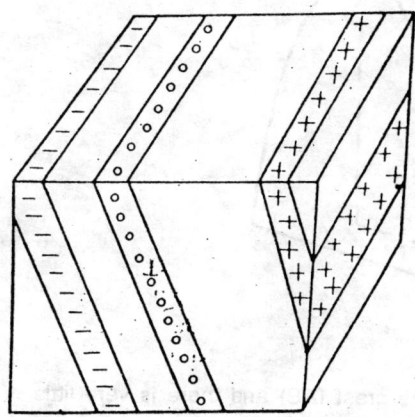

Fig. 97. Homocline.
Beds dip only in one direction and these together form one limb of a very gigantic fold.

(e) **Shape of the fold:** Depending upon the behaviour of the curvature of the fold at or around the axis, four varieties of folds are recognised namely,

 (i) cylindrical,

 (ii) conical,

 (iii) chevron, and

 (iv) box folds.

In a *cylindrical fold,* the two limbs meet into a smooth arc of circle at the axial portion. The form may be described as a half cut circle (Fig. 98). In a *conical fold,* the two limbs culminate into a rather blunt point, instead of having a smooth curvature as is seen in the case of a cylindrical fold. Also the limbs are less curved than that in a cylindrical fold (Fig. 99). The fold is rather "cone" shaped. In a *chevron fold,* the limbs have very little curvature at the axis, and these culminate into a sharp point. The form resembles the letter "V", and therefore the fold is also called as "V" shaped fold (Fig. 100). In the *box fold,* the crests and the troughs of the fold are considerably flat (Fig. 101).

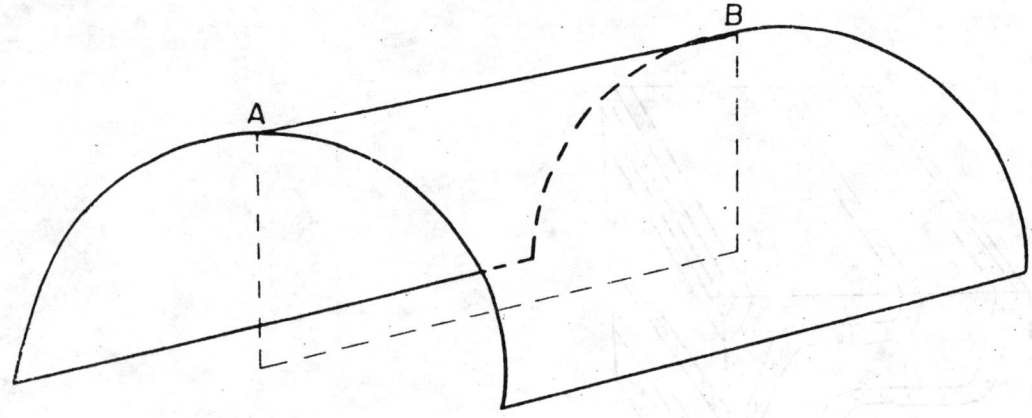

Fig. 98. Cylindrical fold.
Note that even at the crestal part (AB), the fold is perfectly curved without development of any angularity.

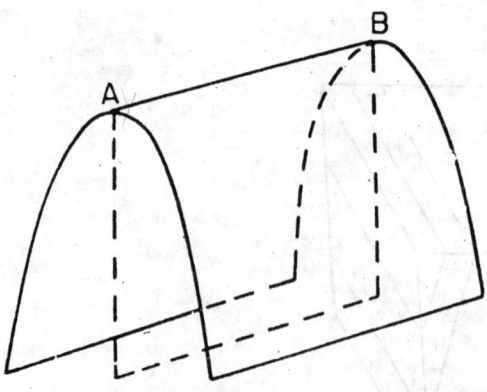

Fig. 99. Conical fold.
Note that the two limbs meet rather sharply at the crest (AB) and there is very little roundness or curvature for the fold.

Excellent example of chevron fold is found in the phyllites exposed 500 meters N 15° W of Kelur village, Shirhatti taluka, Dharwad district, Karnataka state (Gothe 1973), and it is documented in Photo 53. Box fold is described by Desai (1991) from the banded hematite quartzite occurrence from the N.E.B. range of Sandur schist belt, and it is presented in Photo 54.

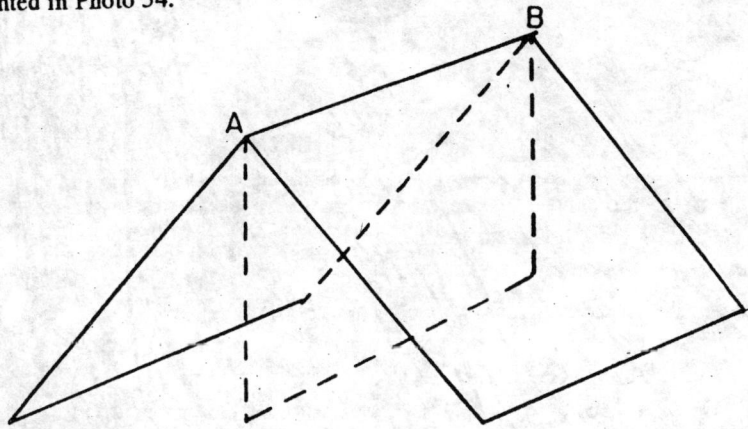

Fig. 100. Chevron fold.
Note that the two limbs meet at a sharp point (AB) at the crest. There is no curvature at all at the crest of the fold.

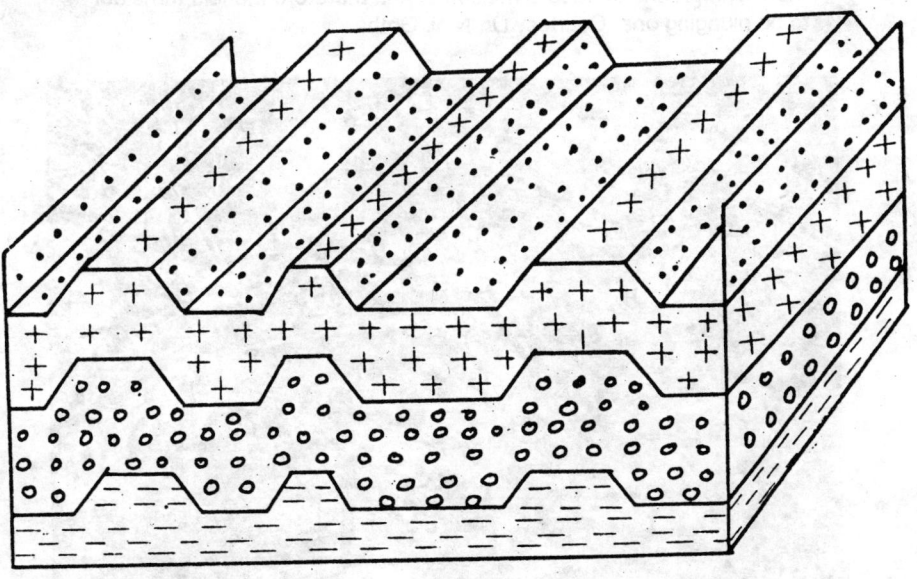

Fig. 101. Box fold.
Note that the crests and the troughs of the fold are flat and not curved or rounded ones.

(f) **Thickness of limbs of fold:** Fold itself being a plastic style of deformation, the flowage or the yielding of the beds (rocks) is a feasible situation. Many times it is found that the thickness of the beds does not remain constant; it is thinner on the limbs, and thicker at the axial portion. Accordingly two varieties are recognised namely,

Photo 53. Field photo of a chevron fold developed in phyllites outcropping 500 meters N15°W of Kelur village, Dharwad district, Karnataka state. The "V"s are pointing due general north (indicated by the hammer). The ground surface is horizontal and therefore the fold turns out to be plunging one. Courtesy Dr. N.N. Gothe.

Photo 54. Field photo of "Box fold" produced in banded hematite quartzite of N.E.B. range, Hospet, Bellary district, Karnataka state. Note the development of bluish coloured layers which are in fact of pseudo-tachylyte, which indicate the action of dislocatory movements (faulting). Courtesy Dr. H.D. Desai.

(i) open fold, and

(ii) closed or tight folk.

In an *open fold,* the thickness of the beds remains constant throughout the limb or the axial part, whereas in a *tight or closed fold,* the thickness is not constant at the limb and the axial part of the fold (Figs. 102, 103).

(g) **Trend of the axis of the fold:** The axis of a fold generally crops out as a straight line and therefore the fold continues to crop out in a linear fashion. The geomorphic expression may be in the form of a hill chain (anticline), or a valley (syncline), and depending upon the number of anticlines and the synclines, there will be developed parallel hill chains and valleys. Many times the individual anticlines or the synclines do not extend in a linear fashion, but appear as numerous folds, the axes of which show a shift. Such folds are called as the "en echelon" folds (Figs. 104, 105).

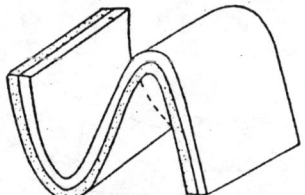

Fig. 102 Open fold.

Note that the thickness of the beds (limbs) is constant at any part of the fold.

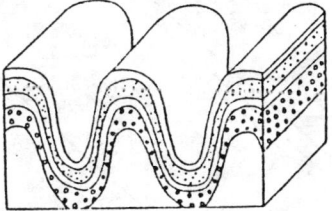

Fig. 103 Closed or Tight Fold

Note that the plastic beds show unequal thickness at the crests, troughs and the limbs of the fold. Beds shown with dots and open circles, have unequal thickness, while others have uniform thickness.

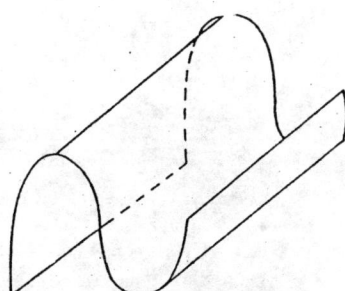

Fig. 104. Note that the fold continues in its extent along the axis without any shift. Therefore it is not an "en echelon fold".

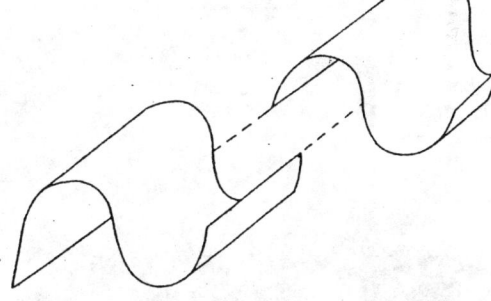

Fig. 105. En echelon fold

Note that the anticline overlaps over the linear extent of syncline. The fold appears to be faulted in its extent along the axis, but actually it is developed through process of deformation and not due to faulting.

B = inclined isoclinical anticline. Axial plane ABCD is inclined. Both the limbs are parallel to each other and dip in the same direction.

C = recumbent fold. Not that the axial plane ABCD is horizontal, so also the two limbs of the fold. Whether it is an anticline or a syncline, can not be told unless the age of the rock at core of the fold is known.

There is another type produced when the axis of the fold is not off set, but is folded again. This is called as a "refolded fold", and it applies both to the anticlines and the synclines. Due to refolding, the fold axis gets bent, and the fold therefore does not continue in a linear fashion, but gets shifted. This is shown in Figs. 106 A,B. Excellent example of this type of fold is offered by the schistose rocks of the Gadag schist belt. The banded hematite quartzites form a part of the belt, and this rock shows a clear change in the trend direction from NNW - SSE through NW - SE, WNW - ESE, W - E, N - S, NNE - SSW, N-S and back to NNW - SSE direction, over a distance of nearly 56 km. (Maclaren, 1906). The banded hematite quartzites are bent at the places of change in the strike direction, thus resulting in the refolding of the axis of the fold. Gothe (1973) and Koppad (1975) who have studied parts of the Gadag schist belt have documented change in the trend of the axis of the fold, and these are presented in Photos 55 and 56.

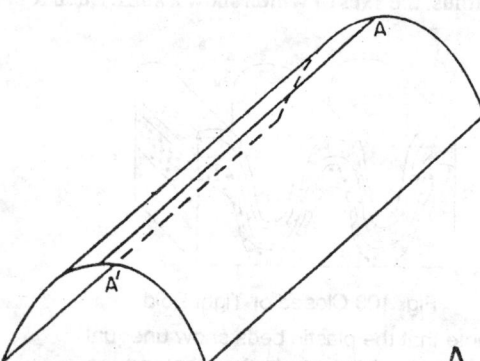

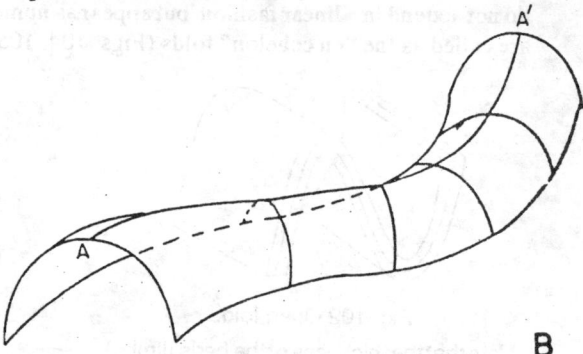

A - anticlinal fold showing continuation of its axis AA' in a straight line when traced along its extent

B - anticlinal fold showing curvi-linear extension of its axis AA' when traced along its extent. This is produced due to refolding of the axis alone.

Figs. 106. A,B. Effect of refolding of axis of the fold.

Photo 57. Field photo showing refolded limb of an isoclinal fold produced in thinnly laminated banded hematite quartzites. Palm of the person indicates horizontal axis of the refolded structure, which is parallel to the main axis of the isoclinal fold. Note that at the axis, the rocks have undergone crushing with the production of breccia. The structure is located in the NE part of the Arpee Iron Ore Mines, Hospet taluka, Bellary district, Karnataka State. Courtesy Dr. H. D. Desai.

Photo 58. Field photo of an unusual knee shaped fold developed in banded hematite quartzite through refolding of the limb of the main fold. This fold is trending N 55° W - S 55° E, the limb being vertical in attitude. Axis of the refolded fold is also vertical. Note the crushing and flowage of the rock at the bent portion (ball pen indicates the knee shaped bend in the rock). This structure is located 3 km N 50° E of Kalahalli village, Sandur taluka, Bellary district, Karnataka State. Courtesy Dr. H.D. Desai.

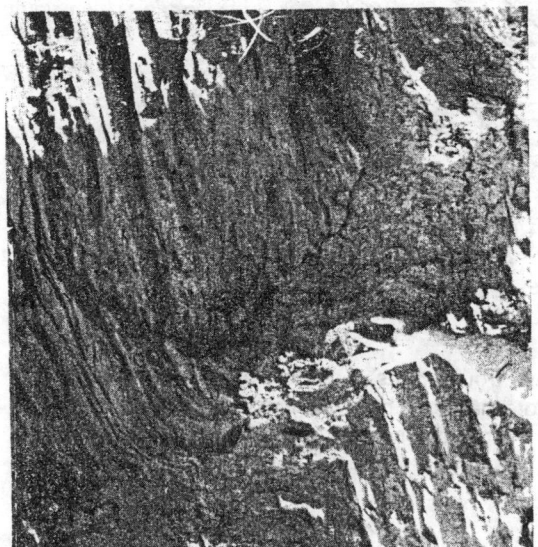

Photo 57. Field photo showing refolded limb of an isoclinal fold produced in thinnly laminated banded hematite quartzites. Palm of the person indicates horizontal axis of the refolded structure, which is parallel to the main axis of the isoclinal fold. Note that at the axis, the rocks have undergone crushing with the production of breccia. The structure is located in the NE part of the Arpee Iron Ore Mines, Hospet taluka, Bellary district, Karnataka State. Courtesy Dr. H. D. Desai.

Photo 58. Field photo of an unusual knee shaped fold developed in banded hematite quartzite through refolding of the limb of the main fold. This fold is trending N 55° W - S 55° E, the limb being vertical in attitude. Axis of the refolded fold is also vertical. Note the crushing and flowage of the rock at the bent portion (ball pen indicates the knee shaped bend in the rock). This structure is located 3 km N 50° E of Kalahalli village, Sandur taluka, Bellary district, Karnataka State. Courtesy Dr. Desai.

Due to refolding, the limb also gets affected, and such types are more common, and are therefore easily detectable. Desai (1991) has described refolded limbs of folds produced in the banded hematite quartzites from the N.E.B. range, Jambunath hills, Bellary district, Karnataka state. In Photo 57, the axis of the refolded limb is horizontal, while in Photo 58, it is vertical. This latter form appears like a "bent knee".

(h) **Depth as basis:** Fold is decidedly a three dimensional structure and therefore it needs to be studied in respect of the continuation of the forms at depth. The data available from the deep mines and that derived from the remote sensing procedures disclose that the dip amount, dip direction and even the thickness of the limbs, do not show uniformity. Accordingly several forms are distinguished. A *similar fold* is one wherein the limbs are thin and the axial parts are thick (Fig. 107). Generally the plastic members (beds) will yield and produce uneven thickness at the limbs and the axes of the folds. A similar fold may be called as a "special kind of tight or a closed fold", wherein all the beds have yielded by the flowage of the material. *A parallel or a concentric fold* is one wherein the shape of the anticline and the syncline goes on changing with depth (Fig. 108). In such folds the thickness of the beds is observed to remain constant. In this regard it resembles the "open folds", but in a parallel fold, the dip increases and then decreases, as the depth increases. Thus the anticline which is having steep dip at the surface, it slowly becomes more and more flat, and again the dip amount increases to resume steepness. This behaviour also will be found in the syncline adjoining to an anticline, and vice versa. *Piercing or diapir folds* are anticlines wherein a plastic member forming the core of the anticline, is found to squeeze its way along the axis of the fold (Fig. 109). *Supratenuous fold* is formed when sedimentation and folding processes operate together. The beds become thickest at the trough of the syncline, and are thinnest at the crest of the anticline (Figs. 110 A,B). This structure is so developed because the anticlinal part was rising whereas the synclinal part was sinking, when the sedimentation was taking place.

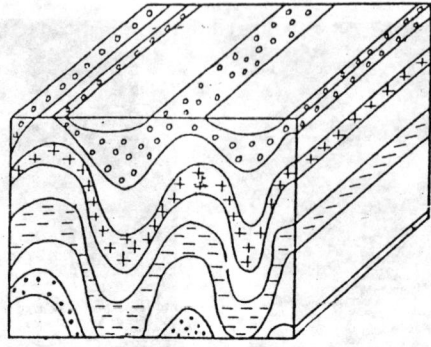

Fig. 107. Similar fold.

Note that the limbs are thin, while the crest and trough parts of the fold are thick. This pattern is transmitted upward and downward. Note that all the beds are affected, unlike that observed in a "closed or tight fold" where only the plastic ones are affected (Fig. 103).

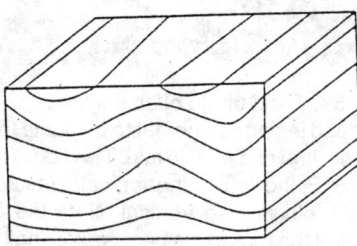

Fig. 108. Parallel or concentric fold.

In this fold, the shape of the anticline or syncline changes when traced along vertical direction or downward direction. The form becomes flat over some depth, and then regains its steeper form.

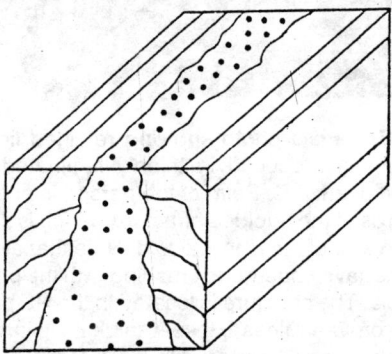

Fig. 109. Piercing or diapir fold.

Due to tight folding, plastic bed from the group gets squeezed along the axial portion of the fold. As the bed forces its way into the fold, it is called as "piercing fold".

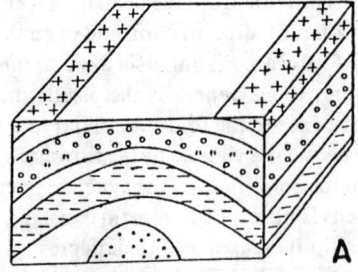

Fig. 110 A. Supratenous fold (anticline).

This structure is produced when upheaval of the basin takes place at the same time of sedimentation. Due to this the beds on the top part of the rising part of the basin, become thin. When such a pile of sediments are folded, the limbs become relatively thicker than the crestral parts.

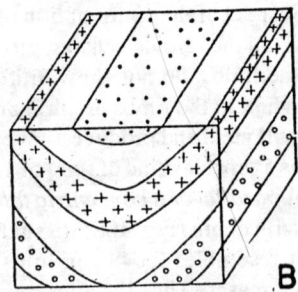

Fig. 110 B. Supratenous fold (synclline)

A syncline on the side of a supratenuous anticline, can be expected to possess different thickness on the limbs and the axial parts. However this is just a suggestion.

is Photos 50, 51, 52, 53, 54, 57, 58, 59, 60 and the frontis piece photo of this chapter, folded forms differ in size, yet these will be designated as anticline or syncline as the case may be.

(j) **Combination of several parameters:** So far the simpler individual bases have been utilised to recognise the different varieties of the folded forms. There are some other folds which combine two or more bases described above. At times some of the folded forms defy their classification on any consideration as those folds are highly "irregular" in nature. Considering these features, the following forms are recognised.

A *fan fold* is one wherein both the limbs are overturned. The "overturned" nature can be established by considering the disposition of the primary sedimentary features like the current bedding, the ripple marks and so on, towards the main fold. Thus in Fig. 111 A, though the two limbs dip towards each other, it is found that the oldest bed is in the centre. The fold therefore becomes an "anticlinal fan fold", though it looks like a syncline. Both the limbs are found to be over turned. In Fig. 111 B, a syncline belonging to the category of a fan fold, is shown. Here though the dip directions are opposing ones, the limbs are found to be over turned, and hence the fold becomes a "synclinal fan fold".

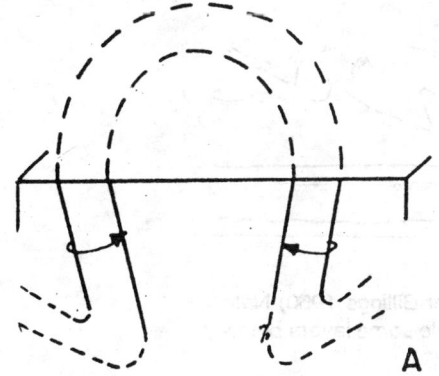
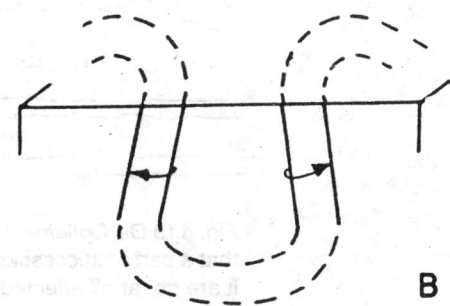

A - Note that both the limbs are overturned. Though both the limbs apparently dip towards each other, it is yet an anticline. It is therefore called as "anticlinal fan fold".

B - Note that both the limbs are overturned. Though the limbs apparently dip away from each other, it is a syncline. It is therefore called as a "synclinal fan fold".

Figs. 111 A,B. Fan folds.

A *disharmonic fold* form is one wherein there is no regularity of dip amount, thickness of the limbs, depth or height of the fold, and so on (Fig. 112). *Decollement* meaning " shearing off "structures is produced when only the top portion of a sedimentary succession gets independently folded, while the underlying formations remain undisturbed. Billings (1960) describes such a fold from the Jura Mountains (Fig. 113). *Drag fold* is found associated with the main anticlinal and the synclinal folds. As the name indicates, it is produced due to the "dragging away" of the beds. An incompetent bed when sandwiched between two competent beds, and this is subjected to folding, then the incompetent bed gets dragged over the competent bed, while the competent beds are folded into major anticline and syncline (Photo 95 and Figs. 114 A,B).

While classifying the folds on the basis of size, some varieties have been already described. There are yet very large sized folds which need to be distinguished. To such a category belong the anticlinorium, the synclinorium, the geosyncline and the geanticline. *Anticlinorium* is a large sized anticline on the two limbs of which many minor anticlines and synclines are produced (Fig. 115). *Synclinorium* is a large sized syncline on the two limbs of which numerous anticlines and synclines are developed (Fig. 116). These two varieties of folds differ from the drag folds in that here all the beds are folded alike, unlike only the incompetent bed in the case of drag folds. *Geosyncline* is a large sized basin in which sedimentation takes place with the sinking of the floor of the

(i) **Size of the fold:** Depending upon the size of the fold, categories like small fold, large fold, minor fold, or major fold, are recognised. However no specifications have been set about the exact dimensions. The parameters like the height and the distance between the two limbs of the fold, are utilised in such a classification. The terms used are therefore relative ones, in the context of the absolute dimensions. Thus

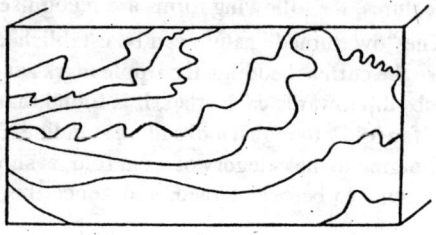

Fig. 112. Disharmonic fold.
Note that the style of each folded form is different and is extremely irregular within its own stretch.

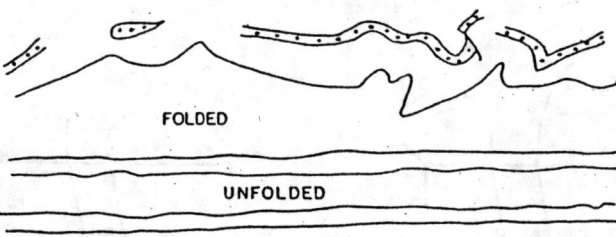

FOLDED

UNFOLDED

Fig. 113 De Collement structure (after Billings 1960) Note that a part of succession is folded while some layers below it are not at all affected.

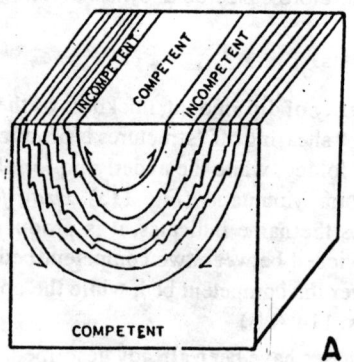

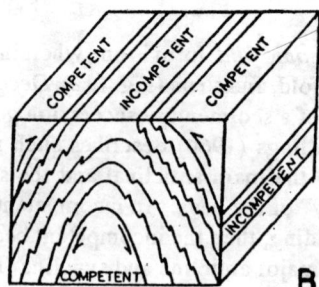

Figs. 114 A,B. Drag folds.
These are produced in the incompetent bed sandwiched between two competent beds. Due to folding, the competent bed develops either a syncline (Fig. 114 A) or an anticline (Fig. 114 B).

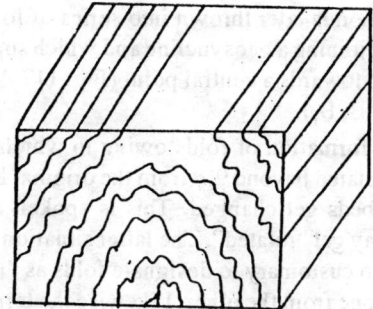

Fig. 115. Anticlinorium.

Note that on the limbs of the major anticline, several minor anticlines and synclines are produced.

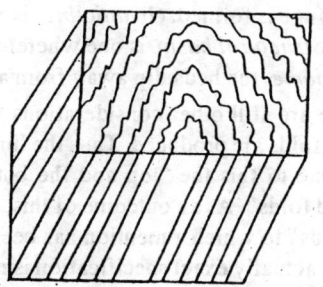

Fig. 116. Synclinorium.

Note that on the limbs of the major syncline, several minor anticlines and synclines are produced.

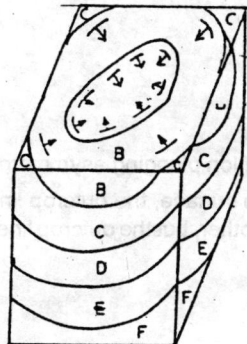

Fig. A - 3 dimensional view.

Note that the beds dip towards the central part from all the sides.

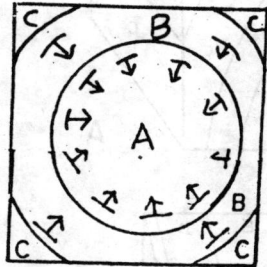

Fig. A₁ - map of Fig. 117 A.

Figs. 117 A,A₁. Structural basin.

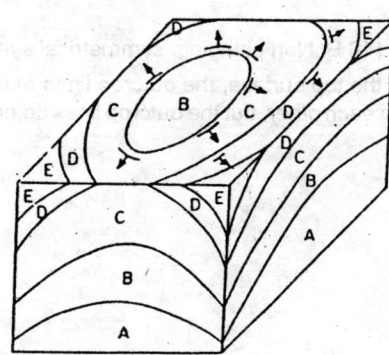

Fig. B - 3 dimensional view.

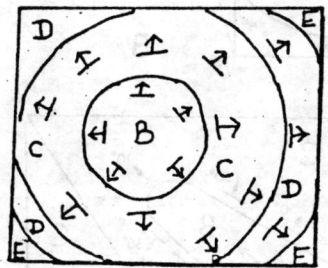

Fig. B₁ - map of Fig. 117 B.

Note that the beds dip away from a central part.

Figs. 117 B,B₁ Structural dome.

basin. Thick pile of sediments accumulates and the whole succession is later thrown into series of folds, and a mountain range is formed. *Geanticline* is a large sized landmass adjoining a geosyncline and which supplied the sediments. *Structural basin* is one wherein the beds are seen to dip towards a central point (Fig. 117 A,A₁). In a *Structural dome* the beds dip away from a central point (Fig. 117 B, B₁).

There are still other considerations that are involved in the formation of folds owing to which different varieties of folds are produced. Thus the limbs of the fold may get rotated beyond 90° from the original horizontal position. Due to this the "top and the bottom positions" of the beds get changed. This is spoken off as the "overturned folds". As an outcome of this, one or both the limbs may get "rotated". The latter situation produces the "fan folds" to which a mention has been already made. It is also customary to designate folds as "major and minor", but actually exact specification is not stated to distinguish one from the other. Likewise the terms" mega and micro folds" are also in vogue. Some such varieties of folds have been shown in Figs. 118 A to F. A complex fold is presented in Photo 60.

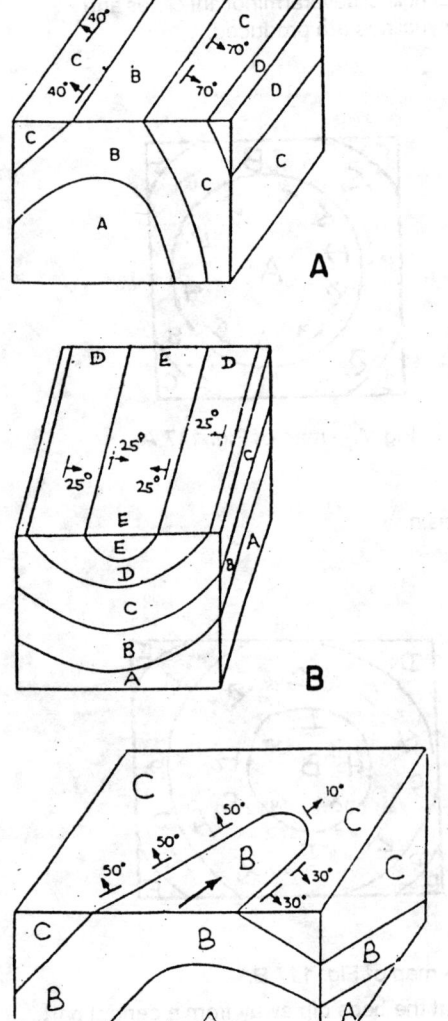

Fig. 118 A. Non plunging asymmetrical anticline.
Note that on the top surface, the outcrop lines of the beds are parallel to each other, but the outcrop lines do not close.

Fig. 118 B. Non plunging, symmetrical syncline.
Note that on the top surface, the outcrop lines of the beds are parallel to each other, but the outcrop lines do not close.

Fig. 118 C. Asymmetrical plunging anticline.
Note that the outcrop lines close on the top surface of the block. It is so because, it is a plunging fold. Direction of plunge is shown by heavy arrow.

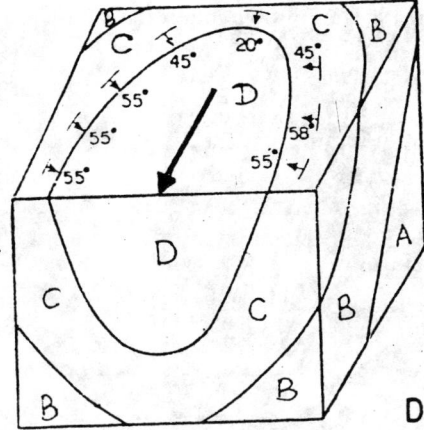

Fig. 118 D. Symmetrical plunging syncline.

Note that the outcrop lines close on the top surface of the block. It is so because, it is a plunging fold. Further the outcrop closes in the opposite direction of plunge of the fold. The direction of plunge is shown by heavy arrow.

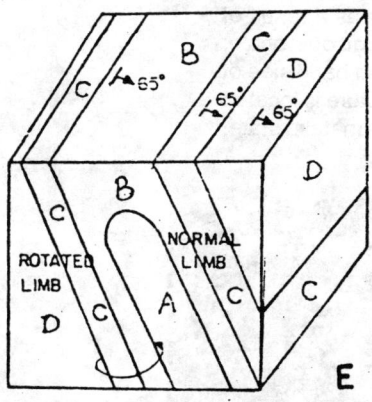

Fig. 118 E. Non plunging, overturned, isoclinal anticline.

Limb on the left hand side of the fold is overturned as indicated by curved arrow. Note that the outcrop lines do not close on the top surface of the block.

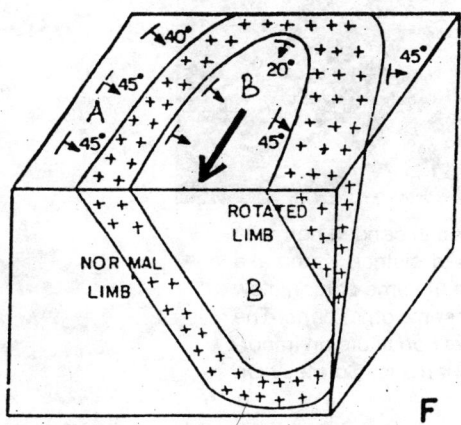

Fig. 118 F. Plunging, overturned, isoclinal syncline.

Note that the limb on the right hand side is overturned. Thick arrow indicates the direction of plunge. Note that the outcrop closes in the opposite direction of the plunge (see top portion of the block).

Photo 59. Field photo showing development of boudins as a result of folding accompanied by refolding. The rock is a ferruginous shale. Development of drag folds is clearly observable on the left hand side of the photo. Ball pen is along the axis of refolding. The struture is located in the Aarpee Iron Ore Mines, Hospet, Bellary district, Karnataka state. Courtesy Dr. H.D. Desai.

Photo 60. Field photo of a complex fold registered in banded hematite quartzite of Nagavi area, Gadag taluka, Dharwad district, Karnataka state. Note that the rocks show V shaped pattern at some portion, flow fold type at the central part, and no folding at all at some other parts. The pattern is therefore complicated one. These rocks constitute an important member of the Gadag schist belt of Dharwarian age. Courtesy Dr. S.C. Puranik.

EXAMPLES OF FOLDS FROM INDIA

So far a detailed technical account of the different types of folds has been given. Krishnan (1968) has noted some examples of folded structures, and these have been given below.

1. The Potwar plateau of western Punjab is a synclinal trough (Soan syncline) having its axis along ENE - WSW direction (p. 52).

2. In Sundargarh district, there is an anticlinorium or geanticline which has an ENE - WSW axial direction (p. 131).

3. The rock exposures around *Jutogh* and *Chor* mountains have been regarded as forming a highly compressed recumbent double anticline, the Chor granites occupying the core of the intervening syncline (p. 138).

4. The Kolar schist belt in which gold field occurs, is 65 km. long and 4 to 6 km. broad. The schists have been folded into synclinorium, the folds being isoclinal along N - S axis and dipping steeply (60° - 80°) to the west. The folds have been refolded along NNW - SSE axis plunging 30° - 40° northwest (p: 149).

RECOGNITTION OF FOLDS

Any structure needs to be proved if it be not exposed completely by way of an outcrop. Folds have the third dimension, namely their continuation at depth. Therefore one needs vertical sections, natural or artificial ones, to unravel the structure. Procedure followed in recognising the folds may be grouped into below noted categories.

  (i) physiographic studies,
 (ii) structural studies,
(iii) outcrop patterns,
 (iv) direct observations,
  (v) photogeological and landsat imageries,
 (vi) geophysical studies, and
(vii) drilling and mining:

The structure (fold) is to be established by taking many evidences together into consideration. Each approach is elaborated in the following pages.

  (i) **Physiographic studies:** According to the geomorphologists, the geomorphic forms are controlled by the structures possessed by the rocks, besides by the other factors. Fold is one such structure which influences the geomorphic forms. In the anticlines, the beds actually get raised up, while in a syncline, these will be depressed or lowered. As a first approximation therefore, the anticlines should give rise to hills, and the synclines to the valleys. However the resistance of the rocks to erosion, also contributes much to the retention as hills (if resistant to weathering) or production of valleys (if not resistant to weathering). Further, conspicuous ridges or the hog-back geomorphic forms are possible only if the rocks possess high degree of inclination. Therefore if chains of hillocks interspersed with valleys be present in a region, then the presence of folded structure is warranted. Further if the rocks constituting the ridges were to display bend or bends in their outcrops, then the possibility of the existence of a refolded fold becomes quite strong. Maclaren (1906) has described such a bend near Kadampur (Dharwad district, Karnataka state) in the banded hematite quartzites and the associated schists. The axis of the fold shows a clear change of trend from the initial NNW - SSE direction through W - E, N - S, NE - SW back to NNW - SSE direction at the other end. These rocks are also possessing high angle of inclination. However prior to drawing the conclusion, a field check is necessary, since faulting accompanied by dragging of the beds may bring about bends in the rocks, and this also will give rise to ridges and the hog-back forms.

 (ii) **Structural studies:** This is by far the most reliable and therefore most sought after method in establishing the existence of folds. This comprises of noting the dip and the strike of the rocks and plotting them

(Figs. 119 A to D). For a successful application of this method, the rocks should be very well exposed and these should extend over a reasonably large area. In the Figs. 119 A to D, simpler cases have been shown. The beds are often overturned and this is a sure indication of the presence of a folded structure. The "rotation or overturning" of the limbs of the fold is established by the use of the primary sedimentary structures like the current bedding, the ripple marks, the graded bedding and so on, and also by the use of the "drag folds". The procedure adopted is described in Chapter 7 of this book. Some illustrations depicting the overturning of the beds and the type of the folds derived therefrom, have been given in Figs. 120 A to D, and 121 A, A₁, B, B₁.

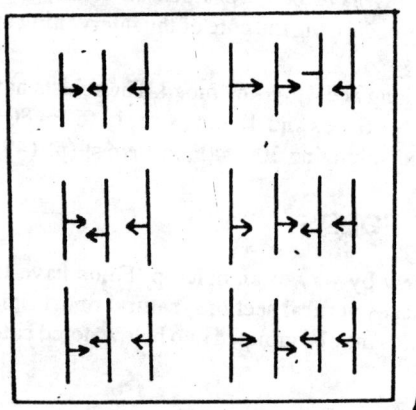

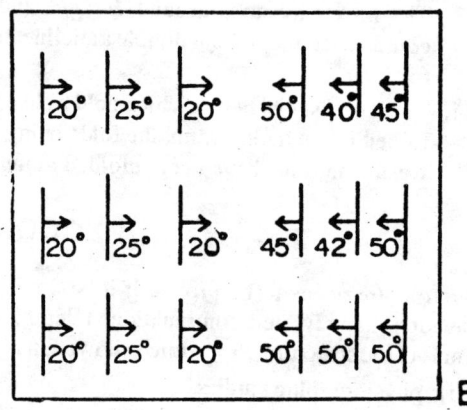

Fig. 119 A. Structural map of folded rocks.
The attitude indicates presence of a syncline, an anticline and again a syncline, on proceeding from the left hand side towards the right hand side of the figure.

Fig. 119 B. Structural map of folded rocks.
The attitude indicates the presence of an asymmetrical syncline, the limb on the left hand side being flatter in inclination.

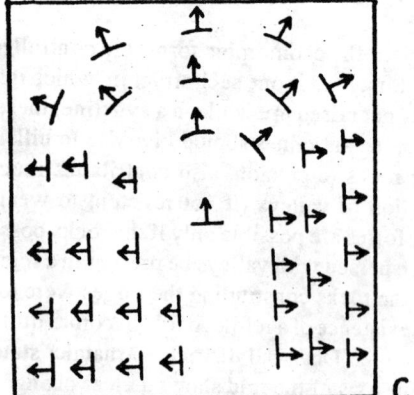

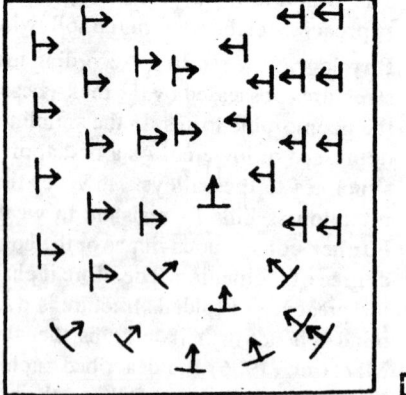

Fig. 119 C. Structural map of folded rocks.
The attitude indicates presence of a plunging anticline, the plunge being away from the observer.

Fig. 119 D. Structural map of folded rocks.
The attitude indicates presence of plunging syncline, the plunge being away from the observer

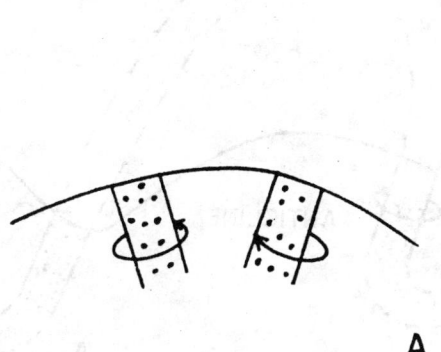

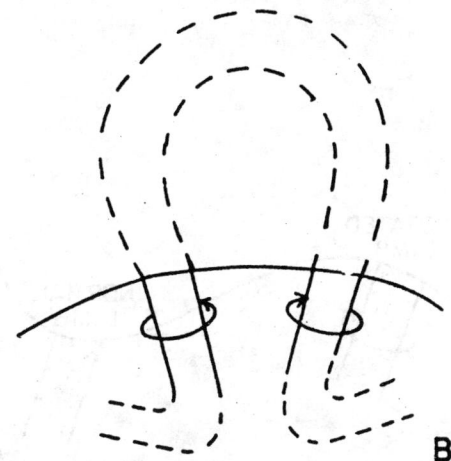

A - limbs dip apparently towards each other. Arrows indicate that both the limbs are overturned.

B - reconstruction of limbs into an "anticlinal fan fold"

Figs. 120 A,B. Vertical section of anticlinal fan fold.

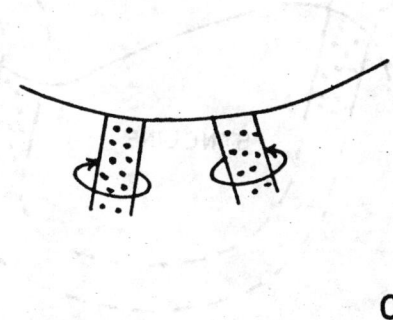

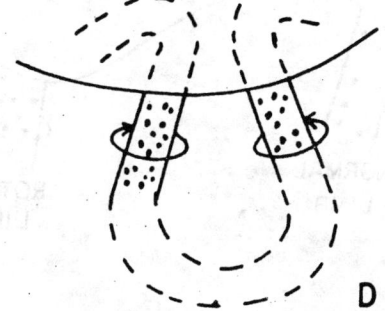

C - limbs apparently dip away from each other. Arrows indicate that both the limbs are overturned.

D - reconstruction of limbs into a "synclinal fan fold".

Figs. 120 C,D. Vertical section of synclinal fan fold.

(iii) **Outcrop patterns:** Nature of the outcrops of the rocks depends upon the attitude (dip and strike) and the topography (slope of the ground surface) a detailed account of which has been given by Gokhale (1987). Outcrop patterns of the horizontal, the dipping and the folded beds are different. Therefore the outcrop patterns are regarded to be quite reliable ones in depicting the folds. The below noted patterns are suggestive.

(a) **Repetition of outcrops:** This may be due to faulting, or due to suitable topography, or also due to the existence of a fold. If evidence for faulting be lacking, then the possibility of favourable topography is to be eliminated. A formation (bed) possessing a low dip, crops out more than once (Figs. 122 A,B). When these two possibilities are ruled out, then the presence of a fold may be inferred. In Figs. 123 A,B recognition of folded structuree is very simple because the dip is either towards or away from a point. When the beds (limbs) are dipping towards or away from each other,

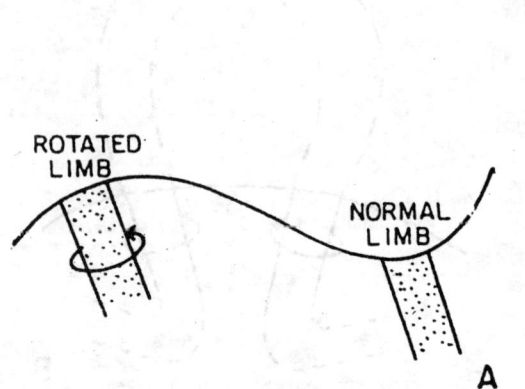

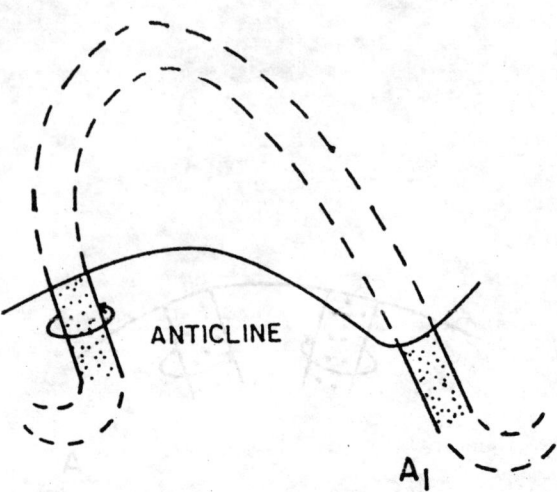

A - both the limbs are dipping to the right hand side, but the limb on the left hand side is overturned as indicated by the arrows.

A₁ - reconstruction of limbs into isoclinal anticlinal fan fold.

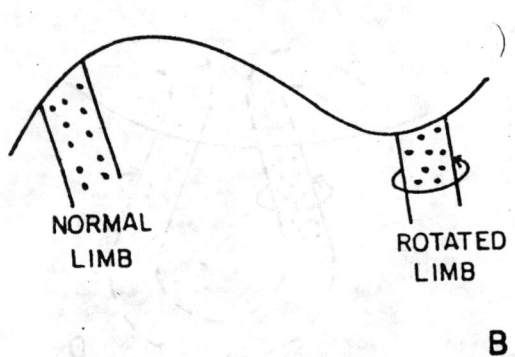

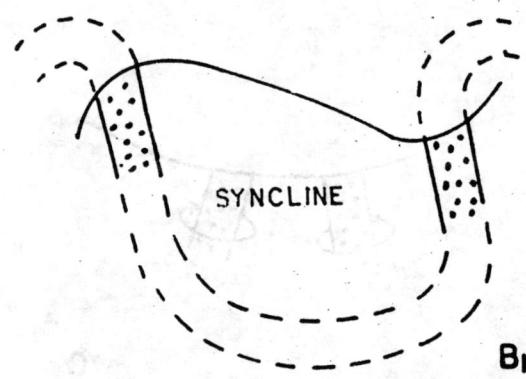

B - both the limbs are dipping to the right hand side, but the limb on the right hand side is overturned as indicated by arrow.

B₁ - reconstruction of limbs into an isoclinal synclinal fan fold.

Figs. 121 A, A₁, B, B₁ Isoclinal fan folds.

such cases clearly and undoubtedly indicate synclinal or anticlinal folds. However it is necessary to ascertain that both the limbs are in their normal position. Thus in Fig. 120 A, though the beds (limbs) are dipping towards each other, the fold is still an anticline, because it is found that the limbs are overturned. In Fig. 120 C, though the beds (limbs) are dipping away from each other, the fold is still a syncline, because it is found that the beds (limbs) are overturned. In Figs. 121 A, B, since the beds are dipping in the same direction, reconstruction of the fold has two possibilities. In the first instance, because the bed (limb) on the left hand side is overturned, the reconstruction is in the form of an anticline. In the second instance, because the bed (limb) on the right hand side is overturned, the reconstruction is in the form of a syncline (Figs. 121 A₁, B₁).

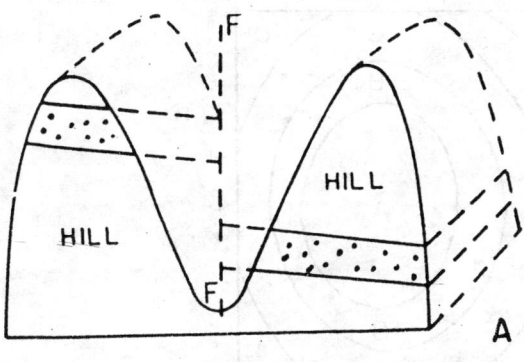

A - due to faulting

B - due to favourable topography and dip direction and amount.

Figs. 122 A,B. Causes of repetition of beds.

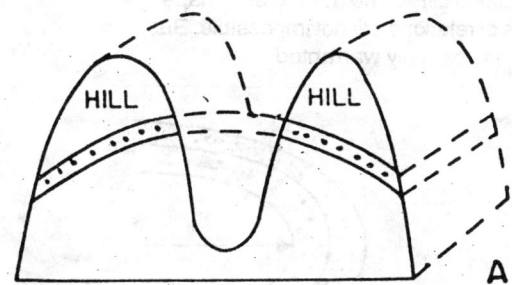

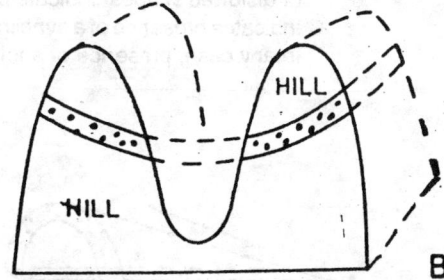

A - due to existence of anticline and favourable topography.

B - due to existence of syncline and favourable topography.

Figs. 123 A,B. Causes of repetition of beds.

(b) **Shape of the outcrop:** In folds, the beds either dip towards or away from each other. Due to this, the outcrops forming the limbs of the fold, are observed to close. The shapes produced by such a closure, are different for the anticline and the syncline. A "diamond or an oval" shaped outcrop is produced for an anticline, while an "hour-glass" shape is produced for a syncline (Fig. 124). Sometimes reversals also are encountered - diamond for a syncline, and an hour-glass, shape for an anticline, However, whether diamond or hour-glass, the shapes definitely indicate the presence of a fold. Distortion of the said shapes are possible in the actual field maps, because the dip amounts are not necessarily uniform throughout the structure. Also the slope of the ground being not uniform, the shapes of the outcrops therefore get distorted. In the isoclinal folds, such outcrop patterns are not possible, because both the limbs dip in the same direction. The outcrop for the two limbs therefore becomes parallel to each other without any closure. If the ground surface be flat, and the outcrops are found to close at the one end, or at both the ends, then singularly plunging and doubly plunging folds, respectively, are to be expected (Figs. 125 A to D and A₁ to D₁).

(iv) **Direct observation:** This comprises of availability of exposure of a complete fold, either an anticline or a syncline, in natural cuttings like valleys, The railway cuttings, the road cuttings also afford excellent opportunities for the exposure of the folds. In such cases no reconstruction is necessary, and there is no ambiguity or necessity for inference to be drawn regarding the existence or otherwise of the fold. Generally small sized folds get exposed in rail or road cuttings. There is the added advantage in this method. The

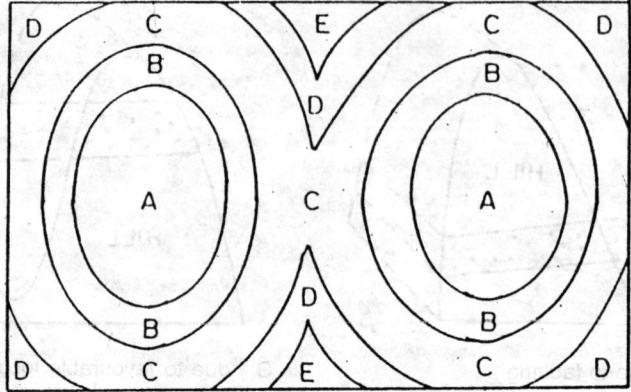

Fig. 124 Shapes of outcrops of fold.

Two shapes are seen. "oval or diamond" and "hourglass". These shapes unquestionably indicate presence of folded structures. Normally "oval, diamond or distorted shapes" indicate presence of an anticline. The "hour glass shape" indicates presence of a syncline. Reversals of relations are not impossible. But in any case, presence of a folded structure is cerainly warranted.

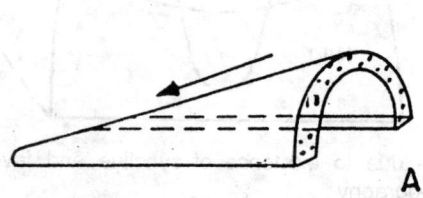

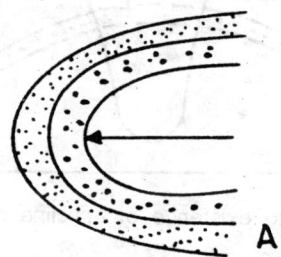

Fig. 125A. Singularly plunging anticline. Arrow indicates the direction of plunge. Note that the "nose" of the fold is in the direction of plunge of the fold.

Fig. 125 A₁. Outcrop pattern of fold shown in Fig. 125A. Note that the outcrops close in the direction of plunge of the fold (indicated by arrow).

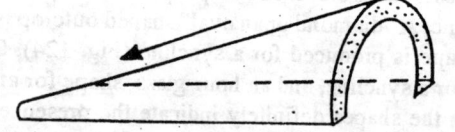

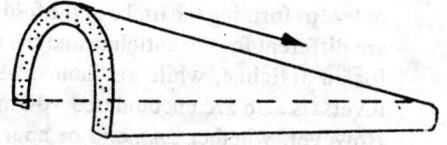

Fig. 125 B. Doubly plunging anticline. Note that the "nose" of the fold is situated in the direction of plunge (indicated by arrows).

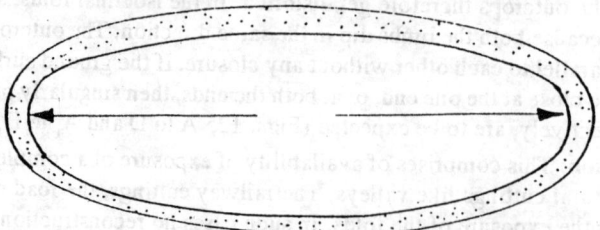

Fig. 125 B₁. Outcrop pattern of fold shown in Fig. 125 B. Note that the outcrops close on both sides, but it is in the direction of plunge. Arrows indicate direction of plunge.

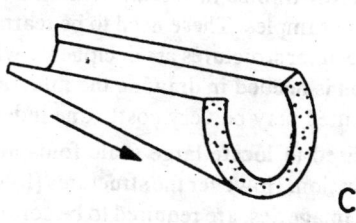

Fig. 125 C. Singuarly plunging syncline. Note that the fold plunges towards right hand side, but the closure (nose) is on the left hand side. Arrow indicates direction of plunge.

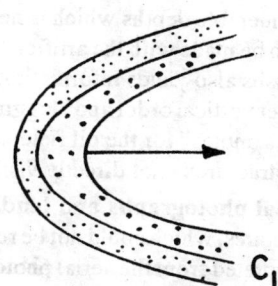

Fig. 125 C$_1$. Outcrop pattern of fold shown in Fig. 125 C. Note that the outcrop closes on the left hand side, while the direction of plunge (indicated by arrow) is on the right hand side.

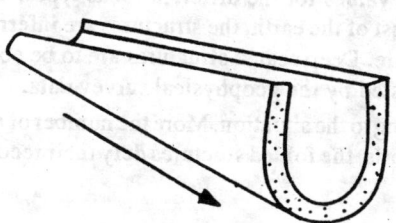

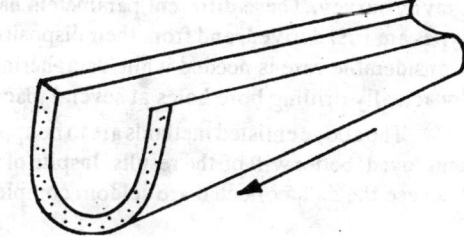

Fig. 125. D. Doubly plunging syncline.

Note that the fold plunges in opposite direction of closure. Arrows indicate direction of plunge.

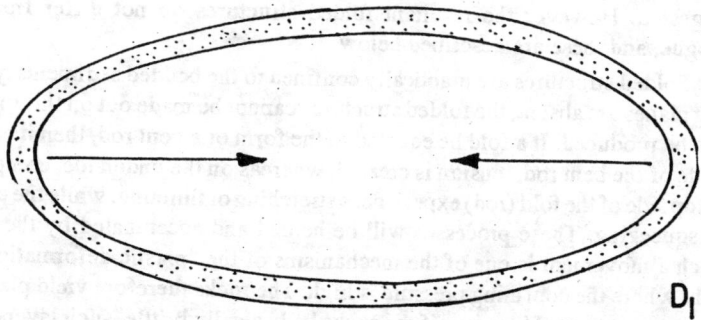

Fig. 125 D$_1$. Outcrop pattern of fold shown in Fig. 125 D. Note that the outcrop closes in the opposite direction of the plunge of the fold. Arrows indicate the direction of plunge of the fold.

depth of folding can be seen directly. Further if two sections perpendicular to each other be available, then the plunging or the non-plunging nature of the fold can be seen directly. But such instances are not frequently available for study. In Photos 50, 51, 52, 53, 54, 55, 56, 57, 58, 59, 60, and in the frontis piece photo of this chapter, different types of folds can be seen directly.

(v) **Quarries, drilling and mining:** The activities of quarrying, drilling and mining yield lot of information. Quarries are better suited, since the exposures are observable directly and that too in the natural sunlight. Also many sections are available in different directions, and these help to study the structures in three dimensions more easily. Further, quarries are available at more number of places unlike the mines. Mining is restricted to the economic mineral deposits, but in this case the information can be obtained from considerable depths which is not possible in the case of quarries. However in the mines the observations are to be made with the artificial lights, and the view is therefore limited in extent. Drilling for the mineral deposits also yields information, but it is in the form of "core samples. These need to be rearranged in the proper vertical order and "logging" is needed. Folds and the other structures are deciphered with the "drill core samples" for the oil fields. However utmost precaution is needed in drawing the inference, because the structure is not directly visible and any mistake committed may be very costly one indeed.

(vi) **Aerial photographs and landsat imageries:** This is utilised to locate large scale folds and the other structures, which could not be reckoned by the direct observation. However the structures (folds included) interpreted from the aerial photographs or from the landsat imageries, are required to be confirmed by the field check. The advantage of this method is that a large area is viewed and the probable site of the structure is narrowed down which avoids the streneous field mapping of area where structures are not developed. Of course the field structures are inferred by studying the dip direction and the amount of dip possessed by the sedimentary rocks, as is inferrable from the aerial photograph or the landsat imageries.

(vii) **Geophysical methods:** These comprise of resistivity survey, seismic survey, magnetic survey, and the gravity survey. These different parameters have different values for the different rocks types. The rock types are first derived, and from their disposition in the crust of the earth, the structures are inferred later. Considerable care is needed while deciphering the structure. Deep seated structures are to be confirmed by actually drilling bore holes at several places, as suggested by the geophysical survey data.

The above enlisted methods are to be applied according to the situation. More the number of methods employed, better will be the results. Inspite of all these efforts, the folded stuctures defy their recognition, because the data collected are seldom complete.

MECHANISM OF FOLDING

The deformative forces no doubt produce the folded structures, but it is desirable to know about the internal adjustments taking place during the process of folding. This is refered to as the mechanism of folding, and several ways have been suggested. However the resultant folded structures do not differ from one another. Four mechanisms are in vogue, and these are described below.

(a) **Flexure folds:** Folded structures are practically confined to the bedded sedimentary rocks. In other rocks where bedding planes are absent, the folded structures cannot be made out on this mechanism, because no bent forms can be produced. If a fold be equated to the form of a bent rod, then it can be appreciated that on the outer side of the bent rod, tension is created, whereas on the underside, compression is developed. Therefore the top side of the fold (rod) experiences stretching or thinning, while the under side experiences thickening or squeezing. These processes will be helped and accentuated by the "intergranular movement", and such a movement is one of the mechanisms of the "plastic deformation". Normally folding occurs at depths where the confining pressure is high. The rocks therefore yield plastically rather than by the development of ruptures. However, if the rocks be basically brittle, such layers or beds may develop fractures, faults etc. These features are shown in Fig. 126.

In the above described procedure, certain internal movement is called into play. Instead of the individual minerals made to move, the indidual layers or the beds themselves are made to move past each other. This is possible because the sedimentary rocks are bedded formations wherein the presence of the several bedding planes is a possibility. The several beds then may be sliding past each other under the influence of tension (top side of the fold), and the compression (underside of the fold). The height of the fold depends on the thickness of the bed. Thinner it is, the height is small, thicker it is, the height is more. All these features have been presented in Figs. 127 A to H.

Thus slipping along the bedding planes thinning, stretching, or thickening, are the mechanisms of the flexure folding. As such wherever thinning, stretching or thickening has been observed, such folds are called as the "flexure folds", and wherever slipping has been observed, those are called as the "flexure slip folds". This mechanism is possible when the rocks are thick and competent ones, so that those could slide past each other or could be thinned, stretched and thickened.

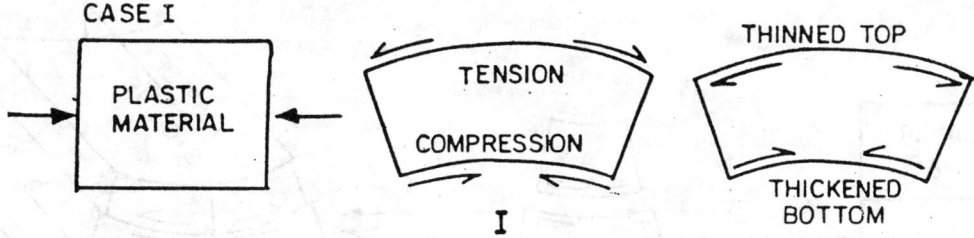

Fig. 126. Case I.

Plastic material subjected to folding. Note that due to compression, the top side is subjected to tension, while the underside is subjected to compression. In between these two zones, shearing situation is created.

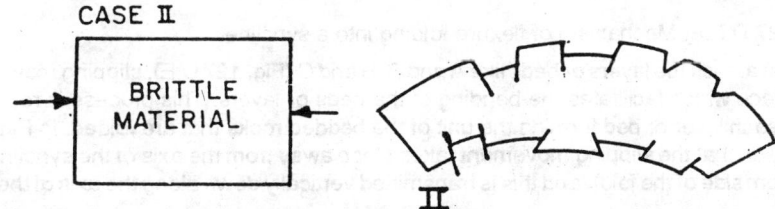

Fig. 126 Case II.

Brittle substance subjected to folding. Note that ruptures, faults, fractures are produced besides bending of the substance into a folded form.

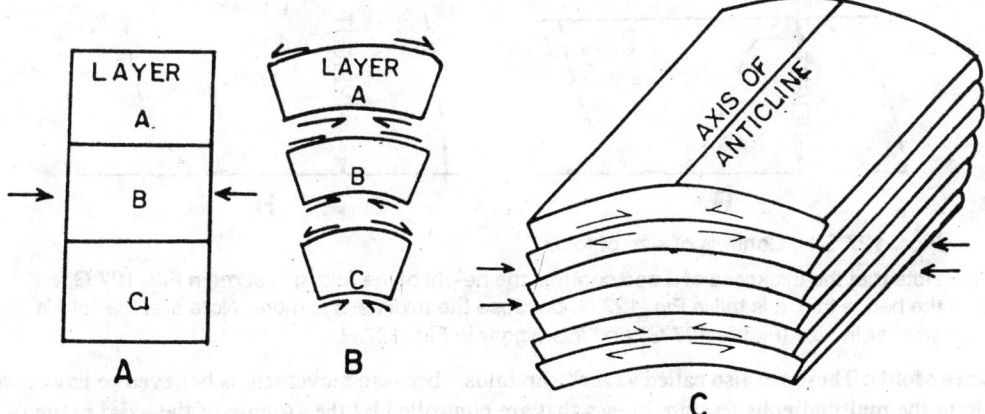

Figs. 127 A,B,C. Mechanism of flexure folding into anticline.

Note that across the layers or beds, like A and B, B and C (Fig. 127 B), slipping movement takes place, which facilitates the bending of the beds or layers. This process is repeated across each layer or bed forming the unit of bedded rocks that are folded. In Fig. 127 C, it is seen that slipping movement takes place towards the axis of the anticline (its under side), and it is transmitted vertically down along the axis of the fold.

(b) **Flow folds:** If the rocks be composed of thinnly bedded, incompetent layers, then the entire unit yields plastically without any slipping movement along the bedding planes. However such folds appear very much alike to those produced by the "flexures". Many minor folds are developed by this mechanism. Flow folds are also called as the "incompetent folds".

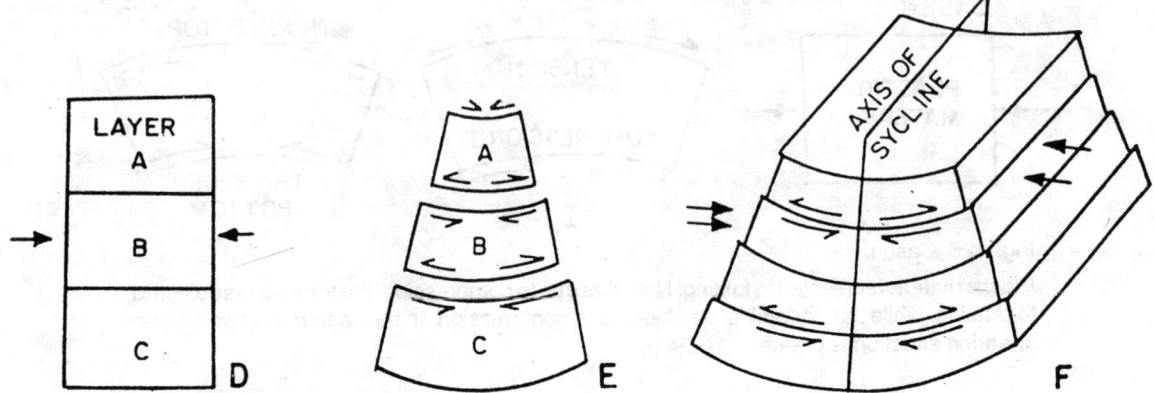

Figs. 127 D,E,F. Mechanism of flexure folding into a syncline.

Note that across the layers or beds like A and B, B and C (Fig. 127 D,E), slipping movement takes place which facilitates the bending of the beds or layers. This process is repeated across each layer or bed forming the unit of the bedded rocks that are folded. In Fig. 127 F, it is seen that the slipping movement takes place away from the axis of the syncline (on the bottom side of the fold), and this is transmitted vertically down along the axis of the fold.

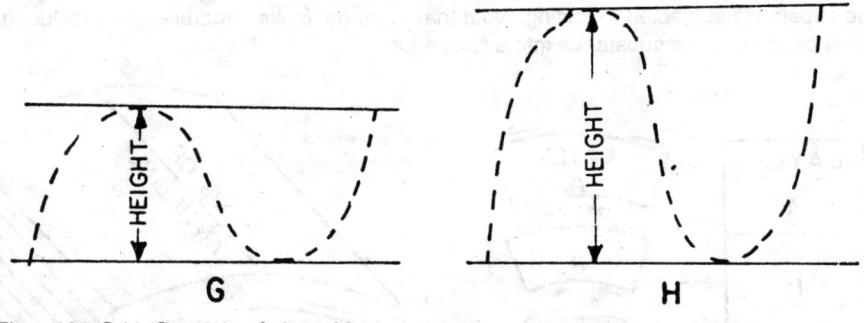

Figs. 127 G,H. Controls of size of fold.

Note that the thickness of a bed controls the height of the fold. It is short in Fig. 127 G, as the bed is thin. It is tall in Fig. 127 G, because the thickness is more. Note that the fold is smaller in size (Figure 127 G) and it is bigger in Fig. 127 H.

(c) **Shear folds:** These are also called as the "slip folds" because movement is believed to have taken place along the multitudinous fracture planes that are controlled by the attitude of the axial plane of the fold (Figs. 128 A to D). The movements are of different magnitudes (Fig. 128 C), and in the end, a folded form is produced. The earlier fracture or the shear planes along which the movement had taken place, those may get healed up. Note that the folds produced by this mechanism have their axial planes inclined i.e., these are neither vertical nor are horizontal in attitude.

(d) **Folds due to vertical movements:** Instead of horizontal compression, vertically directed forces may bring about up-arching of the beds. In fact, the sedimentary basins are uplifted by the vertically acting forces. During such movements, folding can take place. In Fig. 129 the original line ac is up arched to "abc", which obviously needs stretching around "b". Such a process can give rise to a few large sized domes, but not to numerous synclines and anticlines.

The various mechanisms described above might act jointly to bring about folding of a variety of rocks comprising an unit which is affected by the deformative forces. The geoscientist should ascertain as to which one of the mechanisms or a combination there of, was actually operative in the individual cases.

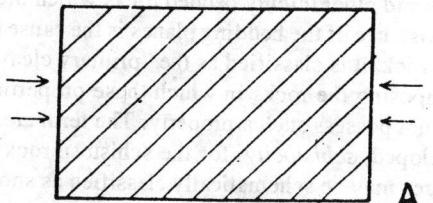

A - bed subjected to shearing forces. Fractures inclined at 60° to the surface are produced.

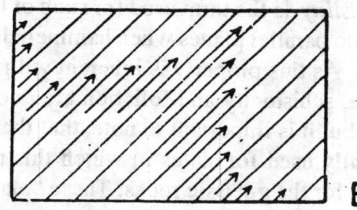

B - differential movement takes place along the fractures due to shearing, and the fractured blocks move over different distances.

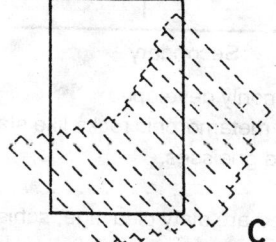

C - actual movement along multitudinous shear fractures, more in the central part, and decreasing on either sides.

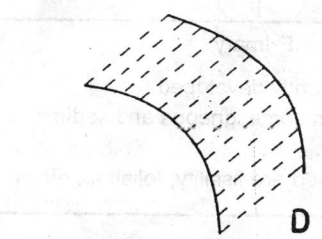

D - Folded form produced ultimately, after healing up of the shear fractures (shown in dashed lines).

Figs. 128 A,B,C,D. Mechanism of shear folding.

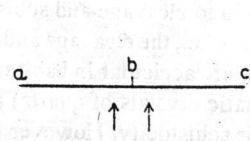

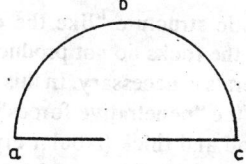

Fig. 129. Mechanism of folding due to vertical movement.

Distance a c, is taken to be same. Therefore stretching of line a c is inevitable at "b", the central part, in order to produce a folded form.

FISSILITY, CLEAVAGE, SCHISTOCITY, FOLIATION, AND LINEATION

These are "plastic" type of deformations and are mainly associated with the metamorphic rocks like the slates, the schists, the granulites, the gneisses and so on. One variety of cleavage called the "axial plane cleavage" is obviously linked with the folds. Moreover these various structures are the result of the action of the deformative forces, but no fracturing or rupturing is observed. Therefore these are grouped under the cateogory of the "plastic structures". Fissility, cleavage, schistocity and foliation will be described together, while the lineation will be described separately.

KINDS OF CLEAVAGE, SCHISTOCITY, AND FOLIATION

Some of the metamorphic rocks, and one or two sedimentary rocks possess the property of breaking into parallel planes when the rock is hammered. This physical property is spoken off as cleavage, schistocity, and foliation in the metamorphic rocks, and as fissility in the sedimentary rocks. Though it is a physical property, it is all the same controlled by the minerals forming such rocks. Further, as it is not possessed by all the rocks, special environment and causes, must be responsible for its development. As it is mostly found in the metamorphic rocks, it is therefore considered as a "secondary" feature. However in some igneous rocks, development of foliation takes place during the process of crystallisation. As such, such foliation is called as a "primary" feature.

Fissility is the term used for such of the sedimentary rocks and other thinly bedded rocks which break or cleave into parallel planes when hammered. In such cases, the existence of the bedding planes is the cause of the cleavage. As this property is inherent with the formation of the rock, it is classified as the "primary cleavage". Cleavage, schistocity and foliation are due to the minerals composing the rocks in which these properties are noticed, but it is important to note, that their parental rocks did not possess such a property. The term *cleavage* is generally used for slates in which this property is best developed, *schistocity* for the schistose rocks, and *foliation* for the gneissic rocks. These "plastic types" of structures may be schematically classified as shown in Fig. 130.

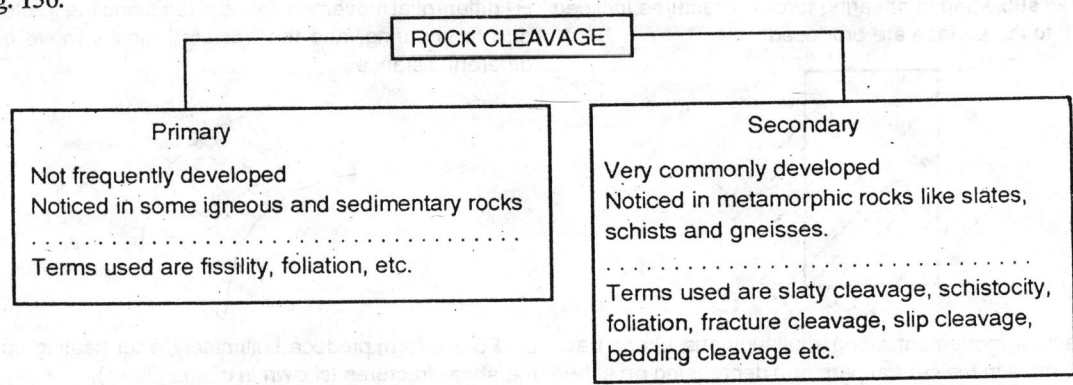

LINEATION

This also is a "plastic structure "like the cleavage and the folds. However this is not readily observable or discernible because the rocks do not produce any planar structure. Akin to cleavage and schistocity, rotation of the constituent minerals is necessary. In this respect, both these structures i.e., the cleavage and the schistocity are due to the action of the "penetrative forces". Obviously, minerals that are accicular in habit or atleast are much more long than broad and thick (tabular crystals of felspars or prismatic crystals of quartz) alone can produce lineation. Very obviously such minerals can not produce cleavage or schistocity. However both cleavage and lineation may be found together though not having equal intensity of development. The commonly occurring minerals in rocks showing lineation are hornblende, kyanite, actinolite and so on. Thus in general, the amphibolic minerals produce lineation more frequently. Followed by these are felspar and quartz. It is also possible that numerous equidimensional minerals may be segregated into linear clusters. Therefore according to the nature of the minerals participating in the development of the lineation, different varieties of lineations are recognised which are described below.

(a) **Minerals oriented lineation:** This is the most common type and is invariably encountered in the hornblende schists and gneisses. Wherever schistocity and lineation are combined, it gives rise to different varieties of forms. The lineation may:

    (i) coincide with the dip of schistocity, or foliation,

    (ii) be at an angle to the schistocity or foliation or,

    (iii) be horizontal while the schistocity or foliation plane is dipping. These different varieties are shown in Figs. 131 A to C. Of course it is not necessary that the rock should possess cleavage or schistose planes. In such cases, the rock can not cleave, but the plunge of lineation could be zero, 90° or any angle between these two values. The appearance of rocks possessing only lineation is shown in Fig. 132.

In the above cases, mineral hornblende has been considered. Mica generally produces schistocity and cleavage, but if the plates be elliptical or a bit elongated in form, then it can also produce lineation in the rocks (Fig. 133). It is also possible that the individual small grains, plates of mica or any other equidimensional mineral, may cluster into oblate bodies and these together may produce lineation (Fig. 134).

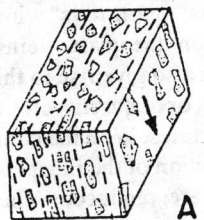 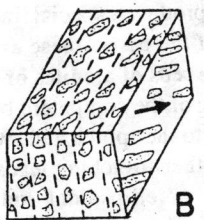 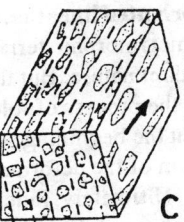

Fig. 131 A. Lineation coinciding with schistocity. Note that the longer axes of hornblende plates, are dipping down the dip of schistocity, Arrow indicates the direction of plunge of lineation. Planes of schistocity are shown by dashed lines.

Fig. 131 B. Lineation at an angle to schistocity. Note that the plunge of lineation (arrow) is at an angle to the dip of schistose plane (dashed lines).

Fig. 131. C Lineation perpendicular to schistocity. Note that the plunge of lineation (arrow) is perpendicular to the dip of schistose planes (dashed lines).

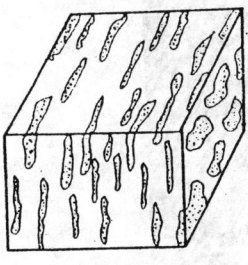

Fig. 132. Lineation without planar structure.
Note that though there is parallelism between the accicular minerals, these are however not confined to planes. Planar structure is therefore not produced, but only lineation is developed.

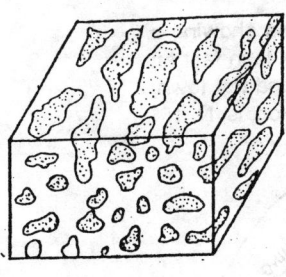

Fig. 133. Lineation due to elongate/ellipsoidal minerals.
Elongate or elliptical plates of mica have given rise lineation. The axes of the plates are parallel to each other, causing lineation structure.

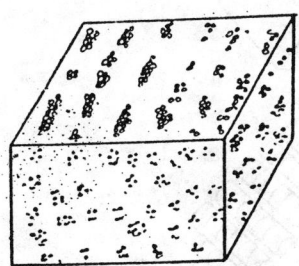

Fig. 134. Lineation due to clustering of minerals.
Aggregation of small, equidimensional grains of minerals may produce clusters. Such clusters may be more long than wide. These may give rise to lineation of their longer axes. The minerals may be quartz, pyroxenes, mica etc. which are not accicular in habit.

Laths of potash felspars are often found to produce lineation in the granitic rocks. Augen gneiss is a good example which has the potential of producing lineation. It is to be noted that only because a mineral is accicular or elongate in form, it need not automatically produce lineation in each case. How many such grains show parallelism between their longer axes, controls the development of lineation. Gothe (1973) describes excellent example of coarse grained augen gneiss that has developed lineation due to the potash felshpars (Photo 61).

(b) **Rock oriented lineation:** Quartz grains produce a special lineation called as "quartz rods" which are found in the metamorphic terranes. In most of the cases, these are produced from the quartz veins that are of concordant nature. But thinner layers or beds of arenites or quartzites can also give rise to this structure. During the process of folding, stretching of the competent beds of arenites or of quartz veins takes place, which in the beginning stages gives rise to the "pinch and swell" like structures. With the continuation of the action of the deformative forces, further stretching leads to the production of the independent lobate bodies. Ultimately independent lobate or ellipsoidal rod like bodies are produced. The stagewise development of the structure is shown in Figs. 135 A to D.

Photo 61. Field photo of a coarse grained granitic gneiss showing a perfect parallelism between laths of K-felspars. The rock is in fact an augen gneiss which is exposed 10 km. N 75° W of Mundargi town, Dharwad district, Karnataka state. The trend of lineation is N 10° W - S 10° E. Courtesy Dr. N.N. Gothe.

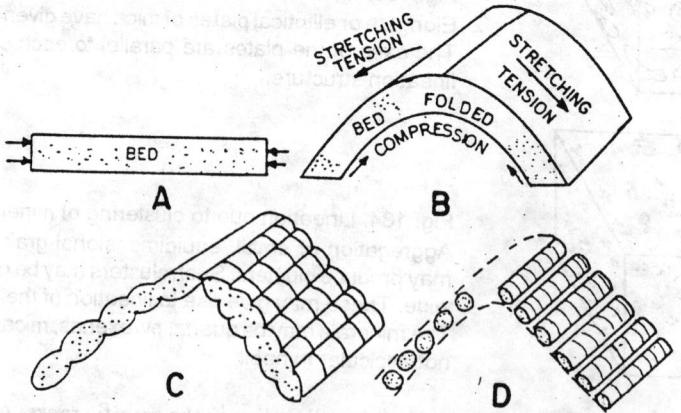

Figs. 135 A,B,C,D. Development of boudins during folding of beds.

A competent bed subjected to compression (A). The bed gets folded into an anticline, developing stretching (tension) on the top side, and compression on the underside of the fold (B). Pinch and swell type of structures are produced due to stretching on the limbs of the fold (C). Independent lobate, ellipsoidal bodies or rods are produced through the continued action of stretching on the limbs of the fold (D).

Gokhale et. al. (1987) have described "fold mullions" from the Halgatti area (Belgaum district, Karnataka state) wherein the quartzarenitic rocks along with the quartz veins traversing them have produced the said structure (Photo 62). At times even the conglomerates develop lineation, if the pebbles and boulders were to be elongate in shape. Koppad (1976) describes such a structure from the Bannipoppa-Bagewadi conglomerate horizon of Dharwarian age (Photo 63). This occurrence is from the Gadag schist belt. Dharwad district, Karnataka state.

Photo 62. Field photo of argillites of Halgatti area (Belgaum district, Karnataka state) documenting development of mullion structure in them. The mullions (rods) dip 30° due N 50° W (towards the observer). The handle of the hammer indicates the plunge of mullions. The rocks are folded and slaty cleavage is developed in them. The mullion structure is attributed to folding and hence it is a plastic deformation. Courtesy Dr. G.V. Hegde.

Photo 63. Field photo of a conglomerate exposed near Bannikoppa village, Shirhatti taluka, Dharwad district, Karnataka state. It forms a part of the Gadag shcist belt. Note the parallelism between the longer axes of the elongated pebbles of the conglomerate. This has given rise the lineation structure. Courtesy Dr. V.B. Koppad.

(c) **Lineation not due to minerals:** Lineation being a linear structure, it can be produced even without the direct participation of the minerals. During faulting striations and grooves are usually produced on the fault planes. As these are lines, these therefore are deemed to produce lineation. In fact from the trend of the striations, varieties like the strike-slip, the dip-slip and the oblique-slip faults are recognised. These structures are not due to any mineral, but are due to the displacement taking place along the fault plane. The line of intersection of the two planar structures can give rise to lineation. Thus the intersection of the bedding plane and the slaty cleavage produces lineation. Cleavage planes being numerous as compared to the bedding planes, these cleavages appear as "lines" on the bedding plane, and thus these lines develop the lineation (Fig. 136).

The conical and the chevron type of folds can produce lineation of their axes, because in these folds, the two limbs meet into sharp lines (Figs. 137 A,B). There can be other situations like the parallelism of the longer axes of the xenoliths occurring in the granitic rocks and in the other igneous rocks. Refolding of the limbs of the main fold will give rise to the secondary folds. These latter folds will have their axes either parallel or perpendicular to the primary or the main fold axis, which will give rise to lineation (Figs. 138 A,B.)

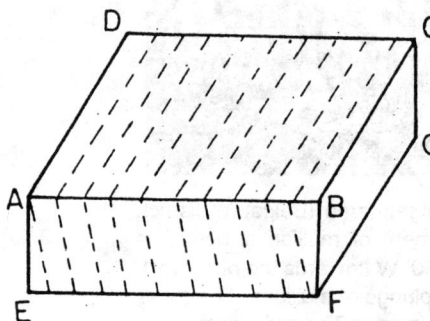

Fig. 136. Lineation due to rock cleavage.
ABCD and EFG are bedding planes. Dashed lines are cleavage planes which are produced in the rock. These lines on the top side of the block, produce lineation on the bedding plane ABCD.

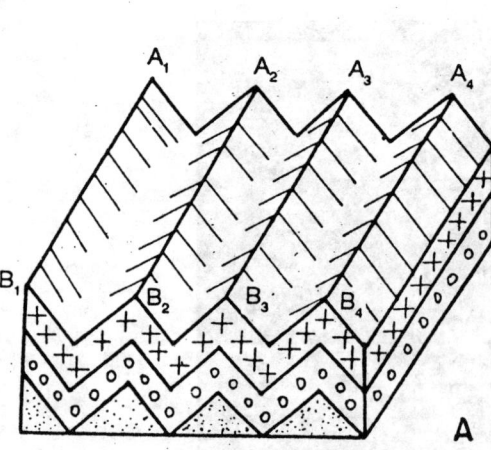

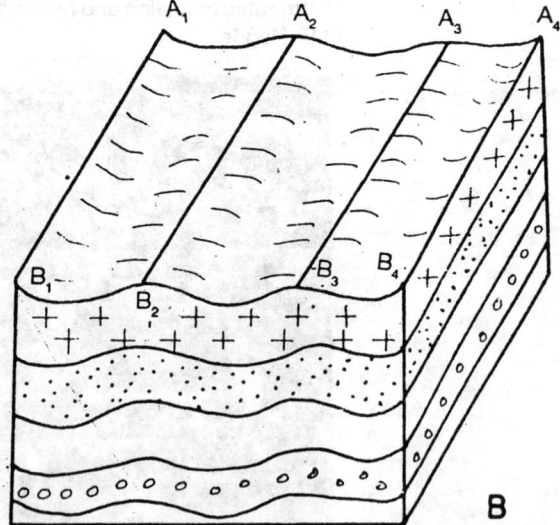

Fig. 137 A. Chevron fold producing lineation.
These folds have sharp axes and thus produce lines. These lines are considered as developing lineation. $A_1 - B_1$, $A_2 - B_2$ etc., are axes of anticlines and these have given rise to lineation.

Fig. 137 B. Conical fold producing lineation.
These folds do not have sharp crests or troughs. The lineation developed by the axes of such folds will be ill-defined. $A_1 - B_1$, $A_2 - B_2$ etc., are the lines of lineation.

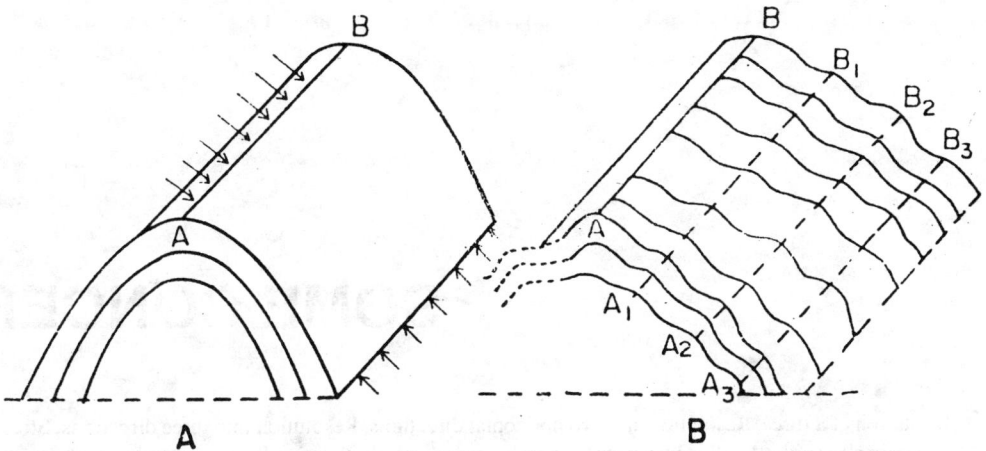

Fig.s 138 A,B.
Anticline affected by a second period of folding of the limbs of the fold. Arrows indicate the direction of refolding. AB is the axis of the main fold (Fig. 138 A). Due to refolding of the limbs, secondary axes A_1 - B_1, A_2 - B_2 etc., are produced which are parallel to the main axis AB. The secondary axes (Fig. 138 B) A_1 - B_1, A_2 - B_2 etc., produce lineation on the limbs of the anticline. Dashed lines stand for the secondary axes of refolding.

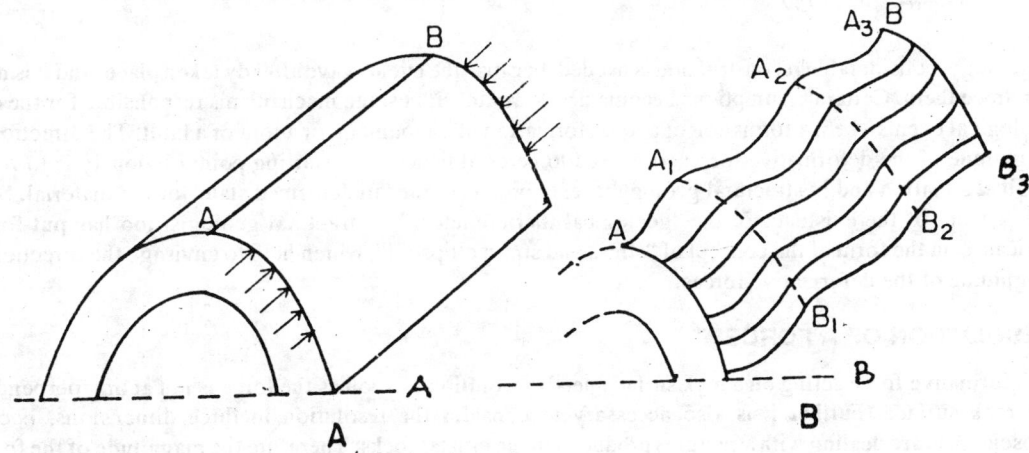

Figs. 139 A,B.
Anticline affected by a second period fo folding of the limbs. Arrows indicate the direction of refolding. AB is the axis of the main fold (Fig. 139 A). Due to refolding of the limbs, secondary fold axes A_1 - B_1, A_2 - B_2 etc., are produced. These are perpendicular to the main fold axis. The secondary axes (dashed lines) produce lineation on the limbs of the anticline (Fig. 139 B).

Combination of the several kinds of the planar structures can give rise to the lineation, but the one produced by the minerals or the rock bodies, can be made out relatively more easily. Compared to cleavage, lineation is not frequently developed, and its recognition is difficult in comparison with the cleavage. It is also noticed that both cleavage and the lineation may be developed together, introducing complexity, because both the planar and the linear structures get superimposed upon one another.

4

SOME CONCEPTS

Resolution of a force, Resolution into two horizontal directions, Resolution into three directions, Stress and strain ellipsoids, Relation between stress and strain ellipsoids, Relation between structures, stress and strain ellipsoids, and direction of deformative forces, Derivation of direction of compressive deformative forces.

Orientation of stress ellipsoids for normal, gravity, reverse, thrust, nappe, sinistral, dextral, tear and wrench faults.

Patterns of stress distribution and resultant structures. Deformative force constant in magnitude. Deformative force not constant in magnitude. Complex type.

Figures 140 to 157.

In geology, considerable reconstruction is needed, because the events have already taken place, and it is necessary to retrace them. Certain assumptions become necessary to suggest the mechanisms responsible for the observed geological events - be it a formation of a rock, formation of a mountain, or a fold or a fault. The direction and the magnitude of the deformative forces also need to be established. The starting point obviously is to restore the original situation and in structural geology it tentamounts to the "undeformed state" of the material. Similar to the concept of the existence of the "geological thermometers", a structural geologist too has put forth some indicators in the form of the concept of "stress and strain ellipsoid", which help to envisage the direction and the magnitude of the deformative forces.

RESOLUTION OF A FORCE

A deformative force acting on a rock surface needs resolution as soon as the force is not acting perpendicular to the rock surface. Further it is also necessary to consider the resolution in three dimensions, because the geoscientists are dealing with structures produced in the crustal rocks. Therefore the magnitude of the force in the two horizontal directions, and in the third vertical direction, is required to be considered. It is a common knowledge that when the forces are acting at a point in two different directions (but not perpendicular to each other), the resultant is produced in an altogether different (third) direction. This analogy is made use of here too, and the resolution of a force at a point into two horizontal directions is considered first. Then the resolution of a force along two horizontal and one vertical direction, is considered next.

RESOLUTION OF A FORCE INTO 2 HORIZONTAL DIRECTIONS

In Figs. 140 A to D, a force OA acting on a surface LM is shown. Depending upon the angle theta (θ) subtended between OA and LM, the force OA gets resolved into a component OB which is perpendicular to LM, and a component OC which is parallel to LM. OB and OC are called the normal and the parallel components, respectively. The nagnitudes of the components OB and OC depend upon the angle theta (θ). It will be observed that as the value of theta becomes less than 90°, the magnitude of OC component increases, but the magnitude of OB component is found to decrease. This feature is appreciable in Figs. 140 B, C and D. The normal component

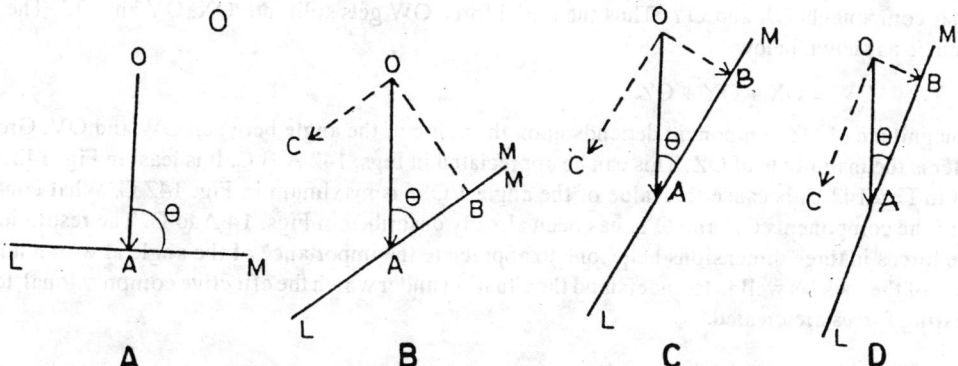

Fig. 140 A,B,C,D. Resolution of a force acting on a rock surface.

OA = direction of deformative force

LM = rock surface on which OA is acting OB = normal component, OC = parallel component

Note that as the value of angle theta decreases from 90° (Fig. 140 A) to lesser values (Figs. 140 B,C,D), the parallel component OC (shear) assumes greater intensity while the normal component OB (compression or tension) assumes smaller values. No resolution is possible in Fig. 140 A, because OA is acting perpendicularly to the surface LM.

assumes the nature of compression or tension. The parallel component (OC in Figs. 140 B,C,D) is called as the shearing force.

Resolution of the compressional and the tensional forces is shown in Figs. 141 A,B, respectively. The magnitude of the normal component OB in Figs. 141 A,B once again depends upon the value of theta. It will be observed that in Fig. 141. A, OB is acting perpendicular to and towards the surface LM. This produces an environment of compression. In Fig. 141 B, OB is perpendicular to but away from the surface LM. This produces an environment of tension.

RESOLUTION OF A FORCE IN THREE DIRECTIONS

This has been shown in Figs. 142 A,B,C. Force OW is acting in a vertical plane OVWZ. The force OW is resolved into two components viz., in a vertical direction OZ, and in a horizontal direction OV. It is to be noted that OZ and OV are perpendicular to each other. The horizontal component OV, is further split into two mutually

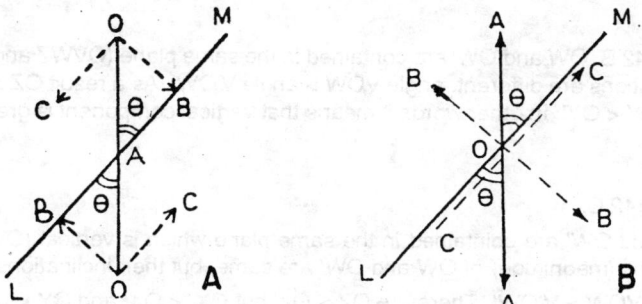

Figs. 141 A,B. Resolution of compressional and tensional forces.

OA = direction of deformative force

LM = rock surface on which deformative force is acting

OB = normal component

OC = parallel component (shear) component

OA becomes compressional force in Figs. 141 A, and tensional force in Fig. 141 B.

perpendicular components OX and OY. Thus the initial force OW gets split into OX, OY and OZ. The relation between them is as shown below.

$$OW = OX + OY + OZ.$$

The magnitude of OZ component depends upon the value of the angle between OW and OV. Greater the value, greater is the magnitude of OZ. This can be appreciated in Figs. 142 A,B,C. It is least in Fig. 142 A, while it is highest in Fig. 142 C, because the value of the angle VOW is maximum in Fig. 142 C. What controls the magnitude of the components OX and OY, has been already described in Figs. 14 A to D. The resolution of the deformative forces in three dimensions helps one to appreciate the importance of the angle at which it is acting on the surface of the rock, as well as to understand the situation under which the effective compressional, tensional and the shearing forces are created.

STRESS AND STRAIN ELLIPSOIDS

According to Newton, to every action, there is an equal and opposite reaction. When the rocks are being deformed, the deformative force constitutes the "action", and the rocks resist the deformative force, this latter sets in the "reaction". Owing to this, the body of the rock is under the effect of the action and the reaction. This situation continues as long as the deformative force acts on the rock. This condition is described by the term "stress". Therefore stress is not a force, but it is a "state or an environment" which is created in the rock. Stress is created due to the external force, and it will be equal to that force upto a certain limit (called as the strength or the competency of the rocks), which is otherwise designated by the term "elastic limit". Beyond that stage, either plastic or ruptural deformation takes place, which has been described in Chapter 1.

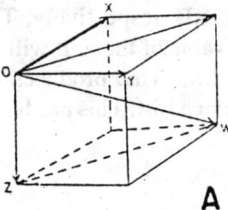

Fig. 142 A. OW = OX + OY + OZ

angle VOW = inclination of deformative force

OW is contained in vertical plane OVWZ.

OX, OY are horizontal components, OZ is vertical component.

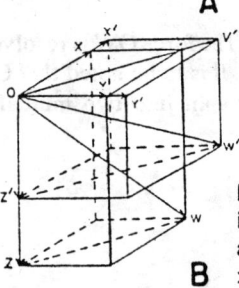

Fig. 142 B. OW and OW' are contained in the same plane (OVWZ and OV'W'Z'), but their inclinations are different. angle VOW > angle V'OW'. As a result OZ > OZ', but OZ < OX' and OY < OY'. In other words it means that vertical component is greater because VOW > V'OW'.

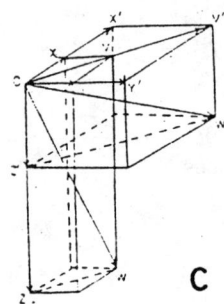

Figs. 142 C.

OW and OW' are cointained in the same plane which is vertical (OVWZ and OV'W'Z'). Lengths (magnitude) of OW and OW' are same, but their inclinations are different.

angle VOW > V'OW'. Therefore OZ > OZ' but OX < OX' and OY < OY'.

Note: The values of components OX, OY, and OZ depend upon the inclination of the force OW in the plane. When it is more (Fig. 142 C), the vertical component becomes more. When it is less (Fig. 142 A) then the vertical component becomes less. Thus OZ is maximum in Fig. 142 C, less in Fig. 142 B and least in Fig. 142 A.

Figs. 142 A,B,C. Resolution of force in three dimensions.

As a measure of deformation, certain initial "body form" is required to be known. To say that stretching or elongation has taken place, originally the material should have been spherical in shape. However a geologist deals with rocks which are composed of minerals. These are not necessarily spherical in shape, and some of them are already more long than broad, owing to their habit of crystallisation. During the deformation of the rocks, the minerals are affected. Therefore the amount of change in the original form of the mineral, is to be taken as a measure of the intensity of deformation. But since the minerals are of mixed shapes, the concept of stress and strain ellipsoids, is therefore put forth. As already said, the resultant structure, be it a joint, a fold, or a fault, is the starting point. And this is considered as the "strain". Thus from the analysis of the strain (meaning the structure), the intensity and the style of the deformation is established. From the strain, the stress conditions are derived, and this latter information helps to orient the direction of the deformative forces. Therefore in structural geology, considerable importance is attached to these conepts.

STRESS ELLIPSOID

Since the stress is a condition which is created due to the external deformative force, its value and nature will be same as that of the external force. Deformative forces acting in the crust of the earth at a point, are resolved into three directions that are mutually perpendicular to each other (see Fig. 142 A,B,C for the resolution of force in three directions). Such a resolution shows that one of the forces is the highest, one is lowest, and the third is intermediate in value. When these are drawn to a scale, the solid body resulting therefrom becomes an ellipsoid, because the intensities of the three forces along the three directions are unequal. Such an ellipsoid is shown in Fig. 143.

STRAIN ELLIPSOID

This has the same significance as that for the stress ellipsoid. In fact this ellipsoid is more useful, because it represents the effect of the deformative forces. Therefore from the "strain ellipsoid" the stress ellipsoid can be derived. The intensity of straining (meaning deformation) will be proportional to that of stress. Hence along the maximum or the greatest stress, the strain will be maximum, and along the minimum or the least stress, the strain will be minimum. The shape of the strain ellipsoid will be similar to that the the stress ellipsoid, but its orientation with respect to the stress ellipsoid will be different. The strain ellipsoid is shown in Fig. 144.

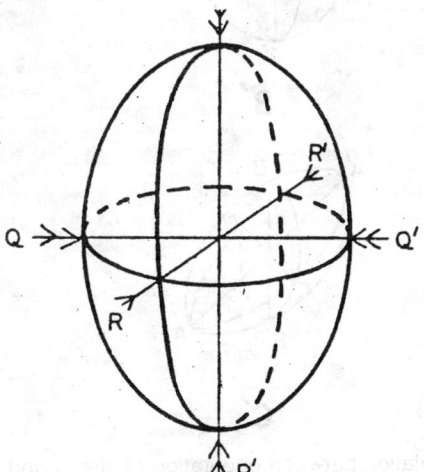
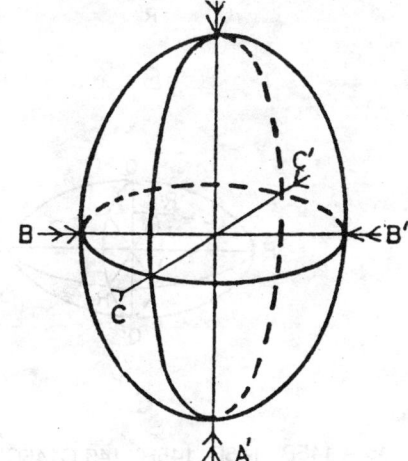

Fig. 143. Stress ellipsoid.

PP = greatest principal stress axis

QQ = intermediate principal stress axis

RR = least principal stress axis.

Fig. 144. Strain ellipsoid

AA' = greatest principal strain axis

BB = intermediate principal strain axis

CC = least principal strain axis

RELATION BETWEEN STRESS AND STRAIN ELLIPSOID

The relation is exactly opposite between these two ellipsoids in respect of the position of the greatest and the least principle axes. Thus the direction of the maximum stress becomes the direction of least strain, and that of the minimum stress, becomes one of the maximum strain. The position of the intermediate stress and the strain axes remains the same. These features are shown in Figs. 145 A to C and A' to C'. From these figures it is apparent that if the strain ellipsoid is horizontal, then the stress ellipsoid becomes vertical and vice versa. Further, from the actual direction of the axes of the ellipsoids as derived in the field, the direction of the deformative forces can be derived.

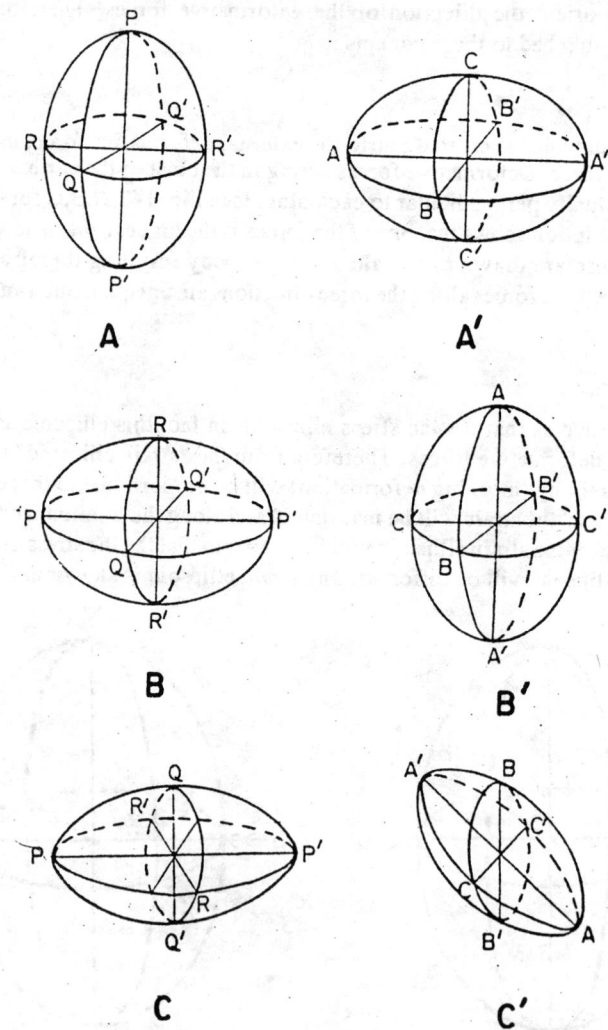

Fig. 145 A 145B', 145B 145B', 145 C145C'. Relation between orientation of stress and strain ellipsoids.

Fig. 145 AA' BB'CC'. Relation between orientation of stress and strain ellipsoids.

When the stress ellipsoid is vertical (meaning PP' is vertical), the strain ellipsoid is horizontal (meaning AA' is horizontal) and vice versa. In the case of stress ellipsoid being horizontal, two situations arise. In one case RR' is vertical, and in the second case, QQ' is vertical.

RELATION BETWEEN STRUCTURES, STRESS AND STRAIN ELLIPSOIDS, AND DIRECTION OF DEFORMATIVE FORCES

The ruptures may be by way of joints, faults or shears, these are produced due to the different deformative forces like the compression, the tension, the shear and so on. The geoscientist desires to derive the kind of the force through the study of the structures. Arguments are to be put forth in support of the force selected - tension, compression or shear. Little difficulty will be experienced in the case of tension or shear, because the direction is either perpendicular or parallel to the trend of the fracture. Thus if the fracture be trending N 40°W - S 40°E direction, then the tensional force must have acted in a N 50°E - S 50°W direction. If it is due to shearing forces, then those must have operated in a N 40° W - S 40°E direction. The problem will be created with the compressive force. The experiment with the square prism however has helped to derive the direction of the forces. The procedure will be described in the following paragraphs, which is applicable to the *compressive forces only.*

While describing the mechanism of the ruptural deformation, reference to the square prism experiment was made. It has been shown that there is a specific relation between the rupture planes and the direction of the maximum deformative force, and the same can be viewed in the light of the stress and the strain ellipsoids. Figs. 146 A,B,C, 147 A,B,C, and 148 A,B,C bring out these relations.

In Figs 146 A to C, 147 A to C, and 148 A to C, the fractures stand for some structures. These actually could be faults or joints. The systematic relation between the strain ellipsoid and the stress ellipsoid, between the stress elipsoid and the square prism, between the square prism and the resultant fractures, and finally that between the deformative forces and the actually developed fracture planes, has been clearly shown in Figs. 146 A to C, 147 A to C and 148 A to C. Thus the position of the fracture planes (meaning the structure) and their orientation in the field, are very important, because only from such a data, the *direction of the actually acted deformative forces, the orientation of the strain ellipsoid, and then that of the stress ellipsoid can be derived.*

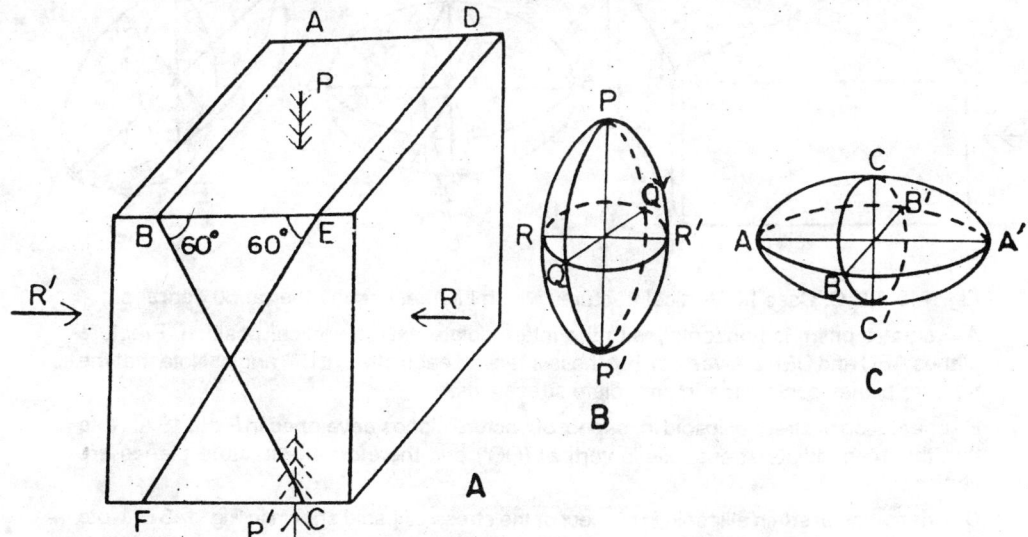

Fig. 146. A, B, C. Case I. Intersecting fractures, the inclination of each fracture plane being around 60 degrees.

A - square prism is vertical. Fracture planes ABC and DEF have inclination of 60°. Note that the fracture planes contain the intermediate stress axis which is horizontal.

B - orientation of stress ellipsoid for fractures developed in Fig. 146 A. Note that the stress ellipsoid is vertical, i.e., PP' is vertical.

C - orientation of strain ellipsoid in respect of stress ellipsoid shown in Fig. 146 B. Note that the strain ellipsoid is horizontal (AA') while CC' is in the vertical position.

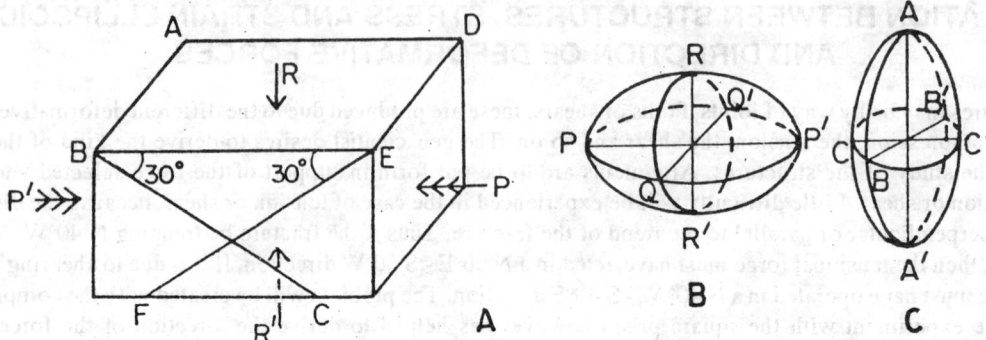

Figs. 147 A,B,C. Case II. Intersecting fractures, the inclination of each plane being around 30 degrees

A - square prism is horizontal. Fracture planes ABC and DEF dip 30 degrees. Note that the fracture planes contain the intermediate stress axis which is horizontal.

B - orientation of stress ellipsoid for fracture planes developed in Fig. 147 A. Note that the stress ellipsoid is horizontal (PP').

C - orientation of strain ellipsoid in respect of stress ellipsoid shown in Fig. 147 B. Note that the strain ellipsoid is vertical (AA' is vertical).

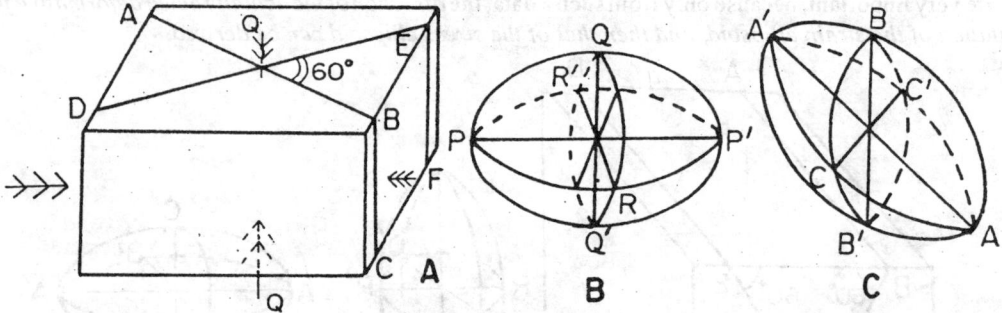

Fig. 148 A,B,C. Case III. Vertical fractures which intersect each other at 60 degrees.

A = square prism is horizontal, with the intermediate axis in vertical position. Fracture planes ABC and DEF are vertical, but these intersect each other at 60° angle. Note that the fracture planes contain the intermediate stress axis.

B - orientation of stress ellipsoid in respect of fracture planes developed in Fig. 148 A. Note that the intermediate stress axis is vertical (QQ') and therefore the fracture planes are vertical.

C - orientation of strain ellipsoid in respect of the stress ellipsoid shown in Fig. 148 B. Note that the strain ellipsoid is horizontal (AA' is horizontal), but the intermediate strain axis is vertical. Hence the fracture planes are also vertical.

DERIVATION OF DIRECTION OF COMPRESSIVE DEFORMATIVE FORCES

The "square prism experiment" is quite useful in achieving this goal. It will be recalled here that the fracture planes bear three types of relations with the deformative forces, namely,

   (i) the fracture planes possess high angle of dip and these planes further intersect at angles of 60° and 120°, then the deformative forces bisect the acute angle of 60° (Figs. 149 A and 14 A).

(ii) the fracture planes are verticlal and are parallel to the direction of the deformative force (Figs. 17, 18).

(iii) the fracture planes are horizontal and are perpendicular to the direction of the deformative force (Fig. 19).

Therefore if the fracture planes intersect each other around an angle of 60°, then the bisectrix of that angle locates the direction of the maximum deformative force (stress). However it is necessary that two fracture planes must develop. If only one plane were to develop, and if such a plane were to dip around 60°, then the deformative force bears an angle of 30° with respect to such a fracture plane (Fig. 149 B). If the fracture planes were to have a low dip of 30°, and further, if these planes were to intersect each other around an angle of 60°, then the deformative force is nearly horizonal and it is obtained by bisecting the angle of 60° (Fig. 150 A). If only one fracture plane having a low dip of 30° were to be present, then the direction of the deformative force can be obtained by constructing an angle of 30° with respect to the fracture plane (Fig. 150 B).

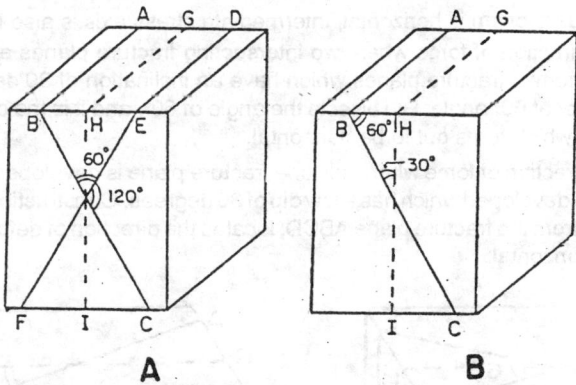

Fig. 149 A. Derivation of the deformative force, when 2 fracture planes are developed. The square prism is vertical.

ABC and DEF are the 2 fracture planes which intersect each other at 60° angle. GHI is the bisectrix of the angle, and it is the direction of the deformative force. It is vertical. The fracture planes ABC and DEF possess an inclination of 60 degrees.

Fig. 149 B. Derivation of the deformative force when only one fracture plane is developed. The square prism is vertical.

ABC is a fracture plane which has a dip of 60 degrees. Construction of a plane GHI which is 30° away from ABC locates the direction of the deformative force. It is once again vertical.

If the fracture planes be vertical and if these intersect each other at an angle of 60°, then the deformative force acts nearly in a horizontal direction, and the plane containing it, bisects the angle of 60° (Fig. 151 A). If only one vertical fracture plane were to be developed, then the direction of the deformative force is obtained by constructing an angle of 30° with respect to such a fracture plane (Fig. 151 B). If the fracture planes be vertical but are parallel to each other, or these intersect each other at an angle of 60°, then the actual direction of deformative force is difficult to ascertain, but it is certain that it is parallel to the fracture planes. It will be recalled here that in the presence of a lubricant, the fracture planes become parallel to the unconfined sides of the prism (Figs. 17, 18). In the natural set up, the lubricant may be the circulating ground water, or the vapours and the gases emanating from a magma and so on. The derivation of the deformative forces under these circumstances is shown in Figs. 152 A,B. The deformative force will be contained in a plane parallel to the vertical fractures. But whether the force acted vertically or horizontally, it is difficult to ascertain. This can be understood better from Figs. 152 A,B, wherein the square prism is shown in the vertical and in the horizontal positions, but the fractures produced in both the cases are similar, in that these are vertical in attitude. If the fracture planes are horizontal, then the deformative force will be perpendicular to it and it will be of the nature of tension, but it is called as the "release fracture" to which a mention has been made in Chapter 1. (Fig. 19).

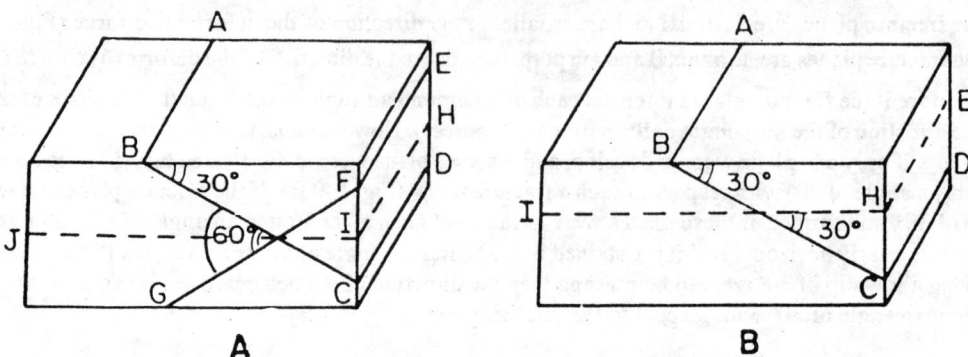

Figs. 150 A,B. Square prism is horizontal, intermediate stress axis is also horizontal.

A - Derivaion of direction of force when two intersecting fracture planes are developed. ABCD and EFG are two fracture planes which have an inclination of 30 degrees. These intersect each other at 60° angle. HIJ bisects the angle of 60°, and it is the direction of the deformative force which turns out to be horizontal.

B - Derivation of direction of force when only one fracture plane is developed. ABCD is the only fracture plane developed which has a low dip of 30 degrees. Construction of plane GHI which is 30° away from the fracture plane ABCD, locates the direction of deformative force. It is found to be horizontal.

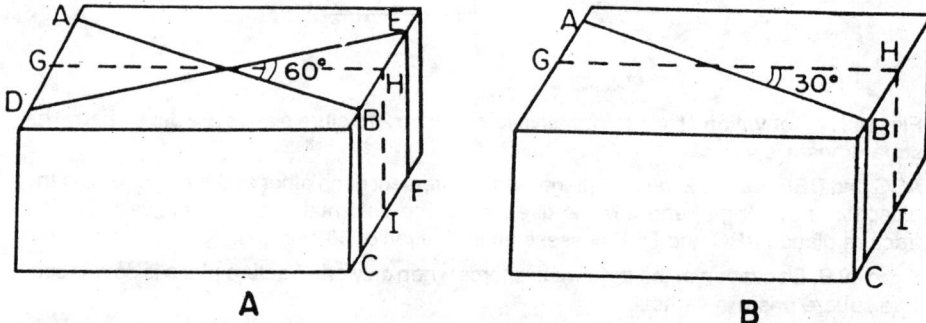

Fig. 151. A,B. Derivation of deformative force when vertical fractures are developed. Square prism is horizontal, the intermediate stress axis is vertical.

A - ABC and DEF are two intersecting but vertical fractures. Angle of intersection is of 60 degrees. Bisectrix of this angle locates (GHI) the direction of the deformative force. It is therefore horizontal in nature, as it is acting along the plane GHI.

B - ABC is the only fracture plane and it is vertical in attitude. The direction of the deformative force is obtained by constructing an angle of 30° with respect to the plane ABC. GHI is such a plane. The deformative force is contained in that plane, and it is horizontal in nature.

These fracture planes may stand for joints, faults, shears and so on. The method of derivation of the direction of the compressive force will be identical in the several rupture structures. However some latitude is to be given for the natural environments. The angular relation of 60° and 120° may not be exactly realised in the crustal rocks. But the values all the same might not be too much different from those of 60° and 120°. The orientation of the stress and the strain ellipsoids therefore become the useful tools in structural analysis.

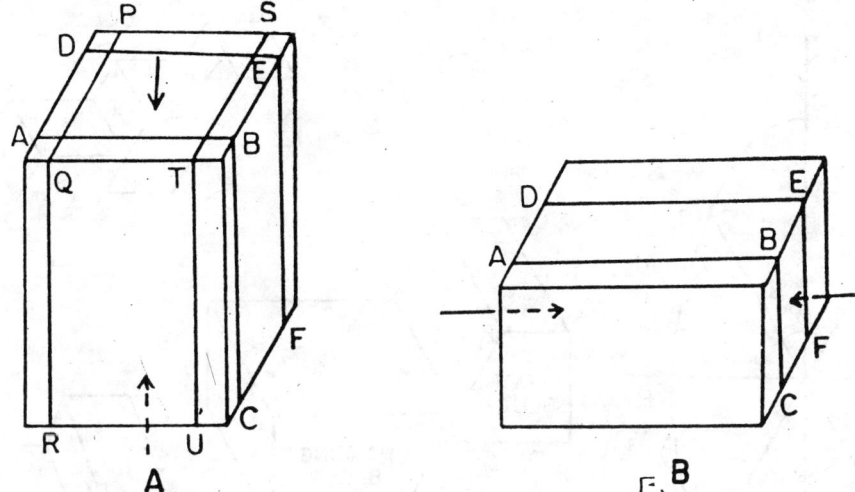

Figs. 152 A, B. Derivation of direction of deformative force when fracture planes are vertical.

A - ABC, DEF, PQR, STU are all vertical planes which intersect each other at 90° degrees. Note that the square prism is vertial and therefore the deformative force also must act in vertical direction.

B - ABC and DEF are vertical planes but these do not intersect each other. Note that direction of deformative force is horizontal because the square prism is horizontal in attitude.

Note that when the fracture planes are vertical, it is possible that these are produced by forces acting in vertical or horizontal direction. Uncertainty therefore exists.

ORIENTATION OF STRESS ELLIPSOIDS FOR SOME STRUCTURES

It is desirable to know the structural environment (orientation of the stress ellipsoid) under which the different structures are produced The orientation of the stress ellipsoid in respect of different types of faults has been described below. It can be extended to the other fractures, ruptures and the joints.

I. NORMAL, GRAVITY FAULTS

These are developed when the stress ellipsoid is vertical i.e., the greatest principle stress axis is vertical. The fault plane will contain the intermediate principle stress axis, and the plane will be perpendicular to the plane containing the greatest principle and the least principle stress axes. The fault plane will be about 30° away from the greatest principle stress axis. These features are shown in Figs. 153 A to D. Thus when the stress ellipsoid is vertical, it is favourable for the development of the normal, the gravity, the step, the graben, the trough and the trench faults. If however the foot block were to move upwards, then it will give rise to the upthrusts. Depending upon the actual geographical location of the least and the intermediate stress axes (R and Q), the actual trend of the fault will be different. Such faults have a high angle dip of about 60 degress.

II. REVERSE, THRUST, NAPPE FAULTS

When the stress ellipsoid is not only horizontal, but the least principle stress axis is vertical, then such fautls are produced. The fault plane has a low dip of about 30°, and it can give rise to the reverse, the overthrust, and the underthrust faults. The orientation of the stress ellipsoid is shown in Figs. 154 A to C. In Figs. 154 B and C, if the foot block moves down, then it can give rise to the underthrust. When the overthrusting is extensive, it leads to the development of a "nappe structure" (Fault).

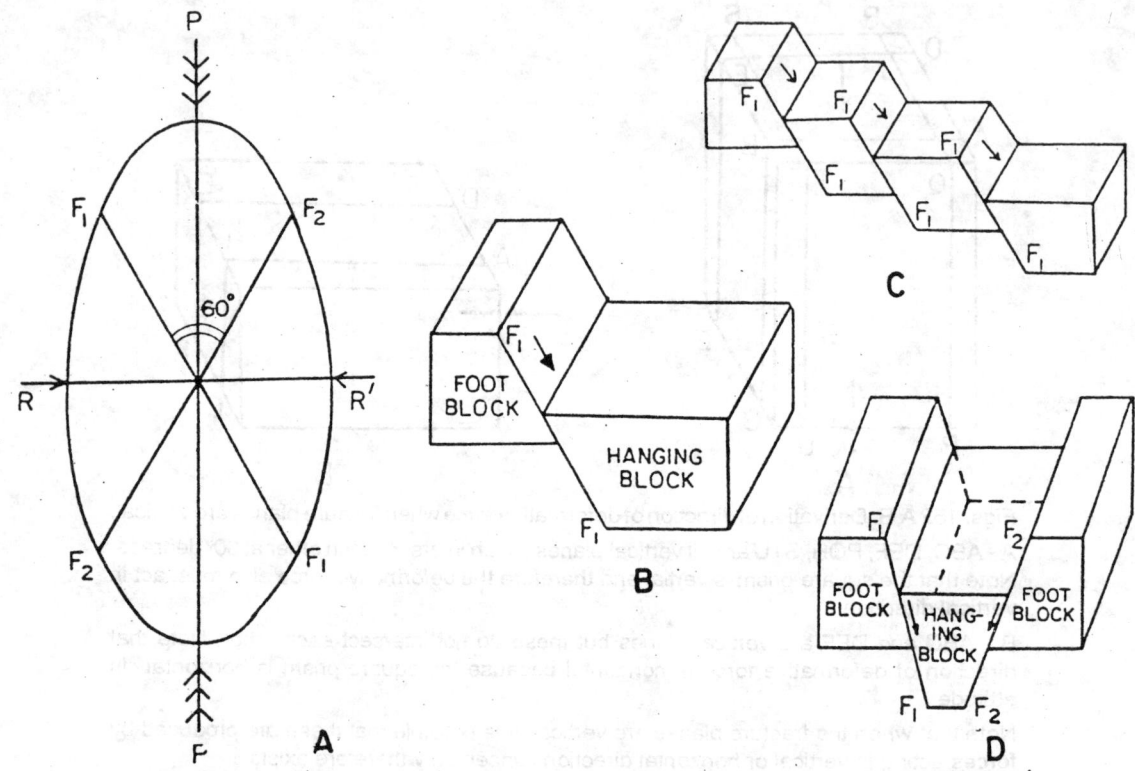

Fig. 153. A,B,C,D. Orientation of stress ellipsoid in respect of normal/gravity/step/graben/ trough/trench faults.

A - stress ellipsoid is vertical i.e. PP' is vertical. F_1 - F_1 and F_2 - F_2 are the traces of the 2 fracture planes, the angle between them being bisected by PP'.

B - Normal/gravity faults. Note that fault F_1 - F_1 is parallel to fracture F_1 - F_1 of Fig. 153 A.

C - Step faults. Note that in this case too faults F_1 - F_1 are parallel to fracture F_1 - F_1 of Fig. 153 A.

D - Graben/trough/trench faults. Note that F_1 - F_1 and F_2 - F_2 fault planes are parallel to F_1 - F_1 and F_2 - F_2 fracture planes of Fig. 153 A

III. SINISTRAL, DEXTRAL, TEAR, WRENCH FAULTS

In this case too, the stress ellipsoid is horizontal, but the intermediate principle stress axis is vertical. The fault is therefore vertical in attitude, but the movement takes places only in horizontal direction (heave) along the fault plane. There is no throw in the or against the direction of gravity. The orientation of the stress ellipsoid is shown in Fig. 155 A, and the faults are shown in Figs. 155 B, C, D, E. In Figs. 155 B,C, dextral faults have been shown. If the direction of the movement be reversed (as compared to that shown in Figs. 155 B,C) it produces sinistral faults (Figs. 155 D,E). If the trend of the faults F_1 - F_1 or F_2 - F_2 shown in the figures be perpendicular to the strike of the rocks affected by the faults, then it gives rise wrench or the tear faults.

PATTERNS OF STRESS DISTRIBUTION AND RESULTANT STRUCTURESS

The relation between the stress ellipsiod and the direction of the deformative force has been elaborated in the earlier section. It has also been noted earlier that the intensity of the deformative forces need not be constant, when traced vertically up and down. If the intensity changes, then the nature of deformation likewise has to change. A few cases will be described in the following paragraphs.

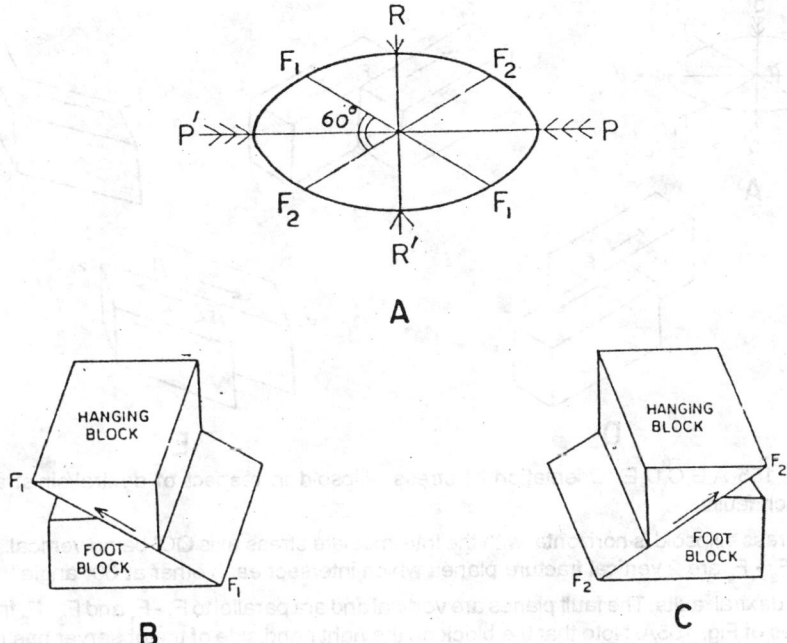

Fig. 154 A,B,C Orientation of stress ellipsoid in respect of thrust/reverse faults.

A - stress ellipsoid is horizontal with least stress axis being vertical. F_1 - F_1 and F_2 - F_2 are 2 fracture planes which bear an anlge of 60° between themselves.

B - thrust fault with a dip of 30° is parallel to fracture plane F_1 - F_1 of Fig. 154 A.

C - thrust fault with a dip of 30° is parallel to fracture plane F_2 - F_2 of Fig. 154 A.

Note that if the foot block were to move down, then it will give rise to underthrust. If the overthrusting be considerable, then it may result in the development of "nappe" structure.

CASE I. DEFORMATIVE FORCE CONSTANT IN MAGNITUDE

The stress distribution and the resultant ruptures have been shown in Figs. 156 A,B. In Fig. 156 B it is seen that the entire block of rock is affected equally, and that the fracture planes bear an angle of 60° or 120° between them. The angle of 60° is bisected by the greatest principle stress axis. Since the stress ellipsoid, is horizontal, the fracture planes have a low inclination of 30° or so, and these may give rise to the thrust faults. The fault planes will further maintain a *uniform inclination* of about 30° at depth.

CASE II. DEFORMATIVE FORCE NOT CONSTANT IN MAGNITUDE

The magnitude of the deformative force may go on increasing with depth. Further, on one side, the force may be more in intensity than that on the other side. This is shown in Fig. 157A. In Fig. 157 B, the orientation of the resultant ruptures is shown. It is seen in Fig. 157 B that the distribution of the fractures is irregular. Further the block on the right hand side i.e., the side on which the intensity of the deformative force is less, *is not at all fractured.* The fractures incline rather more steeply towards the unfractured part of the block of rock (F_1 - F_1), while the other set of fractures maintain a gentle inclination (F_2 - F_2). The most important point to note is that though the entire block of rock is subjected to the deformative forces, a part of it remains unaffected. Such a situation may be interpreted as a case of "unconformity", because some parts of the rock are not at all affected.

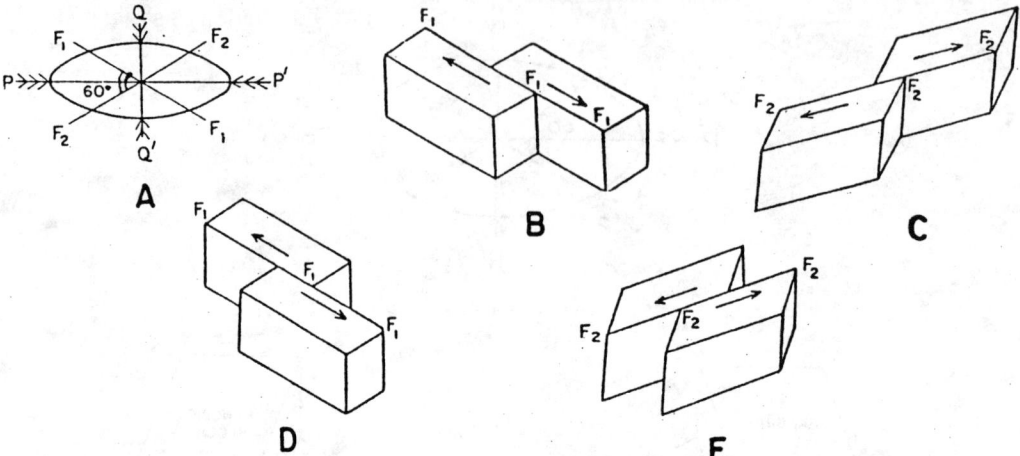

Figs. 155 A,B,C,D,E. Orientation of stress ellipsoid in respect of dextral/sinistral/tear/ wrench faults.

A - stress ellipsoid is horizontal with the intermediate stress axis QQ' being vertical. $F_1 - F_1$ and $F_2 - F_2$ are 2 vertical fracture planes which intersect each other at 60° angle.

B,C - dextral faults. The fault planes are vertical and are parallel to $F_1 - F_1$ and $F_2 - F_2$ fracture planes of Fig. 155A. Note that the block on the right hand side of the observer has moved towards him.

D,E - sinistral faults. The fault planes are vertical and are parallel to fracture planes $F_1 - F_1$ and $F_2 - F_2$ of Fig. 155A. Note that the block on the right hand side of the observer has moved away from him.

‖‖	LEAST PRINCIPLE STRESS AXIS	
≡	GREATEST PRINCIPLE STRESS AXIS	
⟹	DIRECTION OF DEFORMATIVE FORCE	

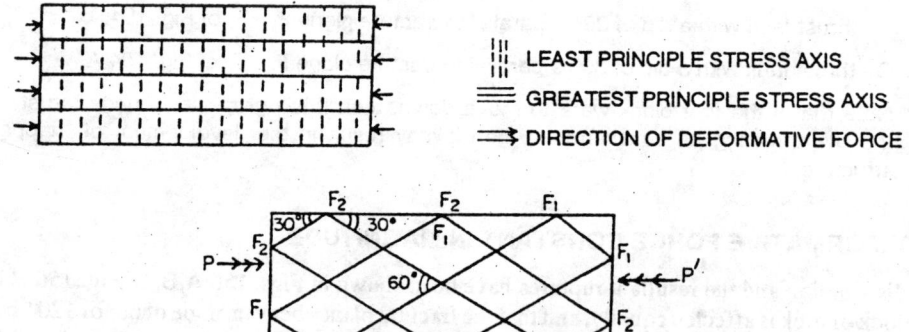

Figs. 156 A,B. (After Billings, 1960)

Stress distribution and orientation of ruptures when stress ellipsoid is horizontal, and least stress axis is vertical.

A - Stress distribution. Greatest principal stress axis is horizontal and least stress axis is vertical. The intensity of deformative force is same in magnitude on both sides as well as depth as is indicated by the constant length of the arrows.

B - orientation of ruptures in the rock.

Note that the angle subtended between the fracture planes $F_1 - F_1$ and $F_2 - F_2$ is of 60° value. Also the slope of these fracture is constant, the inclination being of 30 degrees. Such fractures. can give rise to low angle thrust faults.

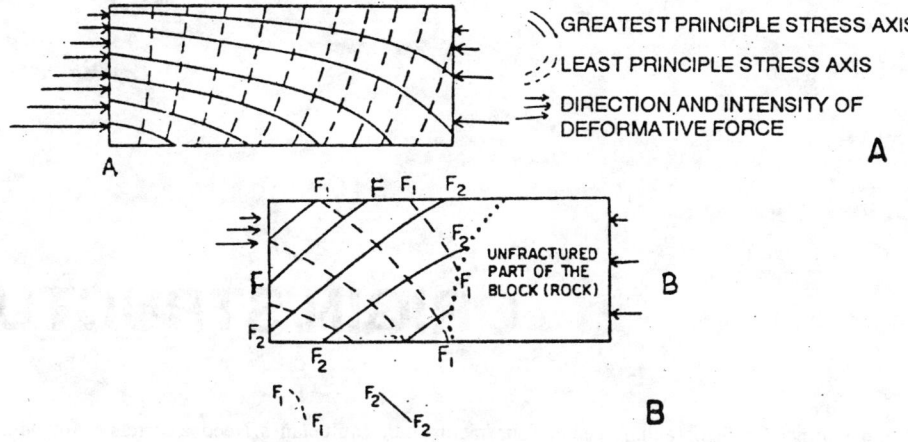

Figs. 157 A,B. (After Billings 1960).

A - stress distribution in the block (rock). The stress ellipsoid is horizontal, the least stress axis being vertical. Intensity of the deformative forces is not same on the two sides. Also the intensity increases with depth and it is more so on the left hand side.

B - orientation of ruptures/fractures in the block (rock).

Note that the development of ruptures/fractures is very irregular. Fractures/ruptures (F_1 F_1) are steeper and are inclined away from the direction of the stronger deformative forces. Fractures/ruptures (F_2 - F_2) are gentle and are inclined away from the direction of weaker deformative forces. Most important part is the development on an area on the right hand side of the block (rock) which is not at all fractured or ruptured.

This is because, that part probably did not exist when the deformative forces had acted. *However the actual fact is that it is a case of the action of the unequal deformative forces in the vertical sense, on the block (rock), and not of the existence of unconformity between the fractured and the unfractured parts of the rock exposure.*

CASE III. THE COMPLEX TYPE

The intensity of the deformative forces may increase for some depth, and it may again decrease for some depth, and again increase. On the other side of the block (rock), the intensity in general may be different as compared to that occurring on the other side of the block (rock). In all such cases, some part of the block (rock) does not get fractured at all. Also the inclination of the fracture planes is also not uniform. However as the complications are many, the actual diagrams are not drawn. Hegde (1984) has noted such a behaviour in the quartzarenitic rocks exposed in the southwestern part of the Kaladgi basin. He has observed that at some places the shears, the fractures, the ruptures are developed, but in between them the rocks are not at all affected.

5

ORIGIN STRUCTURES

Origin of joints, Origin of faults, Need for four fracture cum-fault planes, Need to correlate fault with fold or crushing of rocks at depth, Origin of rotational faults, Faults of translation.

Origin of folds, Tectonic causes, Direction of movement of the deformative forces, Non-tectonic causes.

Origin of fissility, cleavage, schistocity and foliation, Origin of slaty cleavage, fracture cleavage, slip cleavage and bedding cleavage.

Origin of lineation.

Figures 158 to 172.

Photo 64.

So far a detailed account of the different structures developed in the crustal rocks, has been given. One of the aims of the structural studies is to offer explanation about how the different structures were produced. It is true that compressive, tensile and shearing forces bring about the deformation, but not necessarily all of them for the individual structures like the folds, the faults, the joints and so on. No structure exists alone. Thus folds are associated with faults, faults with joints, and so on. For sake of convenience, origin of the individual structures are separately described below.

ORIGIN OF JOINTS

Joints are "rupture structures" and therefore these can develop only when the rocks are in the solid state. In other words it means that the deformative forces can not give rise to the joints in the magmas or the unconsolidated sediments etc. Deformative forces are essential to rupture the rocks. The rocks further should be competent enough. Though this mechanical property is inherent one in most of the rocks, at deeper levels in the crust of the earth, even such rocks behave plastically. As such the ruptures and hence the joints can develop only at shallow depths where the confining pressure will be low.

Though there are a variety of the deformative forces, the resultant joints are not much different from one antother. Yet the structural geologist strives hard to distinguish the tension joints from the shear joints. In the case of fault planes, the movement is decidedly along the fault plane, but it is in the opposite directions, as far as the the two fault blocks are concerned. This situation gives rise to the shearing forces. The joints associated with the fault planes therefore are due shearing forces. These joints are called as the feather joints.

Joints are the major structures, and are therefore found associated with the other major structural events, like the folds, the faults, the emplacement of the large sized plutons and so on. Joints developed in the folds are primarily due to the compression, because the folds are caused by the compressive forces. In the flexure folds, slipping or shearing developes along the bedding planes. Therefore some of the joints noticed in the folds could be due to the shearing forces created during the slippage of one bed over another. The crests and the troughs of the anticlines and the synclines, respectively, experience considerable tension. This culminates into the development of the joints and fractures which are called as the "release joints".

The joints in the igneous and the metamorphic rocks are the outcome of the tension created during the process of cooling of the magmas, and the process of recrystallisation (in the case of metamorphic rocks). Such joints are concentrated more along the contract of the magma with the country rocks. Depending upon the inclination of such a surface, the joints could be horizontal (along the sides of the igneous body), inclined or vertical (top or the roof portion of the igneous body). Further, depending upon the stage of emplacement and consolidation of the large intrusive bodies, joints are produced along the marginal portion of the igneous body (see Chapter 8). These are vertical to highly inclined, moderately dipping ones, and also nearly flat lying joints. Therefore in this case the origin is to be attributed to the stage of consolidation of the magma, and the speed of the upward movement of the magmatic body. This aspect has been described at length in the Chapter on "Granite Tectonics".

When the "spots of tension" are systematically and regularly located with respect to each other, the columnar joints, the rhombic, the rectangular types of joints are developed. Though joints having curviplanar surface are not common, however if these be present, then the torsional deformative forces will be necessary for their development. Thus torsional forces also might be operative in the development of the joints. Joints are primarily produced due to tension and shear. Compression also can give rise to joints if the rocks be resistant ones. The resultant fractures and joints can not be regular in shape and size, because the rocks experience effects of crushing due to the compression. It is therefore argued that if the joint planes be multitudinous and highly irregular, these most likely are due to the action of the compressive forces, and not due to the tension or the torsion. The sheeting structure noticed in some sandstones belongs to the category of "release joints". A detailed account of this has been given in the section dealing with the "square prism experiment.

In conclusion it may be said that the ruptures and hence the joints are caused in the main by the shearing, tensional and compressive forces acting at shallow depths. Torsional forces are at times operative. Cooling and consolidation of the igneous bodies, produces numerous joints which are due to the tension created during the process of cooling. Feature joints are characteristically associated with the faults.

ORIGIN OF FAULTS

Unlike the joints, there is a conspicuous movement of the rocks along the plane of rupture. It is thus quite clear that the shearing forces alone can bring about faulting. All faults are therefore "shear faults". What causes "shear" is debatable. For that purpose even the origin of the compressive, tensile or the torsional forces, is not clearly understood in the context of "which is first".Compression or tension? This aspect has been dealt with in greater details in a separate chapter (Chapter 6). Given the shearing forces, it will be interesting to discuss the development of the different varieties of faults.

NEED FOR THE EXISTENCE OF FOUR FRACTURE-CUM FAULT PLANES

It is customary to draw only one fault plane and then show the displacement. This is feasible only in block diagrams, because it is to be remembered that the crustal rocks build up a contiguous body and not independent blocks. Hence unless a portion of the crust is separated by *four rupture planes*, no movement is at all possible. This feature has been shown in Figs. 158 A,B,C. Though in Fig. 158 A, a fracture plane is formed and shearing forces are generated, either block I or II can not move past each other, because the other side of these blocks are in continuous contact with the crust. In Fig. 158 B another fracture $F_2 - F_2$ is developed, and now the block II alone can move up or down, or in the horizontal direction according to the situation. However even now the movement is not easy, because on the two sides perpendicular to faults $F_1 - F_1$ and $F_2 - F_2$, the rocks are in continuous contact with the crust. Therefore two more fracture/rupture planes PQ and RS are necessary as shown in Fig. 158 C. Owing to these *Four* fracture planes, $F_1 - F_1$, $F_2 - F_2$, PQ and RS, a block of rock is severed on all sides from the crust, and it can now move up or down. Analogogy could be given of cutting a piece from a watermelon. It will be seen that unless *Four* cuts are made in it, the piece cannot be taken out.

It is therefore suggested that if faulting is suspected at some place during the survey, then it will be necessary to examine the area very carefully for the existence of another fault, and two other fractures perpendicular to the two fault planes, in the vicinity. Only then faulting movements are possible.

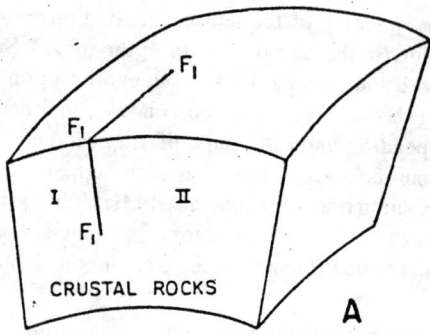

Fig. 158 A. Either block I or II can not move past each other because the rocks are in contact with the crustal rocks.

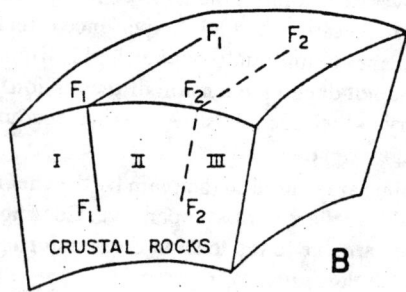

Fig. 158 B. Block II still can not move up or down, because there are no fractures connecting faults F_1 - F_1 and F_2 - F_2.

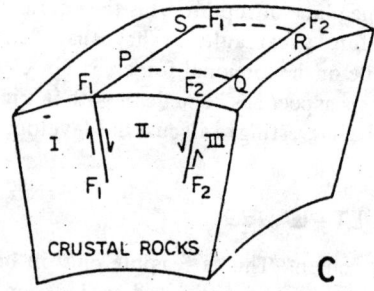

Fig. 158 C. Block II can now move up or down because fractures PQ and RS have now isolated the block II from the crust of the earth, by connecting the faults F_1 -F_1 and F_2 - F_2.

NEED TO CORRELATE FAULT WITH FOLD OR CRUSHING OF ROCKS

Faulting movements in the vertical or horizontal directions are possible only if the rocks in the vicinity of the fault, accomodate the effects of the movement. The rocks are displaced vertically down in the case of the gravity faults. But the downward movement can take place provided the rocks at depth below the fault block, are either folded or are shattered. Owing to these latter structures, room is created for the downward moving rocks due to the faulting. These features are shown in Fig. 159 A,B. In Fig. 159 A, the rocks at depth are folded in the vertical direction, and therefore block II could be displaced downwards. In Fig 159B, the rocks at depth are crushed due to the downward moving rocks. Therefore block II could move down. This is found necessary because there is no open space in the crust of the earth to accomodate the movement of the rocks during faulting. Thus it is evident that there is an intimate association between the faulting and the folding, or between faulting and the development

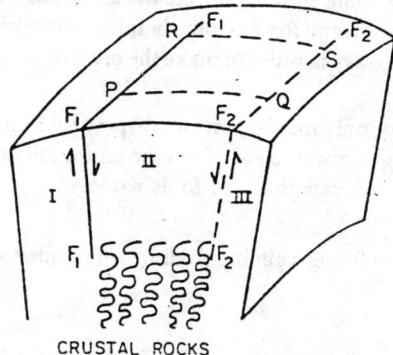

Fig. 159 A. Rocks of block II while moving downwards (faulting) have folded the beds lying at depth In that area.

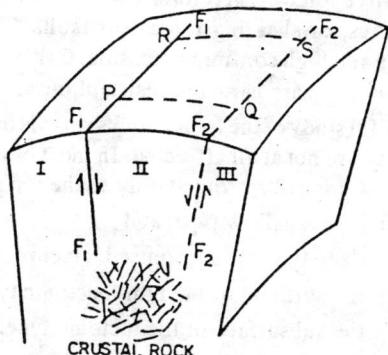

Fig. 159 B. Rocks of block II while moving downwards (faulting) have crushed the rocks at depth occurring at that place.

of crushed rocks. In fact in the homogenous rocks displacement can not be observed, though faulting were to take place. The effect of the shearing forces is expressed by way of the development of the cataclasite. Many faults in the homogeneous rocks go unnoticed, because the development of the cataclasite is not considered as the effect faulting.

ORIGIN OF ROTATIONAL FAULTS

Such faults need a curviplanar surface of rupture. Such a surface can be formed by the torsional or twist type of deformative forces. The movement on such curved surface brings about change in the amount of dip. Rotational movement is possible if a hinge or an axis is developed on the fault plane. This can be understood from the Figs. 49, 50, 51 and 52. In the vicinity of Jamkhandi town, Bijapur district, Karnataka state, excellent rotational fault is exposed. This has been documented in Fig. 83, and Photos 47, 48, and 49.

FAULTS OF TRANSLATION

High angle and vertical faults can be formed only if the deformative forces act vertically on the rocks. In such a situation, the stress ellipsoid is vertical in orientation, and the fault plane or planes assume high angle dip. Normal, gravity, step, graben, trough and trench faults are produced under this condition. This has been shown in Figs. 153 A to D. If the movement be upward along the fault plane, the upthrusts are produced. The reverse, the thrust and the nappe faults are produced when the deformative forces act in the horizontal direction on the rocks. The stress ellipsoid is necessarily horizontal in attitude. Further it is necessary that the least stress axis be vertical. This feature has been shown in Figs. 154 A to C. The overthrusts and the underthrusts can originate in such a situation. Sinistral, dextral, tear and wrench faults are produced when the deformative forces act in horizontal direction on the rocks. The stress ellipsoid is horizontal in attitude, and the intermediate stress axis is vertical. This has been shown in Figs. 155 A to E.

In conclusion it may be said that the type of the fault produced depends upon the direction in which the deformative forces act on the rocks, in the main, and also upon the nature of the fault plane - whether it is vertical or curviplanar one.

ORIGIN OF FOLDS

This is a plastic structure and therefore the deformative forces have to act at great depths in the crust of the earth. Either the rocks should be incompetent ones, or else due to the increase in temperature at depth and also the comfining pressure, the rocks should be made to yield plastically. The layered rocks which are initially horizontal in attitude, these are bent and the limbs so formed are brought closer to each other with the increased action of the deformative forces. Therefore such forces are necessary to bring about folding, unlike the tensional ones for joints, ruptures, or shearing forces for faults. The structural environment for folding therefore comprises of compression and high confining pressure. Considerable depth is necessary in order to make the crustal rocks yield plastically, if these are hard and resistant ones.

A careful study of the folded rocks reveals that in some instances, only the top portions display folding, while at depth these are not at all afffected. In most of the cases the folding is observed to affect the entire thickness of the rocks, and it is not restricted only to the top portion. Therefore two categories of folds namely

  (i) surficial or shallow type, and

  (ii) subsurface type are recognised. Accordingly the deformative forces causing folding are divided into,

    (a) the surficial or the non-tecctonic type, and

    (b) the subsurface or the tectonic type.

The folded crustal rocks constitute one geological event which is accompanying another event. This feature is described as the concept of "paired geological events" and it is elaborated in Chapter 6.

TECTONIC CAUSES

The compressive deformative forces are produced in several ways. A magma can intrude into the crustal rocks only by pushing up or brushing aside the already existing rocks. Therefore the pre-existing rocks undergo decrease in volume because of the compressional forces created by the intruding magma. However it is needed to be proved that the crustal rocks were not folded prior to the act of the intrusion of the magma. Bigger the size of the intruding magma, stronger will be the compressive forces generated by it. Hence the intensity of folding will be likewise more and extensive one. Himalayan mountains, the Aravalli mountains and the other orogenic mountains are observed to have folded rocks within them. Further a core of granitic or the granodioritic rocks is also observed to be present. It is thus quite clear that the granitic intrusions created the compressive forces, and brought about the folding of the crustal rocks. The laccolithic form of the igneous intrusion is due to the uprising of a pile of sediments into which a viscous magma had intruded. The domical shape noticed in the igneous body is in fact due to the overlying sedimentary rocks alone. The sedimentary rocks develop a dome or a fold due to the intrusion of the magma.

Drifting of the continents is an accepted fact, difference of opnion exists only in respect of the type of the motive forces. Westwardly rotation of the earth, the convection currents generated in the mantle, sea floor spreading, the plate tectonics, are some of the active forces that set the continents in motion. As this act drags the crustal rocks, the sediments laid in the basin get folded.

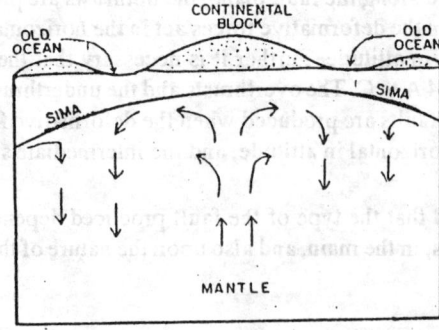

Fig. 160. (After A. Holmes, 1928-29).
Note that convection currents rise below continent and spread out towards the ocean. Currents move upwards first and then move in horizontal direction. At the end of the continent, currents coming from two directions meet and produce compressive forces.

DIRECTION OF MOVEMENT OF THE DEFORMATIVE FORCES

Compressive forces produce the folded structures, but what about the direction in which those had acted on the rocks? In the great a majority of cases, the axial planes of the folds are nearly vertical, or are highly inclined in attitude. The direction of the deformative forces (compression) therefore must be in the horizontal planes. Such horizontal compressive forces might be created by the rotation of the earth on its own axis, drifting of the continents etc. The convection currents generated in the sima, and the intruding igneous bodies, initially produce vertically directed forces which are later deflected in the horizontal direction. This has been documented in Figs. 160 and 161.

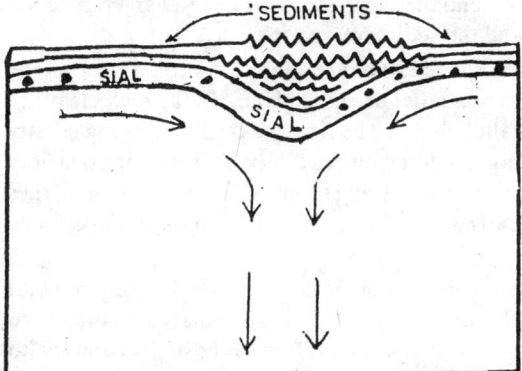

Fig. 161. (After Billings, 1960)
Convection currents (only small part of the whole convection cell is shown) generated in the mantle or substratum converge and produce compressive force. Right above it are seen folded belts. Therefore the currents though these rise upwards and downward directions, these give rise to horizontal forces as well.

As has been noted earlier, the laccolithic form of the igneous body is due to the intrusion of a viscous magma into a pile of sediments. The overlying rocks develop a domical structure due to the vertically directed initial forces, and not to horizontal forces. But in such of the cases where the folded structures are not only developed over the top of the igneous body, but also in the other far off parts, then it is surmised that the vertically rising magma produced horizontal forces, and these latter forces then brought about the folding of the rocks. Some geologists advocate contraction suffered by the crustal rocks owing to which, horizontally directed deformative forces are produced. These are then held responsible for folding encountered in the crustal rocks.

Thus it clear that the *tectonic forces* are due to *vertical* movements of the magmas, and the *horizontal forces* are created later as an outcome of it. Drifting of continents, plate tectonics etc., are in general acting in the horizontal direction. Therefore these lead to the compressive forces, and bring about the folding of the crustal rocks.

NON-TECTONIC CAUSES

Due to this, only the surficial or the exposed parts of the rocks are affected. *Gravity* is called into play and wherever it dominates and the rocks are prone or are susceptible, these may get folded. Thus a bedded formation having a nearly vertical dip when exposed on a steep hill slope, the edges may get bent and produce fold-like structure (Figs. 162 A,B). However the lower or the deeper parts of the same beds are not at all folded. Likewise due to leaching

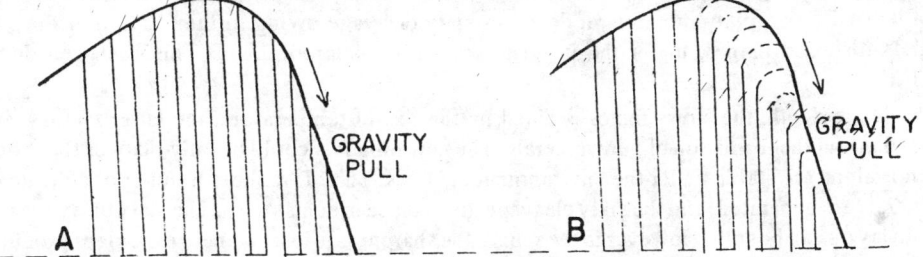

Fig. 162. Beds possessing nearly vertical dip, and exposed on precipitous slopes, may get affected by the gravitative pull (shown by arrow). The ends of the beds get bent and produce anticline like structure.

away of vulnerable rocks that are situated at depth, the overlying rocks might loose support and might sag down, producing a "synclinal like" structure. These are called as the "collapse structures". Landslides, differential compaction of sediments also might produce small scale, local folds.

ORIGIN OF FISSILITY, CLEAVAGE, SCHISTOCITY, AND FOLIATION

These structures also belong to the category of the plastic deformation like the folds. However without developing any external form, the minerals constituting the pre-existing rocks are made to arrange themselves. There fore the confining pressure and considerable compressive forces are essential. These structures are characteristaclly encountered in the metamorphic rocks, less so in the igneous and the sedimentary rocks.

Different rock cleavages are recognised on the basis of their mode of development. Akin to the cleavage noticed in the minerals wherein "loose or poor bond" exists across parallel oriented planes, in the rock cleavage too, " minimum cohesion or poor bond" is developed across parallel planes. The first and the fore most necessity is the presence of or the development of parallely disposed numerous bedding planes easily. This is realised in very thinly laminated rocks like the shales or the sandstones, which may cleave along these numerous bedding planes easily, and produce "fissility''. Obviously thickly bedded rocks can not develop fissility, and this is the experience of the geoscientists.

The cleavage, the schistocity, and the foliation are due to the orientation of the minerals. Not any mineral but only platy and flaky ones can produce the "rock cleavage". Further, the pre-existing minerals are compelled to rearrange themselves into parallel planes. Owing to this, such a cleavage is therefore called as "secondary"to which a mention has been already made. Schists, gneisses, schistose quartzites are metamorphic rocks, and slaty cleavage, foliation are characteristically found in them. Therefore it is apparent that the process of metamorphic recrystallisation and the development of such metamorphic minerals like chlorite, actinolite, hornblende, kyanite, biotite, andalusite, sillimanite are instrumental in the development of cleavage, schistosity and foliation.

Slaty cleavage is produced due to the orientation of the pre-existing platy and flaky minerals like chlorite, sericite etc., into parallel planes. This is brought about by the rotation of these minerals so that their maximum surface is oriented perpendicular to the direction of the maximum deformative force. It is obvious that the deformative force must act only in one direction, and in one line. Otherwise parallelism cannot be brought about. It is also very important to note that the slaty cleavage is developed at an angle (about 60°) to the bedding planes, if the original parnetal rocks were to possess bedding planes. These features have been shown in Figs. 163 A to G.

The directed pressure (stress) which is operative at the stage of the development of the "slaty cleavage", it acts in yet another manner. It is called as the "Riecke's principle". It states that melting occurs at the point of application of maximum stress, and a concurrent deposition at the point of application of minimum stress. Thus a spherical body will get deformed into an ellipsoidal shape (Figl. 164). A sandstone consisting of rounded grains of quartz, gets changed into a foliated quartzite, the foliation being produced by the conversion of the original rounded grains of quartz into ellipsoidal grains. The pre-existing flakes of sericite are rotated into parallel planes and the sericitic sandstone gets changed into a schistose sericite quartzite (Figs. 165 A to D). A shale which consists of detrital flaky, platy minerals, it develops slaty cleavage owing to the rotation of the said minerals into parallely disposed planes, under the influence of stress metamorphism. This has been documented in Figs. 165 E,F.

At moderate depth, the stress factor is aided by the rise of temperature, the latter factor bringing about recrystallisation and the formation of new minerals. The new minerals could be only platy or flaky ones, because such minerals alone are stable under the environment of stress. Therefore more number of chlorite or biotite or similar minerals are generated, and the slaty cleavage gives place to schistocity. The slaty cleavage is quite sharp and very thin layers can be split. However in the schists the sharpness is lost and hence the slabs produced are thick and irregular ones.

With an increase in the grade of metamorphism, the stress starts dwindling, and anti-stress minerals like the

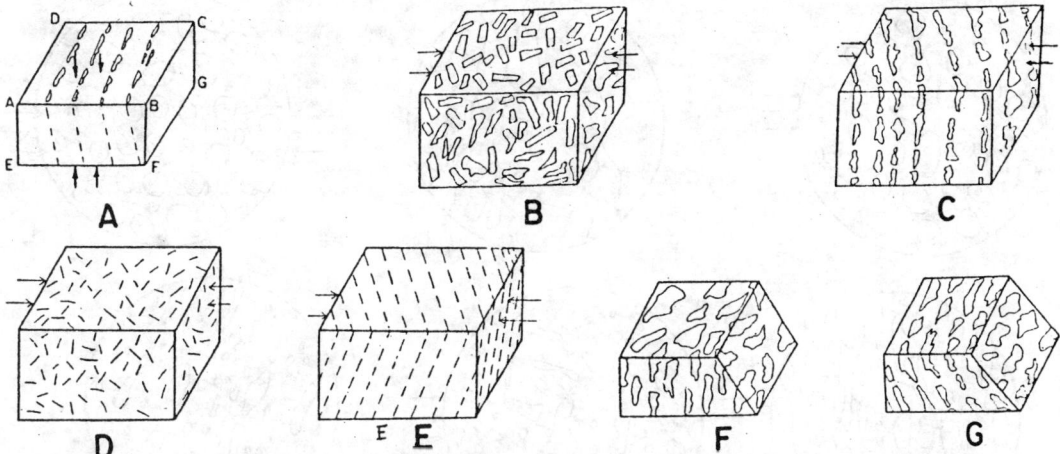

Fig. 163 A. ABCD and EFG are bedding planes. Arrows indicate direction of the action of the deformative forces. The dashed lines represent planes of slaty cleavage which are at an angle (about 60°) to the bedding planes. Platy/flaky minerals are indicated in the ABCD plane, the longer axes of which are parallel to each other.

Fig. 163 B. Note that the platy/flaky minerals are unoriented on any surface. Arrows indicate direction of deformative forces.

Fig. 163 C. Note that the platy/flaky minerals get oriented into parallel positions and thus produce planar structure.

Fig. 163 D. Note that the accicular minerals are unoriented. Arrows indicate direction of deformative force.

Fig. 163 E. Note that the accicular minerals get oriented producing parallelism and a planar structure. The planes are dipping ones.

Fig. 163 F. Though there is a parallelism between the longer axes of the minerals, these are not confined to planes. Therefore it has produced only lineation in the rock.

Fig. 163 G. Longer axes of the minerals are not only parallel to each other but are further confined to parallel planes. The rock therefore has produced lineation together with foliation. The structure is dipping to the right hand side.

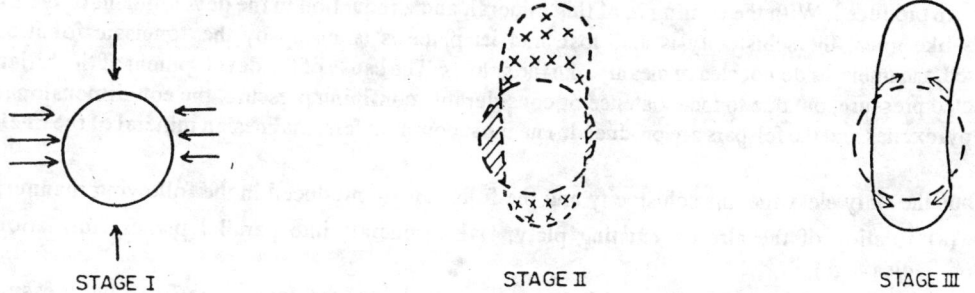

STAGE I   STAGE II   STAGE III

Fig. 164. Riecke's principle.
A spherical body subjected to stress (stage I). Melting (indicated by lines) occurs at the point of maximum stress, while deposition (indicated by crosses) of the melted material takes place at the point of minimum stress (stage II). As a result of this process, the original spherical body gets deformed into an ellipsoidal form (stage !!!).

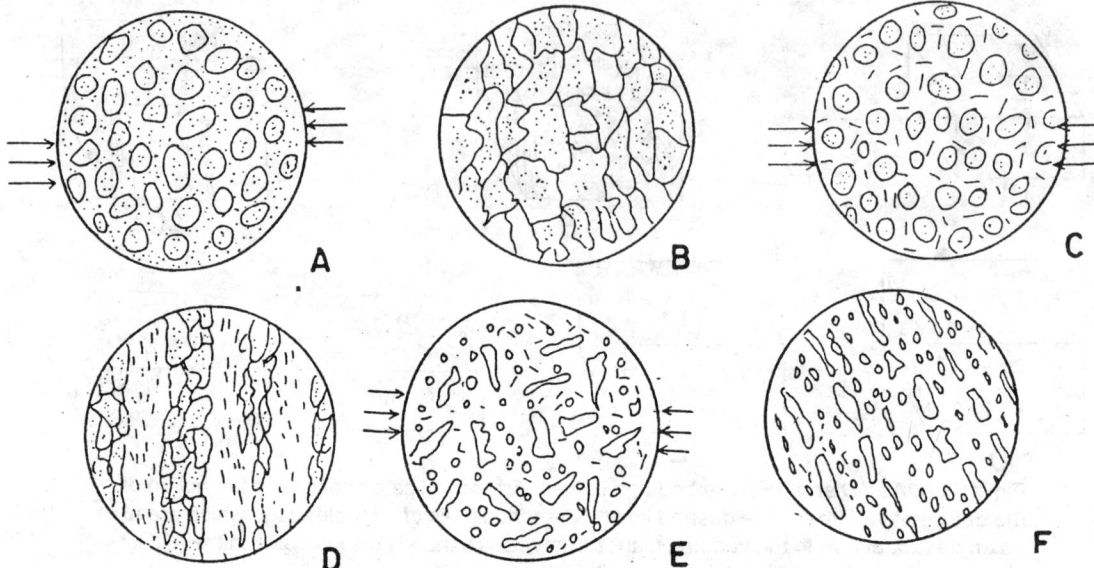

Fig. 165 A. A sandstone acted upon by directed pressure (stress). Note that rounded to sub-rounded grains are cemented in a clayey matrix.

Fig. 166 B. The sandstone shown in Fig. 165 A, is converted into a foliated quartzite. Note that the grains are elongated or elliptical in form, and a mosaic texture is produced.

Fig. 165 C. A sandstone made up of rounded to sub-rounded grains of quartz and cemented in a sericitic matrix, is subjected to directed pressure (stress).

Fig. 165 D. The sandstone shown in Fig. 165 C is converted into a schistose quartzite due to the development of elliptical grains of quartz that are oriented parallel to each other. The flakes of sericite are also confined to parallel planes.

Fig. 165 E. A shale with pre-existing flakes of sericite, chlorite, grains of quartz, is affected by directed pressure (stress). Note the flakes are unoriented.

Fig. 165 F. The rock shown in Fig. 165 E is seen to develop orientation of flakes of sericite, chlorite. Further, the flakes are seen to be confined to several parallel planes. Note that the grains of quartz are not much flattened. The rock becomes a slate.

felspars are produced. With the coming in of this mineral, and a reduction in the development of the platy, flaky minerals like mica, the schistocity is also lost, and its place is taken up by the "gneissic foliation". When hammered, the gneisses do not cleave cleanly like the schists. The cause of the development of the foliation is still the directed pressure, but due to the existence of considerable confining pressure, the equidimensional minerals like the pyroxenes and the felspars are produced. The most common ferromagnesian mineral of the gneissic rocks is mica.

Thus the slaty cleavage, the schistocity and the foliation are produced in the following manner;

    (a) rotation of the already existing platy, flaky minerals into parallel planes (formation of slaty cleavage),

    (b) development of new platy and flaky minerals under the influence of the directed pressure, and the increased heat. These minerals grow along the already developed cleavage planes (formation of schistocity),

    (c) flattening of the competent minerals like quartz, into ellipsoidal bodies, their longer axes getting oriented perpendicular to the direction of the maximum stress (Riecke's principle). This gives rise to the foliated and the schistose fabrics.

The origin of the slaty cleavage, the schistocity and the foliation, are regarded to be "flow type" processes by the majority of the geoscientists. However some others hold the view that it could be a "shear phenomenon". Due to the differential movement, the original body gets deformed into an ellipsoidal one (Figs. 166 A to D). It

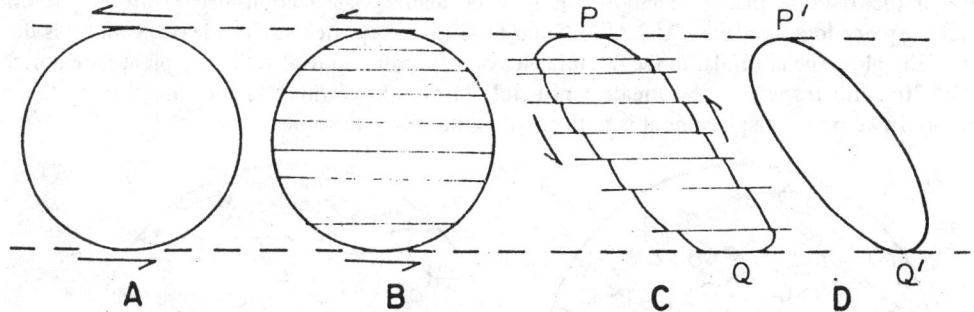

Figs. 166 A,B,C,D. A sphere subjected to shear is shown in Fig. A. Numerous shear planes are developed (Fig. B) and the sphere changes to an elliptical form (Fig. C). Note that the ellipsoidal body (Fig. D) is at an angle to the direction of the shear planes PP' and QQ'

is evident from these figures that the elliptical or the ellipsoidal body will lie at an angle to the shear planes-cum cleavage planes. However the experience is that the flattened or the ellipsoidal bodies (minerals) lie in the plane of cleavage, and not at an angle to it. (Fig. 166 D). The slaty cleavage, the schistocity and the foliation are therefore regarded to be of the "flow type" origin.

Besides the types of cleavages described above, there are three more varieties which differ both as regards their origin and what constitutes such types. These are,

(i) fracture cleavage, (ii) slip cleavage, and (iii) bedding cleavage.

The *fracture cleavage and the slip cleavage* are not influenced by the constituent minerals, and therefore these simulate "fissility" as is developed in some sedimentary rocks. In the fracture cleavage and the slip cleavage, numerous shear planes are produced in the parental rock. It may be recalled here that under the action of the compressive forces, fracture planes dipping 60° or fractures which are vertical, are produced. When such fracture planes are closely spaced, the rock becomes liable to break along them. There is yet another significant point. The slaty cleavage or the cleavage in general, is observed to be developed at an angle to the bedding planes, if the rocks were to possess such planes. In the light of this situation, the fractures or the ruptures created by the compressive forces also will be found to bear similar angle (Fig. 167).

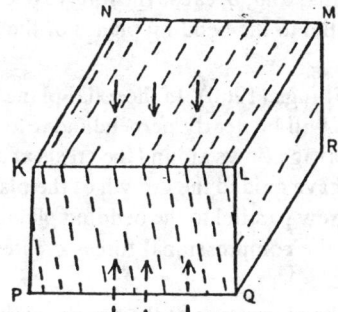

Fig. 167. Development of fracture cleavage. Beds subjected to compressive forces are seen to develop fracture planes (dashed lines) that dip around 60° with respect to the bedding planes. Arrows indicate the direction of compressive forces which are perpendicular to the bedding planes KLMN and PQR.

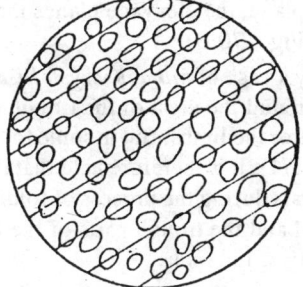

Fig. 168. Note that the minerals also are fractured by the cleavage planes. The minerals do not display any orientation of the grains. The cleavage is therefore due to the fractures alone.

In the *fracture cleavage,* it is further observed that the constituent minerals have not at all contributed to the formation of this structure, because the cleavage planes are seen even to cut through the mineral grains (Fig. 168). In the case of *slip* cleavage, displacement and dragging of the minerals is observed along the cleavage, schistocity or the fracture planes. Therefore it may be deemed as a combined effort of "fracturing and faulting"taking place together (Figs. 169 A,B). Pujar (1989) has reported such a cleavage and it is documented in Photo 64. Slip cleavage is similar to the fracture cleavage, because here also the slip planes are noticed to have an angle of 30° with respect to the greatest principle stress axis (which is the direction of the maximum compression). In addition, displacement is noticed along the fracture planes.

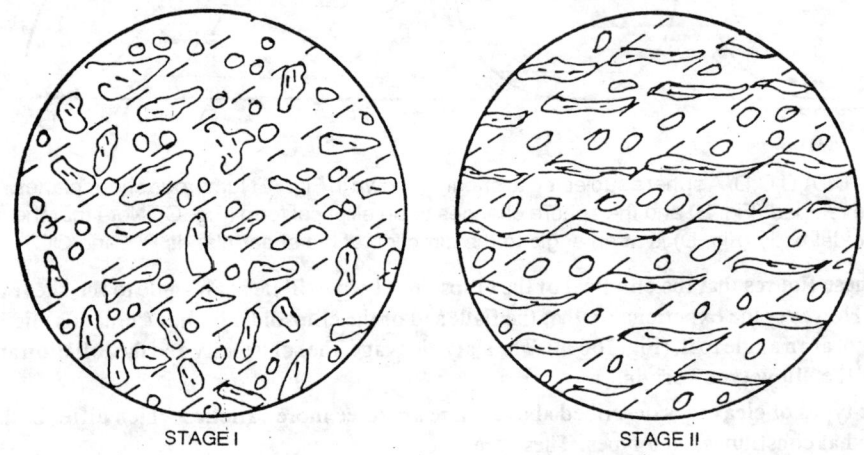

STAGE I   STAGE II

Figs. 169 A,B. Minerals not oriented initially (stage I) are seen to be displaced, elongated and dragged along the fracture-cum-slip planes (stage II). This gives rise to slip-cleavage in the rocks.

BEDDING CLEAVAGE

As noted earlier, cleavage is always produced at angle to the original bedding planes. In some cases however, it is found to be parallel to the bedding planes. Several explanations are tendered to account for its development.

(i) Cleavage is associated with the folds, and it is generally parallel to the axial plane. It is however at an angle to the limbs or the bedding planes (Figs. 170 A). In the isoclinal fold, because the limbs are parallel to the axial plane, the cleavage planes therefore can become parallel to the bedding planes or the limbs of the fold (Fig. 170 B).

(ii) It may a case of *mimetic crystallisation* as suggested by Billings (1960). In the axial plane cleavage, at and around the crest or the trough of the fold, the cleavage will be nearly perpendicular to the bedding planes or the limbs. But in some folds, it is found that the cleavage flows around the crests and the troughs (Fig. 171). It is therefore held that the bedding planes must have guided the growth of the platy and flaky minerals during the process of folding. Minerals therefore grew parallel to the bedding planes even at the crestal and the trough parts of the fold, under the effect of the compressional forces created during the folding of the rocks.

(iii) It may be due to the combined effect of the flattening and the elongation of the grains, perpendicular to the direction of maximum compression (stress), and this is usually perpendicular to the bedding planes. Also according to the mechanism of "flexure folding", stretching or tension occurs on the outer part, while compression or squeezing is created on the underside of the fold. The total effect could be one of stretching, elongation, or lengthening parallel to the bedding planes or the limbs of the fold. Platy and flaky minerals are favoured under this condition, and arrangement of the minerals parallel to the limbs may be produced (Figs. 172 A,B,C).

Photo. 64. Field photo of very well developed fracture cleavage in pink coloured quartzarenites. The locality is 2 km. SE of Magnur village, Belgaum district, Karnataka state. The cleavage is due to the closely spaced shear planes, which are trending in a N 25° E - S 25°W direction. Note that elsewhere the rocks are not at all fractured. Courtesy Dr. G.S. Pujar.

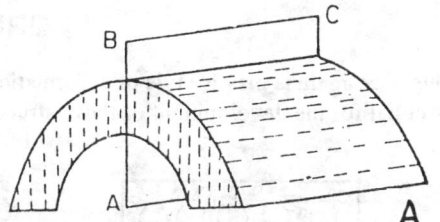

Fig. 170 A. In the axial plane cleavage, the cleavage (dashed lines) is parallel to the axial plane (ABC), but it intersects the limbs or the bedding planes.

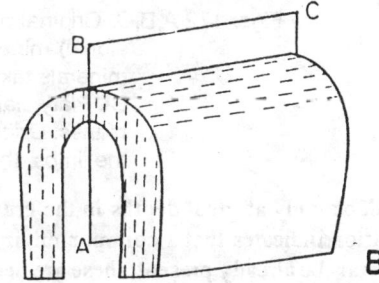

Fig. 170 B. This also is axial plane cleavage, but because the fold is of the "isoclinal" type, the axial plane is parallel to the limbs or the bedding planes, too. Therefore the cleavage (dashed lines) is also parallel to the limbs or the bedding planes.

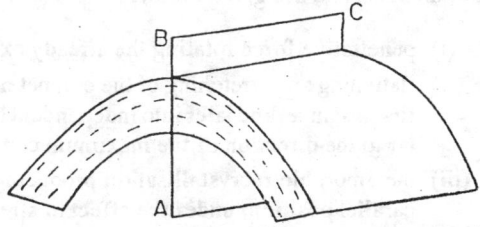

Fig. 171. Mimetic recrystallisation.

Here the cleavage (dashed lines) is not only parallel to the limbs of the fold, or the bedding planes, but even at the crest of the fold, parallelisim is developed. Such is not the case in Figs. 170 A or B.

(iv) It may be due to the load of the superincumbent roks. It is significant to note that the cleavage, schistocity or the foliation is developed in the metamorphic rocks. Folding can occur at great depths. The cleavage therefore may be the outcome of the load metamorphism - vertically directed load of the superincumbent rocks causing the formation of the cleavage.

From the foregoing account, it is apparent that the slaty cleavage, the schistocity and the foliation are characteristically associated with the metamorphic rocks, and that these rocks are further observed to be folded. Thus both folding and the metamorphism have a bearing on the development of the different types of cleavages.

ORIGIN OF LINEATION

This once again is plastic style of deformation witnessed in the crustal rocks. Accicular and tabular minerals are essential for the development of these structures. Besides this, strong confining pressure is necessary which is

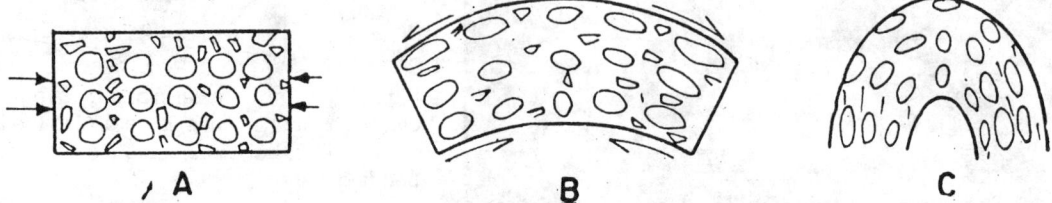

Figs. 172 A,B,C. Original platy, flaky minerals are unoriented, spherical (equidimensional) minerals are undeformed (Fig. A). Stretching of the spherical minerals takes place and new platy, flaky minerals are produced (Fig. B). Platy, flaky minerals get completely oriented parallel to the bedding planes. Original spherical minerals are elongated (Fig. C). Note that on the limbs, the minerals are more elongated.

available only at great depths in the crust of the earth. Association of lineation with cleavage, schistocity and foliation indicates that a common environment is needed for the development of these structures. If accicular minerals be already present, these are needed to be rotated into parallel dispositions. This necessiates the role of the "penetrative deformative force". If accicular minerals are not present, and yet lineation is noticed, in such cases stretching and flattening of the competent minerals of the rocks, is required. This is feasible under the directed pressure which results in the development of the foliation fabric and lineation. Akin to cleavage, the lineation also is commonly encountered in the metamorphic rocks. Thus metamorphic recrystallisation is a causative factor in the production of lineation. In the igneous rocks, laths of felspars often develop lineation. This is obviously due to the flowage of the magma or the lava. The different modes of origin of the lineation are summarised and are given below.

(i) penetrative force rotating the already existing accicular minerals into parallel position,

(ii) flattening and stretching of the competent minerals or rock bodies into "pinch and swell" structure in the first instance, and later into independent boudins or ellipsoidal bodies or rods. This is effected perpendicular to the direction of the maximum compression,

(iii) metamorphic recrystallisation producing more accicular or prismatic minerals which are oriented into parallel position, under the effect of stress,

(iv) flowage of magma or lava which brings about the orientation of the tabular minerals or the xenoliths into parallel position,

(v) intersection of two planar structures producing lineation.

6

ORIGIN AND CLASSIFICATION OF DEFORMATIVE FORCES AND CONCEPT OF PAIRED GEOLOGICAL EVENTS

Origin of deformative forces, Classification and sources of deformative forces, Schematic diagram of classification of deformative forces, Tectonic type of deformative forces, Vertical uplift as a source, Convection currents as a source, Igneous intrusions as a source.
Concept of paired geological events.

INTRODUCTION

So far the various structures encountered in the crustal rocks have been described at length. The systematics of their development also has been discussed. The close relation existing between the various deformative forces, the different kinds of rocks (mechanical properties) and the different structures developed in them, has been elaborated. Structures can be formed only if the deformative forces are created. But what causes the initiation of these forces also should deserve a special consideration. Several speculations alone could be made regarding the sources of the deformative forces, since it is beyond the scope of demonstrating their initiation, and the development of folding, faulting, shearing and the other structures in the crustal rocks. This is so because whatever experimentation that can be carried out in the laboratory, these very much fall short of in imitating them on the geological scales. Especially the factor of time can hardly be imitated in the laboratory experiments, because the geological (structural included) events enact over periods of millions of years. However the sources of the deformative forces which are actually needed, these can be inferred only through the process of careful observations of the structural features as are developed and disposed in the crustal rocks, and the process of logical thinking. The deformative forces also need to be classified. These aspects are described in the following paragraphs.

ORIGIN OF THE DEFORMATIVE FORCES

The rising and the sinking of the landmasses has been proved by the geologists. Earlier sites of seas and the oceans, are now seen to be occupied by the mountains that are composed of folded and faulted rocks. Rocks adjoining the intrusive magmatic bodies, are many times found to be folded, faulted, fractured and so on. It is therefore evident, that the vertically acting forces are operative in the crust of the earth. A rising intrusive magmatic body, requires space in the crust of the earth. Therefore the rocks existing in that place are required to be pushed aside. During this process, the rocks must be getting folded and or faulted.

Thus the cause of the deformative forces, is to be reckoned in the upheaval and the sinking of the landmasses, and the intrusive magmatic rocks. But whatcauses the upheaval or the sinking of the landmasses, is a question in itself. The fact is that the crust of the earth is being built up in stages, and one of the outcomes is the observed deformation produced in the crustal rocks. One of the mechanisms postulated for the development of the folded mountains, is the convection currents generated in the sima. Due to this, the sediments laid in the geosynclines are heaved up. Folding and faulting are some of the results of this movement. Likewise, the continents are set in motion through the mechanism of the plate tectonics. Thus the movement of the Gondwana landmass towards the Chinese landmass, gave rise to the Himalayan folded mountains which are a store house of various structures like those developed in the Alps. But at the same time, there are plutonic igneous rocks like the granites, the gabbros etc., in the Himalayan mountains, which are intrusive into the sedimentary rocks. Therefore even these igneous plutons could have been responsible for the development of the deformative forces.

The Cuddaphs and the Vindhyans are the instances of the upheaval of the basins from the sea bottom. Obviously, vertically directed forces must have been operative during the said upheavals. The Aravalli mountains, the Dharwarian schistose rocks are the examples of the the fold mountains which have undergone extensive erosion. In the past, these mountains also must have been formed due to the movements of the landmasses towards each other, and in the process, folding and squeezing of the sediments laid in the then existing geosynclinal basins, must have taken place. To look for the deformative forces, the main geological event is to be reckoned, and the answer is readily obtainable. Generally events occur in pairs. An upheaval is to be paired by sinking of a landmass, nearby to it. This aspect is elaborated in a separate section of this chapter. It is thus clear that regarding the origin of the deformative forces, there are many conjectures made on the basis of the available structural data.

CLASSIFICATION OF SOURCES OF DEFROMATIVE FORCES

The distribution of the several structures encountered in the crustal rocks reveals that some of them are located at shallow depth and these do not extend much in depth. Some others are located at comparatively deeper levels in the crust of the earth, and such structures are found to continue considerably at depth. Based on these observations, the sources of the deformative forces are split into two major types namely.

  I. surficial or non-tectonic type, and

  II. sub-surface or tectonic type.

The "non-tectonic type" of forces are regarded to act on the crustal rocks from "without inwards" whereas the "tectonic type" of forces are regarded to have acted on the crustal rocks from "within outwards". In other words it means that in the former case, the forces are generated at or near the surface of the earth, while in the latter instant, these are generated in the deeper parts of the earth.

There is yet another way of effecting the classification of the deformative forces. After all the forces act on the rocks.

Therefore two situations are possible, namely forces acting on,

  (a) the entire rocks as one unit, and

  (b) the unit (mineral or minerals) of the rock.

The technical terms used for the above two varieties of the forces are non-penetrative and penetrative, respectively. The effects of the penetrative forces have been elaborated in Chapter 8.

As has been noted in the earlier chapters, the non-tectonic, the tectonic, the penetrative and the non-penetrative forces are further distinguishable into compression, tension and so on. The several distinctions made amongst the deformative forces have been presented in a schematic diagram which is given below.

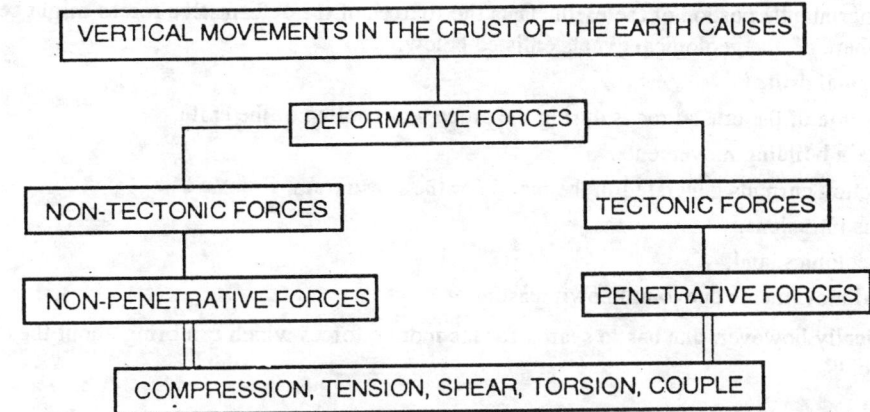

From the above given schematic diagram of the classification of the deformative forces, it is quite apparent that probably the vertical movements created in the earth are responsible for the initial deformative forces, and these later take up different characters like the penetrative, the non-penetrative, the compression and so on. However nothing can be said with certainty regarding what caused the initial deformative forces. It is therefore held as a matter of conjecture.

NON-TECTONIC TYPE OF DEFORMATIVE FORCES

Gravity is the main stay of the non-tectonic deformative forces. Wherever gravity is dominant and rocks likewise are prone to or susceptible to its action, deformation is brought about. Thus outcrops of vertical beds display folds only at their top portions, because these bend under the pull of gravity (Fig. 162 A,B). Leaching away of chemically susceptible rocks like limestones might bring about collapse of the overlying beds. This situation might give rise syncline like structure, or a breccia that resembles a fault breccia. Landslides, differential compaction of sedimentas also can give rise to the deformative forces which are of the non-tectonic type. As already noted, such structures do not continue much at depth.

TECTONIC TYPE OF DEFORMATIVE FORCES

These are also called as the orogenic forces, since these are found to be associated with the major geological events like the upheaval of a landmass, sinking of a basin and so on. Most of the structural features encountered in the crustal rocks are produced due to these forces. As has been already noted, the deformative forces are given different names like compression, tension and so on. But it is difficult to pin point which one of them was the initial force. Ashas been described in Chapter 1, compression can give rise to shear, even tension can give rise to shear. Further in order to cause compression at one place or spot in the crust of the earth, it is felt logiocal to expect the creation of spot of tension in the near-vicinity ofsuch a spot of compression, and vice versa. Granting for a moment that both compression and tension are independently created, the question still remains unanswered as to what caused the compression and the tension. Uncertainty thus prevails in exactly ascertaining the nature of the initial deformative force.

It is of great importance to note that the geological events seldom take place alone, and generally two or more events participate together. Thus the Himalayan, the Aravalli, the Alps and the other orogenic mountains are observed to be comprised of folded and faulted bedded rocks. In addition, these orogenic mountains are also seen to have a core of the intrusive plutonic rocks like granites, granodiorites, gabbros and so on. Such igneous bodies are of batholithic dimensions. If the earth came into being through the process of cooling of the original gaseous and molten state, then contraction and a reduction in volume of the outermost part of the earth (crust) is inevitable. Such a contraction also can create deformative forces because the contracted outer crust has to fit on to a bigger inner mantle portion of the earth. Thus the sources of the deformative forces might be any one or a combination there of, the geological events enlisted below.

(a) continental drift,

(b) contraction of the crustal rocks during the process of cooling of the earth,

(c) mountain building movements,

(d) convection currents generated in the mantle or the substratum,

(e) igneous intrusions,

(f) plate tectonics, and,

(g) vertical movements due to unknown reasons.

Sarcastically however, one has to search for the motive forces which can bring about the above enlisted geological events.

DIRECTION OF MOVEMENT OF THE DEFORMATIVE FORCES

The direction in which the deformative forces had operated in the geological past, can be established only through a careful study of the diposition of the structures developed in the crust of the earth. In a very generalised manner it may be said that the deformative foreces can act in the horizontal and the vertical directions. But how to establish these two directions? Proof is forthcoming from the attitude of the folded structures. In the great majority of cases, the axial planes of the folds are nearly vertical, or are highly inclined in attitude. The direction of the movements of the deformative forces that had caused such folds must be in the horizontal planes. Such horizontal compressive (in the case of folds) and tensile (in the case of fracture planes) forces might be created by the rotation of the earth on its own axis, the drifting of the continents towards each other and so on. Thus the folded mountains of Himalayas are due to the movement of the Gondwana landmass towards the Chinese landmass to which a mention has been already made.

VERTICAL UPLIFT AS A SOURCE

The rising and the sinking of the landmasses has been proved by the geologists. Thus the Himalayan mountains, the Cuddaphs and the Vindhyans are the clear instances of the upheaval of the basins from the sea bottom. Obviously vertically directed forces must have been operative in the said upheavals, to which a mention has been already made.

CONVECTION CURRENTS IN THE MANTLE AS A SOURCE

The convection currents generated in the sima or in the mantle of the earth, move vertically upwards. As these reach the bottom of the sialic layer, these move along the sial-sima contact i.e., convection currents then start moving in horizontal directions (Fig. 160, 161). Thus the vertically rising convection currents might later give rise to the horizontally directed forces.

IGNEOUS INTRUSION AS SOURCE

As has been noted while explaining the origin of folds, the laccolithic form of the igneous intrusion is due to the vertical movement of the magma into the sediments, and the latter rocks get upwarped (folded). Therefore the deformative forces created by the uprising magma are vertical in nature. The uparched rocks are not produced through the horizontal compression in this case. A magma can intrude into the crustal rocks only by pushing up or brushing aside the pre-existing rocks. Therefore the crustal rocks in the vicinity of the magma have to undergo a decrease in volume because of the compressional forces created by the intruding magmas. However it is necessaary to prove that the pre-existing crustal rocks were not folded prior to the act of the intrusion of the magma.

CONCEPT OF PAIRED GEOLOGICAL EVENTS

It is clear from the foregoing account that the deformative forces required for the production of the various structural features, these are the outcome of an another geological event that had enacted in the crust of the earth. An upheval of a part of the crust of the earth, is required to be accompanied by the sinking of a nearby landmass. It might be due also to the intrusion of an igneous body at depth. The domical or the anticlinal fold developed in the rocks found on the top of the laccolith, is due to the igneous intrusion. Thus the domical structre is paired by the geological event of the igneous intrusion. In such of the igneous intrusions where the folded structures are not only developed at the top of the igneous body (like that in a laccolith) but away from it too, then it is due to the pushing aside of the crustal rocks. But all the same, here too the igneous intrusion has produced the folded structures in the crustal rocks. Magma itself needs to be generated by yet another geological process like the melting of the lower parts of a sinking sedimentary basin. A sinking body (due to gravity) might be pushing the rocks aside thus creating room for itself. It is therefore argued by some geologists that faulting might lead to folding. Opposite view also is tenable namely, folding leading to faulting of the rocks. The fact remains that the two geological episodes are paired - sinking of landmass and the formation of folded or faulted structures or both.

Drifting of continents is an accepted fact. Eastwardly directed rotation of the earth, convection currents generated in the mantle, sea floor spreading, subduction, plate teconics are some of the motive forces that can set the continents in motion. As this movement drags the crustal rocks, horizontally directed deformative forces might be then generated which can produce several structures in the crustal rocks. Thus again folding and the other structures developed in the rocks get paired or connected with the geological event of the drifting of the continents.

Krishnan (1968) has noted several instances of the pairing of the geological events, though he has not called it as "paired events". Some of these are quoted below.

(a) The Gangetic plains owe their origin to a sag in the crust probably formed contemporaneously with the upliftment of the Himalayas (p. 2),

(b) The chief coal fields of India owe their preservation to block faulting. It is possible that block faulting was partly contemporaneous with the erruption of Rajmahal lavas (p. 56),

(c) The rock exposures around *Jutogh* and *Chor* mountains have been regarded as forming a highly compressed recumbent double anticline, the *Chor* granite occupying the core of the intervening syncline (p. 138).

Thus it is logical to infer that the tectonic deformative forces are due to the horizontal and the vertical movements in the crust of the earth. Thus geological event is associated with some other events as are described above. The fact thus becomes quite apparent that no geological event is enacted alone, it has to be paired or accompanied by another one.

7

IMPORTANCE OF STRUCTURES

Economic importance of joints, fractures and ruptures, Importance of faults and folds. Examples of mineralisation associated with faults and folds from India.

Academic importance of joints, ruptures, drag folds, axial plane cleavage, fracture cleavage, oscillation ripples, current bedding and graded bedding.

Figures - 173 to 213.

Photo - 65.

Studies no doubt are carried out to obtain a complete knowledge of any subject. Afterwards the usefulness or the utility, is to be considered. The various structures described in the previous chapters help to know the vastness and the varieties of the structures developed in the crustal rocks. This will form the academic side of the studies. But is there anything else to which the structures could be utilised? The geoscientists are looking for the deposits of the economic minerals. Many times there is a considerable coincidence of mineralisation and the presence of a particular structure. Thus the presence of a structure raises the possibility of the occurrence of a mineral deposit. Keeping this feature in view, the academic and the economic utility of the various structures will be described in the following pages.

ECONOMIC IMPORTANCE

IMPORTANCE OF JOINTS, FRACTURES AND RUPTURES

As noted earlier, ruptures in general and the joints in particular play an important role. The movement of groundwater, the mineralising solutions and the like, will be governed by the structures. Structurewise considered, the fractures, the joints and the ruptures will offer opportunity for the groundwater and the mineralising solutions to reach the interior parts of the rocks. Ruptures, fractures and the joints are designated as the secondary porosity of the rocks. Thus the seepage of the groundwater will be maximum wherever the rocks are intricately fractured and ruptured. In the search for the groundwater potentials, the nature and the intensity of the fracturing of the rocks, therefore holds the clue. However groundwater also could be lost, if impervious layers are not present at depth to prevent any further movement. Thus the porosity and the permeability will be enhanced by the presence of the joints, the fractures, the ruptures and so on. Ladder vein deposits are the outcome of the mineralisation along the tension joints developed in the dykes. In general then, the fractures and the ruptures are good receptacles for storing groundwater as well as to bring about the deposition of the valuable minerals.

IMPORTANCE OF FAULTS AND FOLDS

Faults may bring pervious and the impervious rocks in juxta position with each other. Such places mark the accumulation of groundwater. Springs are often found at the site of faults, because of this reason. Minor faults act as the distributors of the mineralising solutions, but the fault planes as such are not actively mineralsed. Folding creates reservoirs for oil and the natural gas. Most of the oil fields are located at the sites of anticlines. Formations are rendered shallow at the anticlines, and if there be a horizon containing economic minerals, then the cost over sinking vertical shafts, gets reduced (Figs. 173 A,B). If it is thrown into isoclinal folds, the horizon is encountered twice in one shaft (Figs. 174 A,B,C).

Alongside with the development of the folds, tension joints are created which may get mineralised. Not only the joints are mineralised, but also mineralisation is noticed along the bedding planes. These are collectively called as the "pitches and flats (Fig. 175). Koppad (1976) has recorded ladder veins of quartz in the refolded banded hematite quartzites (Photo 65). The world famous Saddle Reef of Bendigo, Australia, exemplifies the role of folds in the process of mineralisation. Due to tight folding, openings were created between the beds which were later filled by gold bearing quartz geins (Fig. 176). According to Bateman (1951), one of the causes of the "ore shoots" is the availability of suitable "structure". Therefore the presence of a structure raises the possibility of mineralisation to accompany too.

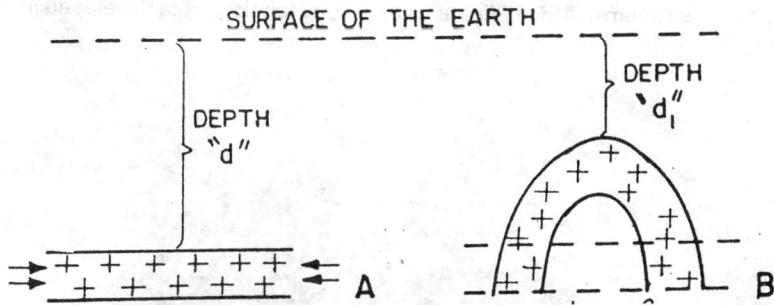

Figs. 173 A,B. A horizontally reposing body which is at a depth "d" from the surface, is affected by folding (Fig. A). Due to this (anticline) the body is raised to a shallower depth "d₁" (Fig. B).

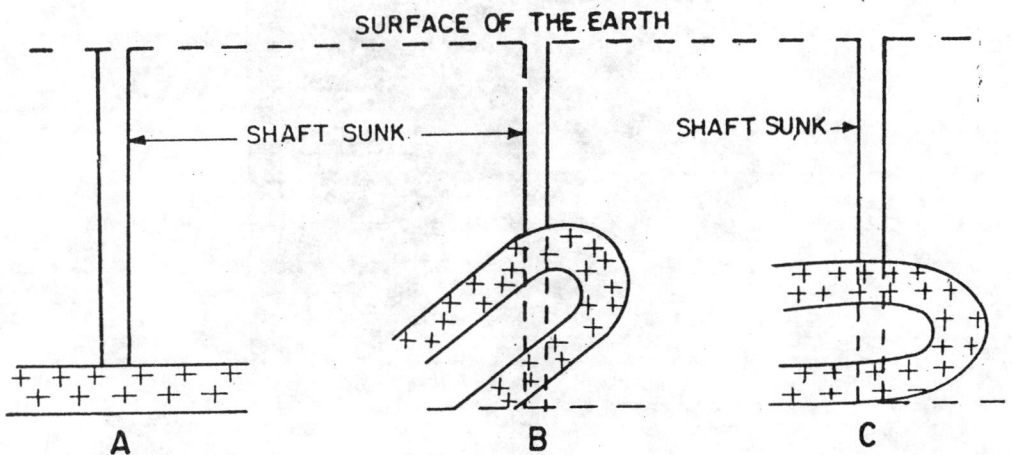

Figs. 174 A,B,C. A horizontally reposing mineralised body is usually encountered only once in a vertical shaft (Fig. A). Due to isoclinal folding, the same body is rendered shallow, and it occurs twice in a vertical shaft (Fig. 174 B). In the case of a recumbent fold, the body is reached twice in a vertical shaft, besides it being rendered shallower in depth.

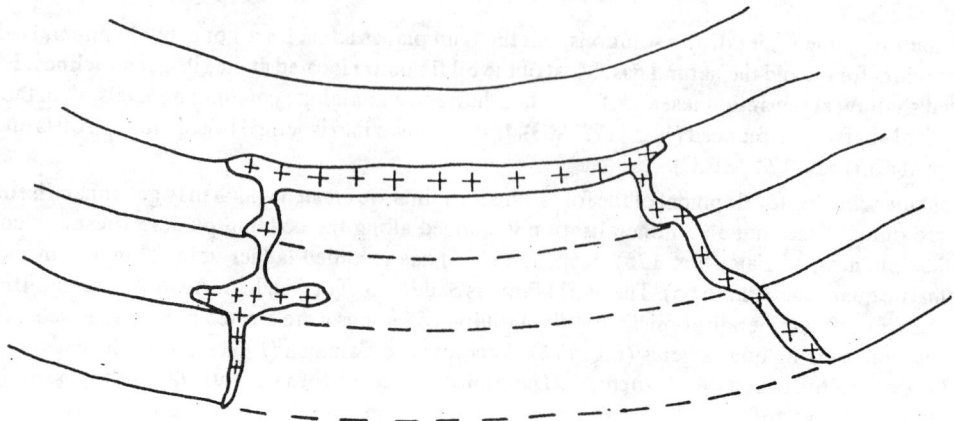

Fig. 175 (after Bateman 1951) Mineralisation along tension joints developed in a sycline.

Photo. 65. Field photo of quartz veins filling up joints trending perpendicular to the layers of hematite from banded hematite quartzite. This has given rise to ladder veins. The locality is Tarikoppa hills, Shirhatti taluka Dharwad district, Karnataka state. Courtesy Dr. V.B. Koppad.

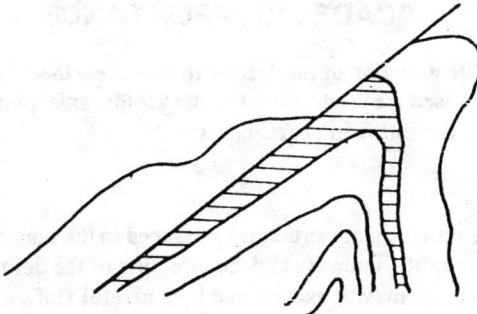

Fig. 176. (After Bateman, 1951). Saddle Reef of Bendigo. Quartz vein with gold (cross line.

EXAMPLES OF MINERALISATION ASSOCIATED WITH FAULTS, FOLDS FROM INDIA

Krishnan (1968) has described several instances of mineralisation associated with structures. Some of them have been described below.

1. Talc bearing serpentine rocks occur along a NNE - SSW zone to the northeast of Udaipur city and also in Sirohi, Dungarpur and Idar. Their distribution generally follows the trend of the axis of folding of Aravalli and Delhi system of rocks in the region (p. 68).

2. The manganese belt of Nagpur-Chhindwara region is structurally very complex. It is a combination of isoclinal, recumbent folds, numerous steep dipping strike faults, thrust faults. The folds and faults produce en echelon pattern (p. 114).

3. The Singhbhum thrust zone extends from Porhat in western Singhbhum through Chakradharpur, Amda, Rakha mines, Mosabani and Sunrgi into Mayurbhanj over a distance of 160 km. It has an E - W course in the western part and turns to the SE in the eastern part. The thrust zone marks the overfolded limb of a geanticline (p. 125). It is 2 to 5 km. wide and is bordered on its north and south by well marked shearing (p. 126).

   According to Banerji (1962) mineralisation took place in three stages. Earliest led to the formation of apatite-magnetite lenses. Uranium mineralisation in the form disseminated uraninite, torbernite and autunite was the second. Copper sulphide mineralisation is the last (p. 126-127).

4. The Pb-Zn deposits of Zawar near Udaipur, occur as replacement veins and fissure fillings, occupying fault and fracture zones in dolomites of Aravalli age associated with phyllites and quartzites (p. 151).

5. The manganese ores associated with kodurites of Eastern Ghat, occur in the axial parts of the main folds and some cross folds (p. 473).

6. The main oil fields of Burma are situated on the first anticline east of the main syncline (p. 473).

7. The Digboi oil field is about 16 km. long and the structures narrow, being cut off on the northwest by the Naga thrust which bring the Tipams against the alluvium (p. 474).

8. Geophysical prospecting conducted in the Brahmaputra valley to the northwest of Naga thrust indicated presence of structure at depth. Drill holes at Nahorkatiya, Moran, Lakwa and Rudrasagar have struck oil (p. 474).

9. The Cambay area is a NNW trending faulted trough, and it extends beyond Mehsana. Oil and gas has been found in this area, as well as in Ankleshwar-Olpad region south of Narmada river (p. 475).

ACADEMIC IMPORTANCE

Some of the structures indicate the direction of the deformative forces, the "right side up" of the beds, anticline, syncline and so on. The structures used are joints, ruptures, drag folds, axial plane cleavage, fracture cleavage and so on. These will be described in the following paragraphs.

JOINTS AND RUPTURES

These have a systematic relation with the main structures produced in the crustal rocks. A researcher could utilise the pattern of joints to derive the quality, intensity and the direction of the deformative forces. Magmatic or non-magmatic processes of rock evolution may be ascertained by a careful and a detailed study of the disposition of the ruptures and the joints. It is often noticed that the joints or the ruptures are irregularly distributed in the rocks. Some areas are devoid of joints or the ruptures, or a contrast in the pattern of jointing is noticed. The significance of this varies from rock to rock, and from place to place. It may be inferred that perhaps the rocks did not exist at that time when the ruptures and the joints were produced in the adjacent rocks. However the actual cause may be that the intensity of the deformative forces was not constant and uniform, and it varied with depth. This produces inhomogeneity in the disposition of the stresses, and this inturn gives rise to unequal fracturing of the rocks to such an extent that some parts are not at all ruptured or fractured. Thus one can comment on the nature of the deformative forces from a careful and a detailed study of the disposition and the pattern of the joints. Columnar joints in particular are very useful. The long columns bear a systematic relation with the attitude of the rocks. If the columns be vertical, then the igneous body containing them is horizontal in attitude, and if the columns be horizontal, then the igneous body is vertical. Hegde (1984) describes horizontal columns in a vertical dyke (Photo 13).

DRAG FOLDS

While describing the kinds of folds, mention was made to "minor folds". Drag folds belong to that category. These are produced due to the dragging of an incompetent bed sandwiched between two competent beds that are subjected to folding. While dealing with the flexure folds, mention was made to the differential movement taking place across the top side and the underside of the anticline (Figs. 126, 127 A to H). Relation between the drag folds and the major fold is shown in Fig. 177.

Study of Fig. 177 reveals that,

(i) the axial planes of the drag folds are parallel to the axial plane of the main fold,

(ii) the axial planes of the drag folds meet the limbs of the main fold at acute and abtuse angles,

(iii) the direction of dragging is in the direction of the acute angles,

(iv) in the case of *anticline,* the direction of dragging is up and towards the axial plane, considering the top side of the main fold. For the underside of the same fold, the direction of dragging is down and away from the axial plane of the main fold,

(v) in the case of a *syncline,* the direction of dragging is up but away from the axial plane, considering the top side of the main fold. For the underside of the same fold, the direction of dragging is down and towards the axial plane of the main fold.

The above noted features are utilised to derive,

(a) the kind of the fold i.e., whether an anticline or a syncline, and

(b) the nature of the limb or the limbs of the fold i.e., whether normal, rotated or overturned ones.

The procedure is as follows.

Presuming that the drag folds are congruous ones, then the axial planes of the drag folds are established, first. Then the acute angle between these and the bedding planes (the limbs) are derived. From the disposition of the acute angles, the direction of dragging (shearing) is established. From the direction of dragging, the top and the bottom of the bed (limb), is established. All these data are utilised to comment on the type of the fold. Some hypothetical cases will now be described to illustrate the methodology.

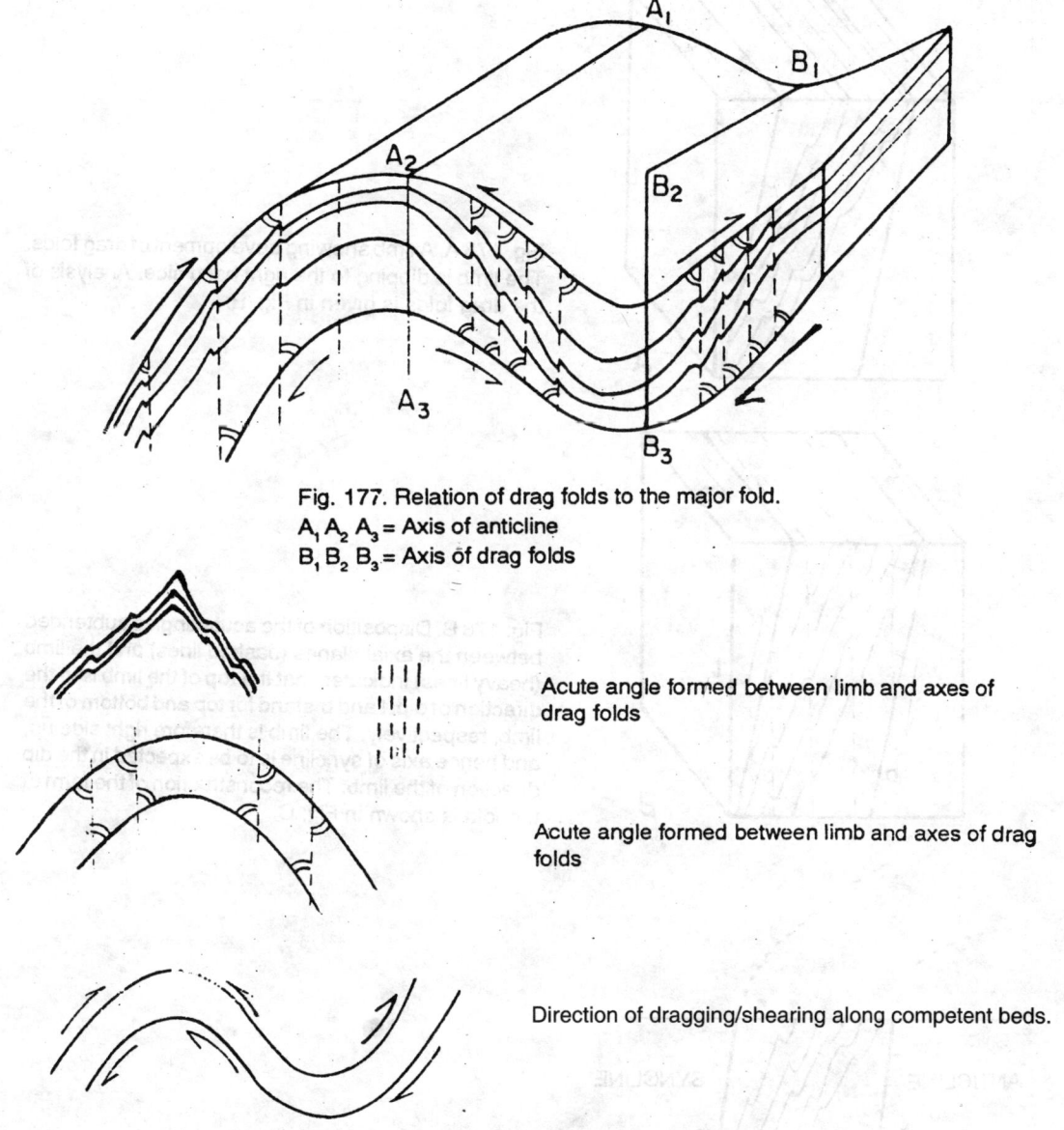

Fig. 177. Relation of drag folds to the major fold.
A₁ A₂ A₃ = Axis of anticline
B₁ B₂ B₃ = Axis of drag folds

Acute angle formed between limb and axes of drag folds

Acute angle formed between limb and axes of drag folds

Direction of dragging/shearing along competent beds.

Fig. 177. Relation of drag folds to the major fold.

CASE I LIMBS NOT ROTATED

In Fig. 178 **A**, the exposure of a limb of a fold in which drag folds are developed is shown. From the disposition of the acute angles subtended by the axial planes of the drag folds with the limb of the fold (Fig. 178 B), the direction of dragging (shearing) is derived to be upwards for the top of the limb, and it is downwards for the bottom of the limb. Therefore, the top of the limb (bed) coincides with the dip direction of the limb (bed), and hence the limb is in the "normal" position, i.e., it is not rotated or overturned. Further, as the direction of dragging is upwards, it is surmised that it points towards the axis of an anticline. Therefore, the limb shown in Fig. 178 **A** marks the normal limb of an anticline, and the axis of that fold is to be expected towards the left hand side of the outcrop. The reconstruction of the form of the fold is shown in Fig. 178 C.

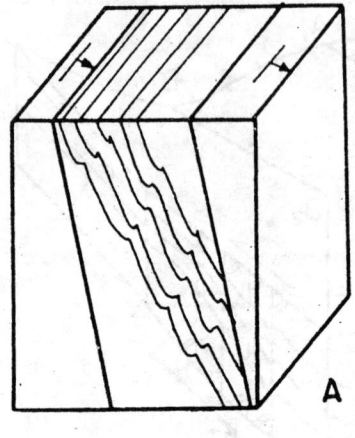

Fig. 178 A. A limb showing development of drag folds. The limb is dipping to the right hand side. Analysis of the drag folds is given in Fig. B.

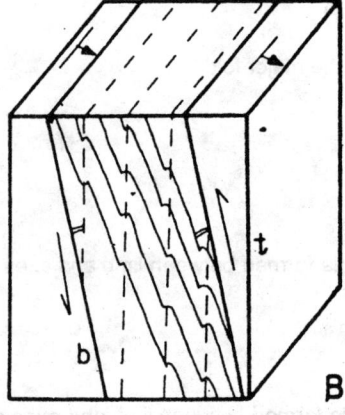

Fig. 178 B. Disposition of the acute angles subtended between the axial planes (dashed lines) and the limb (heavy lines) indicates that the top of the limb is in the direction of dip. t and b stand for top and bottom of the limb, respectively. The limb is therefore right side up, and hence axis of syncline is to be expected in the dip direction of the limb. The reconstruction of the form of the fold is shown in Fig. C.

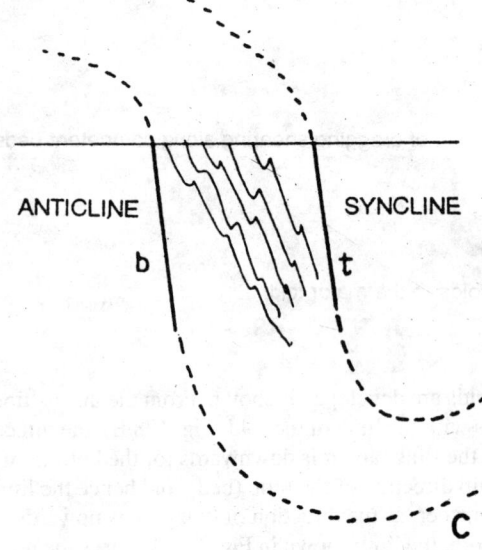

ANTICLINE   SYNCLINE

Fig. 178 C. Reconstruction into a normal fold because the limb is not overturned.

Fig. 178 A, B, C.

Another situation will be considered wherein though the limb is not overturned, it has however been rotated to a vertical attitude. This has been shown in Fig. 179 A. Analysis of the drag folds reveals (Fig. 179 B) that the limb (bed) is not overturned. When the bed is vertical, the top of the limb (bed) could be on the right hand or on the left hand side. In the present case, because the dragging is upwards on the right hand side, it marks the top of the limb (bed). Therefore the axis of the anticline is to be expected on the left hand side of the exposure. The reconstruction of the form of the fold is shown in Fig. 179 C.

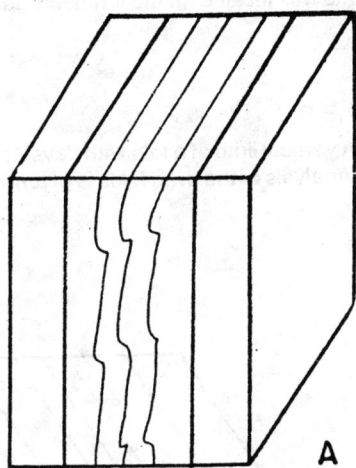

Fig. 179 A. A vertical limb of a fold with development of drag folds. Analysis of the drag folds is given in Fig. B.

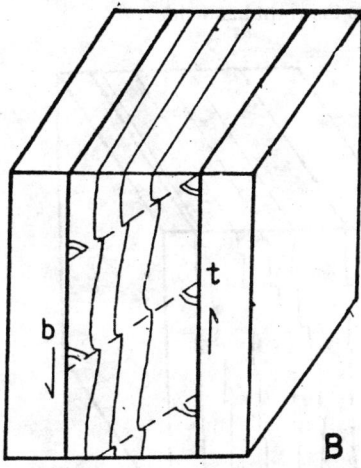

Fig. 179 B. Disposition of the acute angles subtended between the axial planes and the limbs, indicates that dragging is upwards on the right hand side of the limb, and it is downwards on the left hand side of the limb. Therefore top is located on the right hand side. Hence axis of syncline is to be expected on the same side. The reconstruction of the form of the fold is shown in Fig. C.

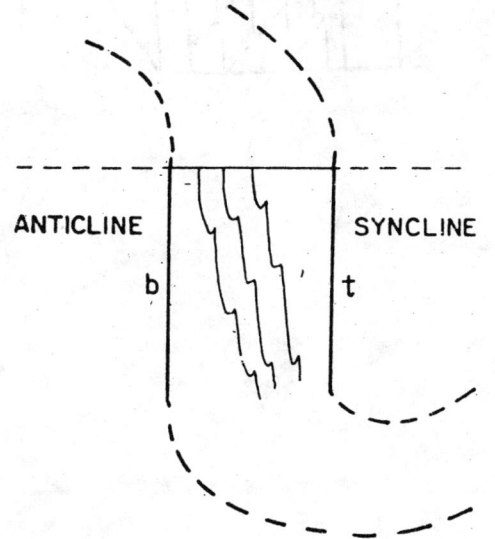

Fig. 179 C. Limb is rotated but is not overturned. It is therefore a rotated fold and not an overturned one.

Fig. 179 A,B,C.

Yet another case will be considered wherein again a vertical limb is exposed, but the disposition of the drag folds is different as compared to that shown in the earlier case (Fig. 179 A). This has been shown in Fig. 180 A. Analysis of the drag folds indicates that the dragging (shearing) of the bed is upwards on the left hand side of the limb, while it is downwards on the right hand side of the limb (Fig. 180 B). Therefore, the top of the limb is on the left hand side, and the bottom is on the right hand side. The axis of the anticline is, therefore, to be expected on the right hand side, and the axis of the syncline on the left hand side. The reconstruction of the form of the fold is shown in Fig. 180 C. Note that in the previous case (Fig. 179 C) anticline was located on the left hand side and the syncline on the right hand side.

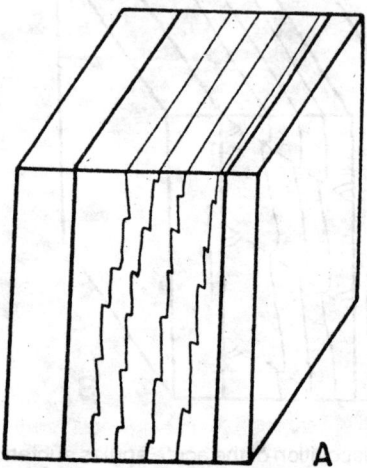

Figs. 180 A,B,C.

Fig. 180 A A vertical limb of a fold with development of drag folds. Analysis of the drag folds is given in Fig. B.

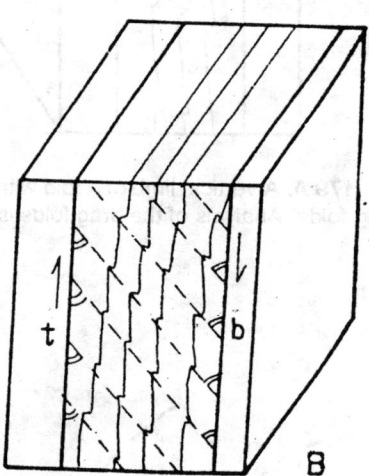

Fig. 180 B Disposition of the acute angles suntended between the axial planes and the limb, indicates that dragging is upwards on the left hand side, and it is downwards on the right hand side. Therefore top of the limb is to be expected on the left hand side. Hence the axis of the syncline is to be expected on the left hand side. Reconstruction of the form of the fold is shown in Fig. C.

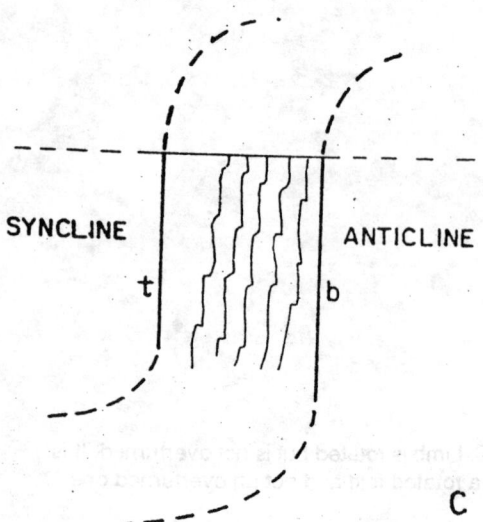

Fig. 180 C. The fold is rotated but it is not overturned.

CASE II LIMB OVERTURNED

This raises the possibility of one or both the limbs being overturned. Several situations arise and these are described further, In Fig. 181 A, is shown a dipping limb of a fold which has developed drag folds. Analysis of the disposition of the axial planes of the drag folds shows that the dragging of the beds is downwards on the right hand side, and it is upwards on the left hand side (Fig. 181 B). This shows that the top of the limb (bed) is against the direction of the dip of the limb (bed) as seen at present. The inference is hence drawn that this limb is rotated from its original position when it was dipping towards left. As a consequence of this, the nature of the axis of the fold gets changed. In the direction of the dip of the limb (bed), axis of anticline is to be expected, while to the left hand side, axis of syncline is to be located. The reconstruction of the form of the fold is shown in Fig. 181 C.

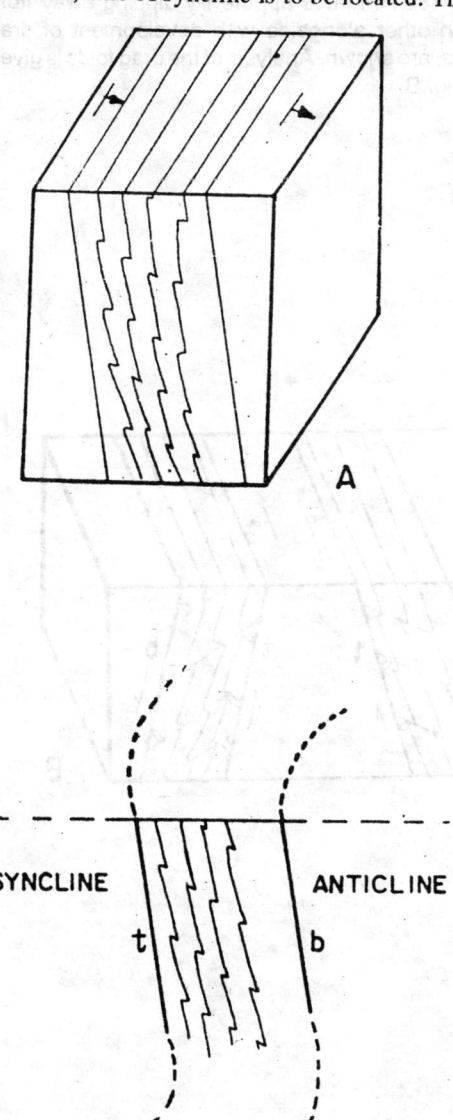

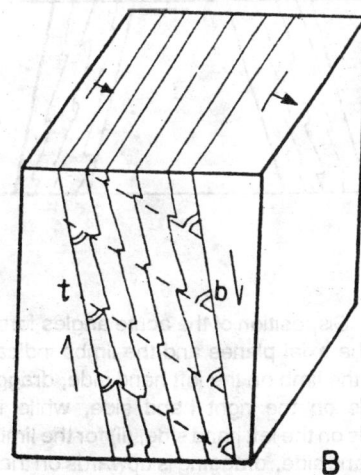

Fig. 181 A. A dipping limb with development of drag folds. Analysis of drag folds is given in Fig. B.

Fig. 181 B. Disposition of the acute angles subtended between the axial planes and the limb indicates that the dragging is upwards on the left hand side, and it is downwards on the right hand side. Therefore top is to be expected against the direction of the dip of the limb. The limb is therefore overturned. Axis of syncline is to be expected on left hand side. Reconstruction of the form of the fold is shown in Fig. C.

SYNCLINE   ANTICLINE

Fig. 181 C. It is an overturned fold

Figs. 181 A,B,C.

So far only one limb has been shown to be rotated and overturned. The development of the "fan fold" is through the rotation of both the limbs of the fold. This has been shown in Fig. 182 A. The top and the bottom of the limb (bed) are inferred by the disposition of the drag folds as has been described earlier. Analysis of the drag folds indicates that though the limb on the left hand side dips towards left, the top is located to the right hand side (Fig. 182 B). Similarly, , the limb on the right hand side dips towards right hand side, but the top is on the left hand side. Thus both the limbs are overturned, and though the fold apparently looks like an anticline, it is a synclinal fan fold. The reconstruction of the form of the fold is shown in Fig. 182 C.

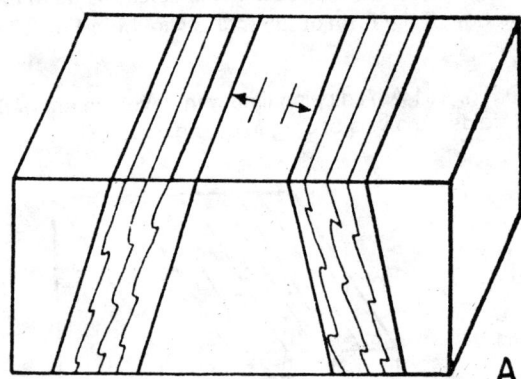

Fig. 182 A. Limbs apparently dipping away from each other alongside with development of drag folds, are shown. Analysis of the drag folds is given in Fig. B.

Fig. 182 B. Disposition of the acute angles formed between the axial planes and the limbs indicates that (i) for the limb on the left hand side, dragging is upwards on the right hand side, while it is downwards on the left hand side, (ii) for the limb on the right hand side, dragging is upwards on the left hand side, while it is downwards on the right hand side. Therefore both the limbs are overturned. Hence axis of syncline is to be expected inbetween the two limbs. Reconstruction of the form of the fold is shown in Fig. C.

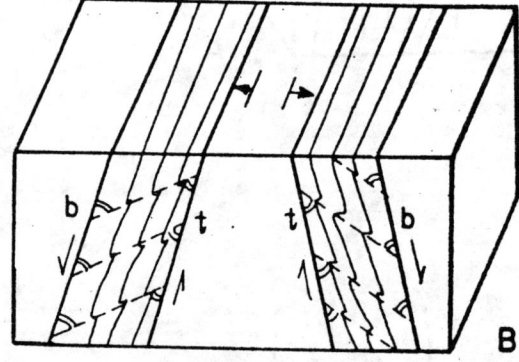

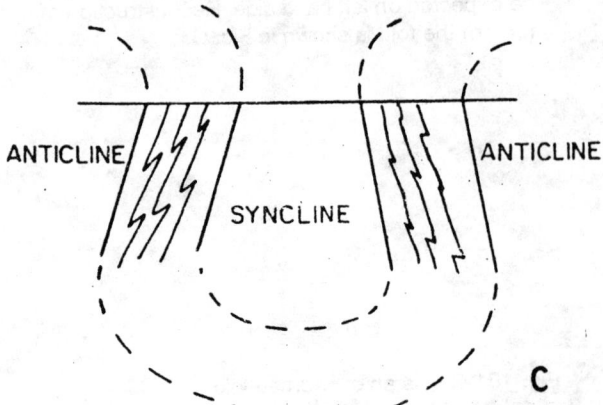

Fig. 182 C. The fold belongs to the category of a fan fold. Thus though it apparently looks like an anticline, it is a "synclinal fan fold".

Figs. 182 A,B,C.

The case of an anticlinal fan fold has been presented in Fig. 183 A. Analysis of the drag folds indicates that both the limbs are overturned (Fig. 183 B). Dragging (shearing) is upwards for the left hand side of the left limb, and the right hand side for the right limb. Thus in both the limbs, bottom of the limb (bed), is to be expected in the direction of the dip of the limb. Though the limbs apparently dip towards each other (Fig. 183 A), the analysis of the drag folds has shown that it is not a syncline, but it is an "anticlinal fan fold". The reconstruction of the form of the fold is shown in Fig. 183 C.

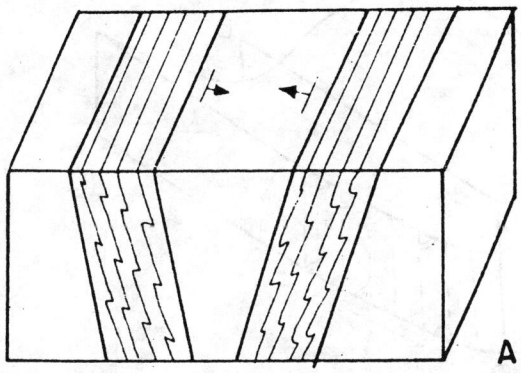

Fig. 183 A. Two limbs apparently dipping towards each other alongside with the development of drag folds are shown. Analysis of the drag folds is given in Fig. B.

Fig. 183 B. Disposition of the acute angles formed between the axial planes and the limbs indicates that (i) for the limb on the left hand side, dragging is upwards on the left hand side, while it is downwards on the right hand side, (ii) for the limb on the right hand side, dragging is upwards on the right hand side, while it is downwards on the left hand side. Therefore both the limbs are overturned. Hence axis of anticline is to be expected in between the two limbs. Reconstruction of the form of the fold is shown in Fig. C.

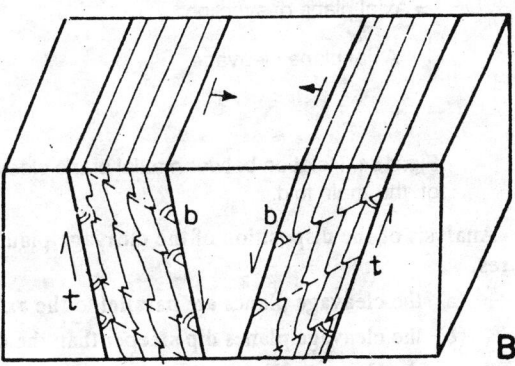

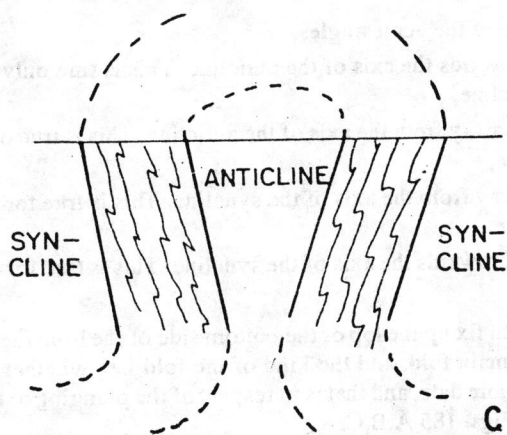

Fig. 183 C. Fold belongs to the category of a fan fold. Though the limbs apparently dip towards each other it turns out to be an "anticlinal fan fold".

Figs. 183 A,B,C.

AXIAL PLANE CLEAVAGE

This structure is more often used in ascertaining the "right side up" position of the limb of the folds. The arguments and the reasoning put forth are similar to that done for the drag folds. This cleavage is systematically related to the main fold, and indicates the attitude of the axial plane of the main fold, as well as the direction of dragging (shearing). This is shown in Fig. 184.

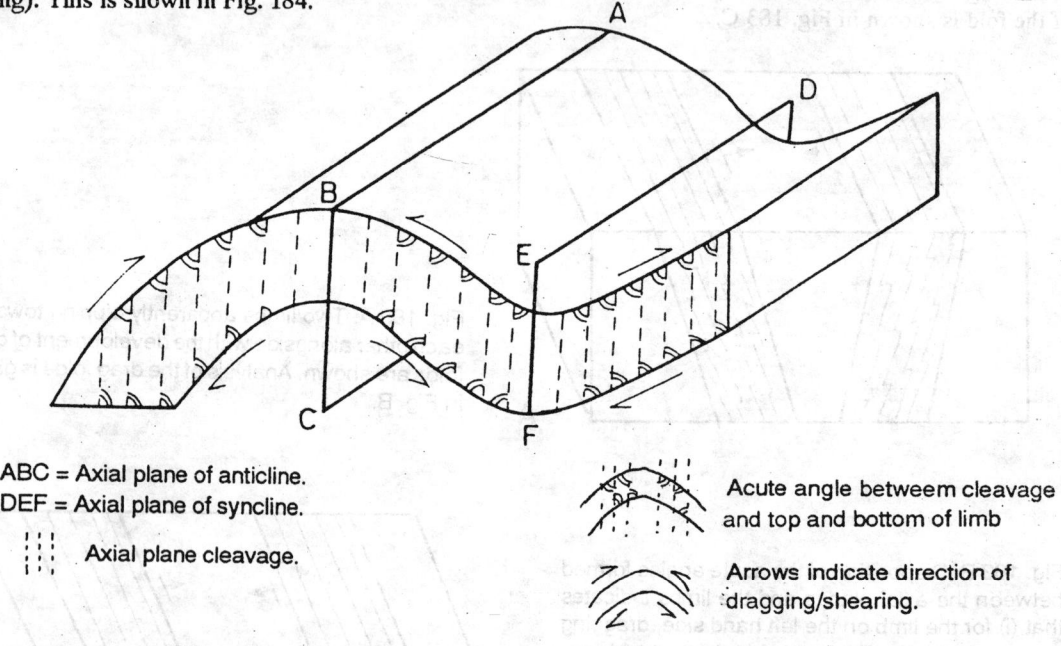

ABC = Axial plane of anticline.
DEF = Axial plane of syncline.

||| Axial plane cleavage.

Acute angle betweem cleavage and top and bottom of limb

Arrows indicate direction of dragging/shearing.

Fig. 184. Relation between axial plane cleavage, the limbs of the fold, and the axial plane of the main fold.

Analysis of the disposition of the cleavage planes in relation to the fold (Fig. 184) reveals the following features.

(a) the cleavage planes are parallel to the axial plane of the main fold,

(b) the cleavage planes dip steeper than the dip of the limbs of the fold,

(c) the cleavage planes subtend acute angles with the top side, and the bottom side of the limbs (beds) of the fold,

(d) the dragging or the shearing is in the direction of the acute angles,

(e) the shearing or the dragging is upwards and towards the axis of the anticline. This is true only for the top side or the top of the limbs of the anticline,

(f) the shearing or the dragging is downwards but away from the axis of the anticline. This is true only for the underside or the bottom of the anticline,

(g) the dragging or the shearing is upwards but away from the axis of the syncline. This is true for the top side or the top of the Limbs of the syncline.

(h) the dragging or the shearing is downwards and towards the axis of the syncline. This is true for the underside or the bottom of the syncline.

Thus the disposition of the axial plane cleavage helps to fix up the top or the bottom side of the limb (bed). It also helps to know the attitude of the axial plane of the main fold, and the kind of the fold i.e., whether an anticline or a syncline. The axial plane cleavage yields one more data, and that is in respect of the plunging or the non-plunging nature of the fold. This aspect is depicted in Figs. 185 A,B,C.

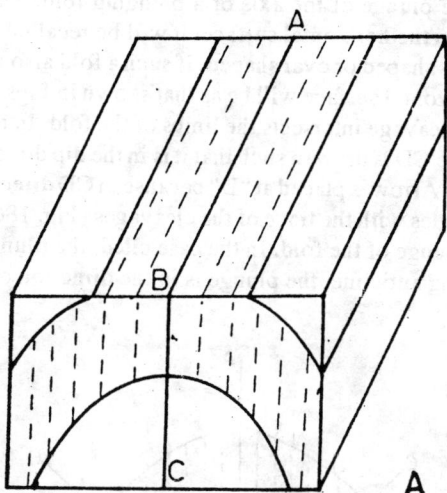

Fig. 185 A. Disposition of axial plane cleagage in a non plunging fold.

Note that the outcrop lines of cleavages are parallel to the axis of the fold (AB) as well as to BC which is a vertical section.

Fig. 185 B. Plunging anticline.

The plunge is from B to A. Cleavages are not only at angle to the limbs in the vertical section, but these are also at an angle to the limbs (outcrop) on the horizontal surface (in Fig. A it was not so). Important point to note is that outcrop closes in the direction of the plunge of the axis of the fold.

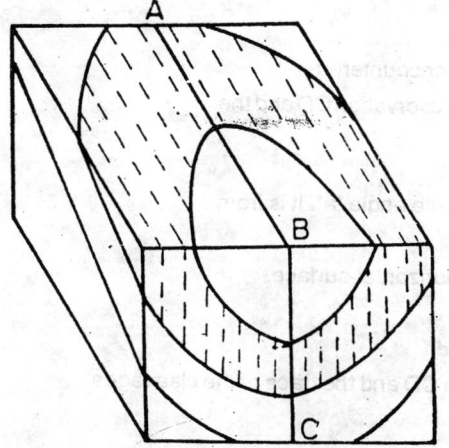

Fig. 185 C. Plunging syncline.

The plunge is from A to B as against that found in Fig. 185 B. In this fold also the cleavages are at an angle to the limbs of the fold, both in the vertical and horizontal sections. Important point is that the closure of the fold is in the opposite direction of the plunge i.e., fold closes towards A, while plunge is towards B.

In the Figs. 185 A,B,C the disposition of the axial plane cleavage in respect of non-plunging anticline, plunging anticline, and a plunging syncline, is shown. From these fighures it is apparent that the trace of the axial plane cleavage on the horizontal surface (outcrop), is parallel to the limbs of the fold in the case of a non-plunging fold (Fig. 185 A). However in the case of a plunging fold, the trace of the axial plane cleavage, intersects the limbs of the fold (Figs. 185 B,C). The inference to be drawan therefore is as follows.

    (i) if the axial plane cleavage be parallel to the limbs of the fold when projected on a horizontal surface, then it is a non-plunging fold,

    (ii) if the axial plane cleavage intersects the limbs of the fold when projected on a horizontal surface, then it is a plunging fold.

It becomes further necessary to derive the direction of the plunge of the axis of a plunging fold. This is achieved with the help of the axial plane cleavage as projected on the horizontal surface. It will be recalled here that the outcrop pattern of a plunging fold is characteristically V-shaped or oval shaped. If such a fold also were to develop the axial plane cleavage, the disposition of it on a horizontal surface will be as that shown in Figs. 186 A,B. Study of Fig. 186 A shows that the trace of the axial plane cleavage intersects the limbs of the fold. In order to ascertain the direction of the plunge of the axis of the fold, a line CD is drawan such that it is in the dip direction of the beds, meaning that it points to the younger beds of the series. Arrow is placed at "D" because in CD direction, younger beds are encountered. Such a line (CD) creates acute angles with the trace of the cleavages (Fig. 186 A). The direction of the acute angles points in the direction of the plunge of the fold. In the case cited, the plunge is from A to B. It may be recalled here that in the case of a plunging anticline, the plunge is in the direction of the closure of the fold. called, the "nose" of the fold.

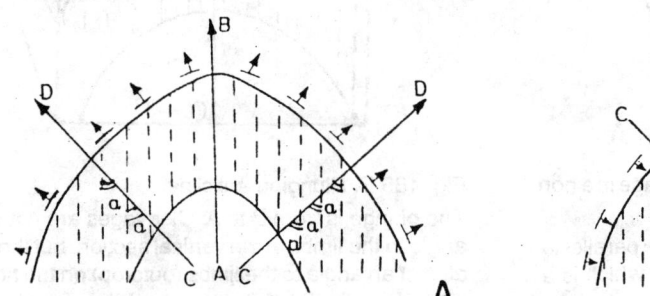
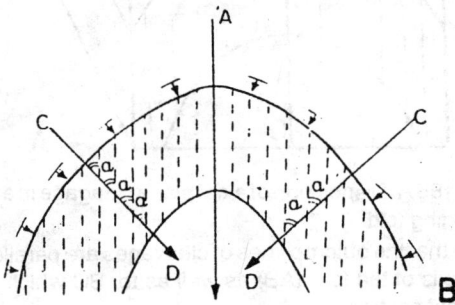

Fig. 186 A. Outcrop of a plunging anticlinal fold on a horizontal surface.

AB = trace of axis of plunging anticline.

CD = direction in which younger beds are encountered

a = acute angle formed between the line of observation CD and the trace of cleavages (dashed lines).

→ = dip direction of the limbs of the fold

Note that the plunge is in the direction of acute angle "a". It is from A to B, outcrop of the fold closes at B.

Fig. 186 B. Outcrop of a plunging synclinal fold on a horizontal-surface.

AB = trace of axis of plunging syncline.

CD = direction in which younger beds are encountered.

a = acute angle formed between the line of observation CD and the trace of the cleavages (dashed lines).

|→ = dip direction of the limbs of the fold.

Note that the plunge is in the direction of the acute angle "a". it is from A to B, but the outcrop of the fold closes at A i.e., in the opposite direction of plunge of the fold.

The disposition of the axial plane cleavage in a plunging synclinal fold is shown in Fig. 186 B. Study of this figure shows that in this case also the cleavage traces intersect the limbs (beds) of the fold. The line CD which is drawn in the direction of the younger beds, creates acute angles "a" with the cleavage. These acute angles point towards "B" end of the axis AB. It may be recalled here that in the case of a plunging syncline, the fold closes in the the opposite direction of the plunge. In Fig. 186 B, the closure is at "A" end, while the plunge is towards "B".

The relation between the axial plane cleavage and the attitude of the fold has been so far described. A few cases will be considered to illustrate the deciphering of the type of the fold from the disposition of the axial plane cleavage.

CASE I LIMB ROTATED BUT NOT OVERTURNED

In Fig. 187 A, is shown a limb of a fold which is dipping but is not vertical. Clavages are also developed. Analysis of the disposition of the cleavages with respect to the limbs of the fold shows (Fig. 187 B) that,

    (i) the cleavages are vertical,

    (ii) the shearing is upwards on the right hand side of the limb,

    (iii) the trace of the cleavages on the horizontal surface is parallel to the limbs (bedding planes).

    Therefore it is inferred that,

        (a) the axial plane of the fold is vertical and hence the overturning of the fold has not taken place,

        (b) the fold is a non-plunging one,

        (c) the top of the limb (bed) is in the direction of the dip of the limb of the fold. Therefore the limb is not rotated, and

        (d) the axis of the anticline is to be expected to the left hand side of the limb.

    The reconstruction of the form of the fold is shown in Fig. 187 C.

    In Figs. 188 A, a vertical limb of a fold alongside with the development of the axial plane cleavages, is shown. In such cases the establishment of the top and the bottom of the limb (bed) gains importance. In the absence of the axial plane cleavage, it becomes difficult to locate the axis of the fold, because the true, meaning the original direction of the dip cannot be established. Two sub-types are noted within the category of fold with a vertical limb, and these are shown in two different figures. In one case, the top turns out to be on the left hand side (Fig. 188 A), while in the other case, it turns out to be on the right hand side (Fig. 189 A).

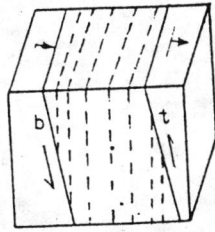

Fig. 187 A. A limb dipping to the right hand side along side with the development of axial plane cleavage is shown. Analysis of the cleavages is given in Fig. B

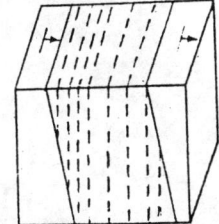

Fig. 187 B. Disposition of the acute angles formed between the cleavages and the limb of the fold indicates that dragging is upwards on the right hand side, it is downwards on the left hand side. The limb is right side up and top is in the direction of dip of the limb. Axis of syncline is to be expected on the right hand side of the limb. Reconstruction of the form of the fold is shown in Fig. C.

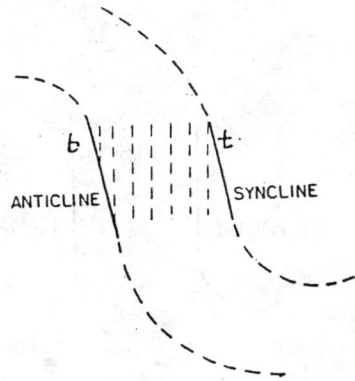

Fig. 187 C. The limb is not overturned, therefore it is a normal fold.

Figs. 187 A,B,C.

TOP ON THE LEFT HAND SIDE (FIGS. 188 A,B,C)

Analysis of the disposition of the cleavages with respect to the limb of the fold reveals that the top is located on the left hand side, because the shearing is upwards on that side, and the bottom is on the right hand side, because the shearing is downwards on that side (Fig. 188 B). The shear planes are nearly vertical which indicates that the axial plane of the fold is also vertical, and hence the dip of the cleavages is gentler than that of the limb of the fold. Therefore considerable rotation leading to verticality is effected. Axis of the anticline is to be expected to the right hand side of the limb. The reconstruction of the form of the fold is shown in Fig. 188 C.

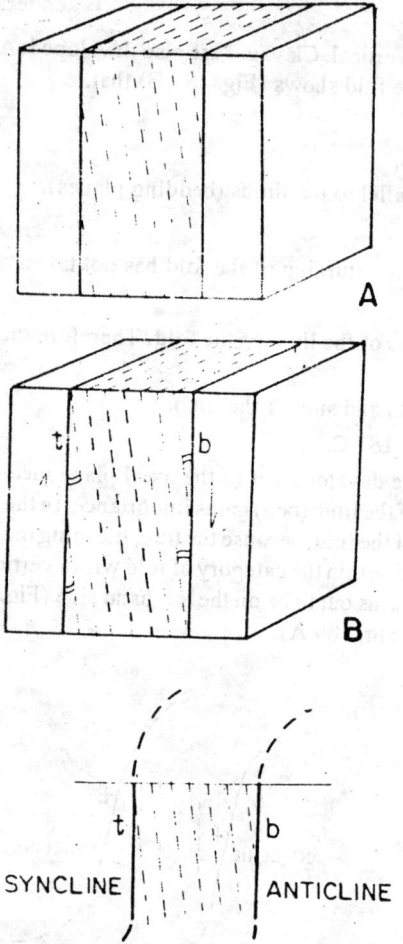

Fig. 188 A. A vertical limb with development of axial plane cleavages is shown. Analysis of the cleavages is given in Fig. B.

Fig. 188 B. Disposition of the acute angles formed between the cleavages and the limb indicates that the dragging is upwards on the left hand side, it is downwards on the right hand side. Therefore the top of the limb is on the left hand side, and the axis of the syncline is to be expected on the left hand side. Reconstruction of the form of the fold is shown in Fig. C.

Fig. 188 C. The limb is not overturned.

Figs. 188 A,B,C.

TOP IS ON THE RIGHT HAND SIDE (FIGS. 189 A,B,C)

Analysis of the disposition of the cleavages with respect to the limbs reveals that the top is located on the right hand side, because the shearing is upwards on that side (Fig. 189 B), and the bottom is located on the left hand side, because the shearing is downwards on that side (in the previous case, Fig. 188 B the situation was exactly opposite). The cleavages are nearly vertical indicating thereby that the axial plane of the fold is also nearly vertical. It is again significant to note that the dip of the cleavages is gentler than the dip of the limb of the fold (Fig. 189 B). Hence the limb is rotated, but it is not overturned. Axis of the anticline is to be expected to the left hand side of the limb. The reconstruction of the form of the fold is shown in Fig. 189 C).

CASE II LIMB OR LIMBS OVERTURNED

Overturning may affect one or both the limbs, the latter variety being called as the "fan fold". In Fig. 190 A, a limb of a fold with the development of the axial plane cleavages, is shown. Analysis of the disposition of the cleavages with respect to the limb (bed) of the fold shows that the shearing is downwards on the right hand side of the limb, and it is upwards on the left hand side of the limb (Fig. 190 B). Therefore the top of the limb is on the left hand side, while the bottom is on the right hand side of the limb. Further the dip amount of the cleavages is gentler than the dip of the limb. Therefore it is to be inferred that the limb is not only rotated, but it is also overturned. Axis of the anticline is to be expected to the right hand side i.e., in the direction of the dip of the limb. The reconstruction of the form of the fold is shown in Fig. 190 C.

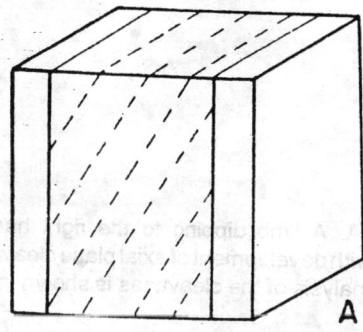

Fig. 189 A. A vertical limb with development of axial plane cleavage is shown. Analysis of cleavages is given in Fig. B.

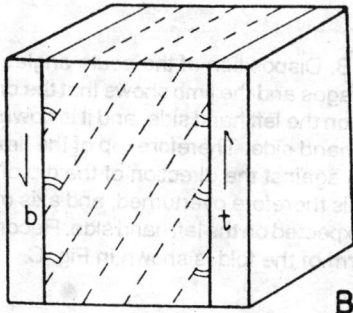

Fig. 189 B. Disposition of the acute angles formed between the cleavages and the limb indicates that the dragging is upwards on the right hand side, and it is downwards on the left hand side. Therefore top of the limb is to be expected on the right hand side, and the axis of the syncline likewise is to be expected on the right hand side. Reconstruction of the form of the fold is shown in Fig. C.

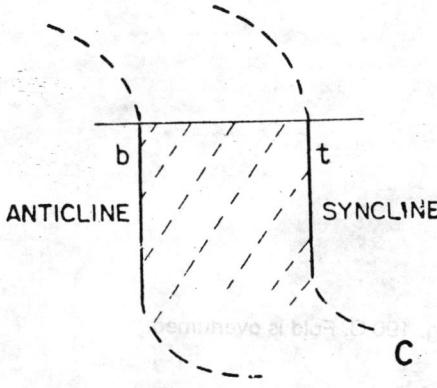

Fig. 189 C. The limb is overturned.

Figs. 189 A,B,C.

BOTH THE LIMBS OVERTURNED

In Fig. 191 A, two limbs are shown alongside with the development of the cleavages. In this case the two limbs are apparently dipping away from each other. Analysis of the disposition of the direction of shearing shows that in the dip direction, bottom of the limb (bed) of the fold is located (Fig. 191 B). It can be observed that the amount of dip of the cleavages is gentler than that of the dip of the limbs (beds). This indicates that not only both the limbs are rotated, but these are further overturned (bottom in the direction of the dip of the limbs). Therefore the fold belongs to the category of a "fan fold", and in the direction of the dip of the limbs, axis of anticline is to be expected. Though the limbs apparently dip away from each other, yet the fold turns out to be a syncline, and it is therefore called as a "synclinal fan fold". The reconstruction of the form of the fold is shown in Fig. 191 C.

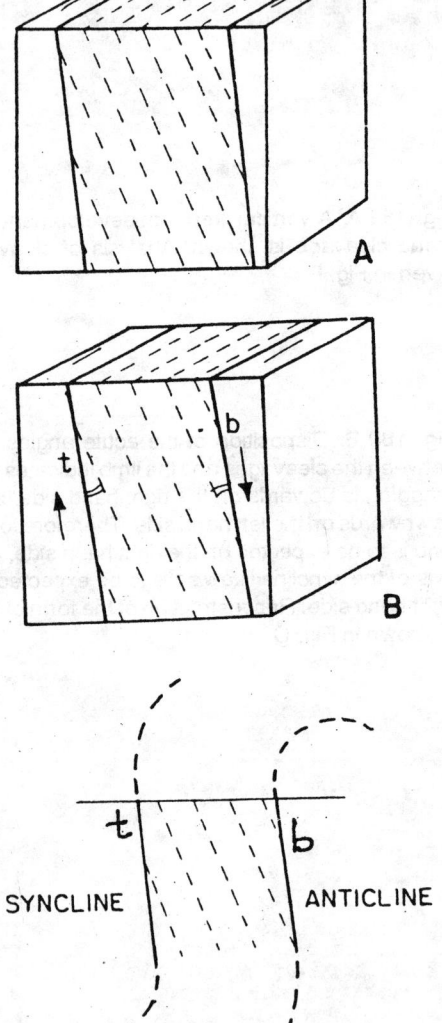

Fig. 190 A. A limb dipping to the right hand side together with development of axial plane cleavages, is shown. Analysis of the cleavages is shown in Fig. B.

Fig. 190 B. Disposition of the acute angles between the cleavages and the limb shows that the dragging is upwards on the left hand side, and it is downwards on the right hand side. Therefore top of the limb is to be expected against the direction of the dip of the limb. The limb is therefore overturned, and axis of syncline is to be expected on the left hand side. Reconstruction of the form of the fold is shown in Fig. C.

Fig. 190 C. Fold is overturned.

Fig. 190 A,B,C.

Another case is shown in Fig. 192 A, wherein the limbs apparently dip towards each other. Analysis of the disposition of the direction of shearing shows that in the direction of the dip of the limbs, bottom of the bed is located (Fig. 192 B). Further the cleavage planes have a dip flatter than that of the limbs (Fig. 192 B). This indicates that the limbs are not only rotated, but these are further overturned (bottom in the direction of the dip of the limbs). Therefore the fold belongs to the category of a "fan fold", and in the direction of the dip of the limbs, axis of anticline is to be expected. Though the limbs dip towards each other, the fold turns out to be an anticline, and therefore it is an "anticlinal fan fold". The reconstruction of the form of the fold is shown in Fig. 192 C.

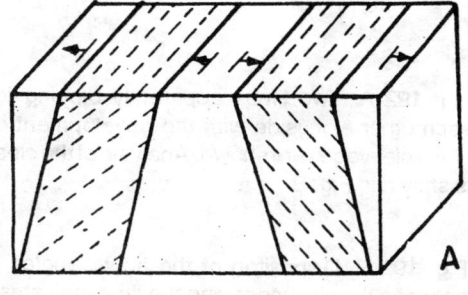

Fig. 191 A. Two limbs apparently dipping away from each other alongside with the development of axial plane cleavages are shown. Analysis of the cleavages is given in Fig. C.

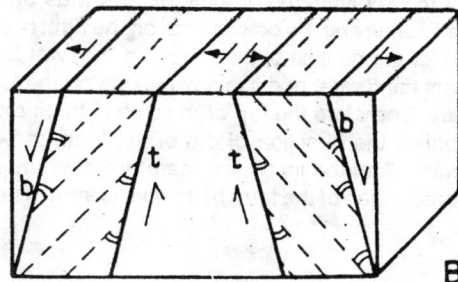

Fig. 191 B. Disposition of the acute angles formed between the cleavages and the limbs shows that (i) for the left limb, dragging is upwards on the right hand side, and it is downwards on the left hand side, (ii) for the right limb, dragging is upwards on the left hand side, and it is downwards on the right hand side. Therefore top of the limb is to be expected against the direction of dip of the limbs. Therefore both the limbs are overturned, and axis of the syncline is to be expected in between the two limbs. Reconstruction of the form of the fold is shown in Fig. C.

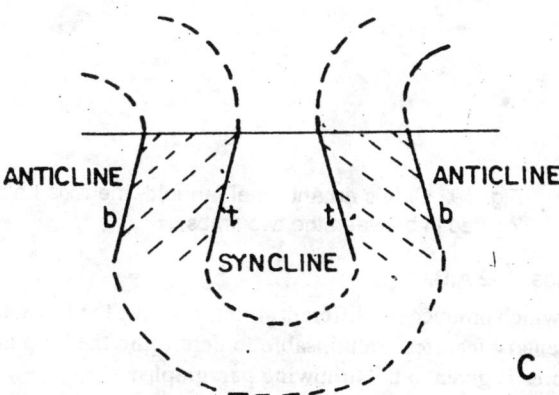

Fig. 191 C. Both the limbs are overturned. It is synclinal fan fold.

Figs. 191 A,B,C.

FRACTURE CLEAVAGE

This is also utilised in the same way as the axial plane cleavage, if it is a "congrous one" i.e., it has been produced alongside with the development of the main folded structures. Since such a cleavage is not frequently developed, it has not been dealt with in details.

OTHER STRUCTURES

There are some other structures which are used to ascertain the "top and bottom" of the beds. These are the oscillation ripples, the current bedding, the graded bedding and so on. These are in fact primary sedimentary

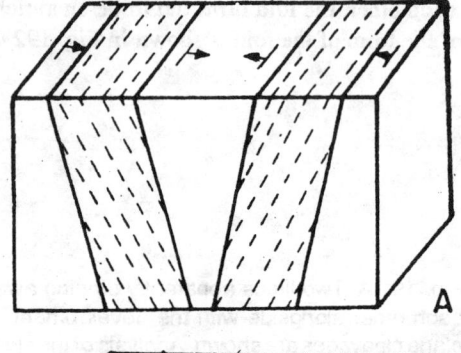

Fig. 192 A. Two limbs apparently dipping towards each other alongside with the development of axial plane cleavages are shown. Analysis of the cleavages is shown in Fig. B

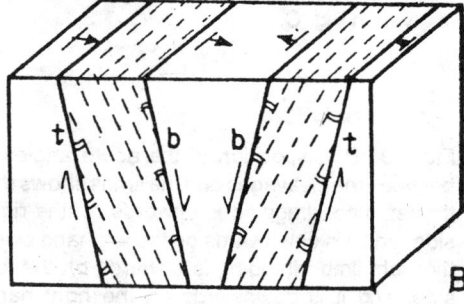

Fig. 192 B. Dispositon of the acute angles formed between the cleavages and the limb indicates that (i) for the left limb, the dragging is upwards on the left hand side, and it is downwards on the right hand side, (ii) for the right limb, the dragging is upwards on the right hand side, and it is downwards on the left hand side. Therefore the top of the limb is to be expected against the direction of dip of the limb, in both the limbs. Both the limbs are therefore overturned. Reconstruction of the form of the fold is shown in Fig. C.

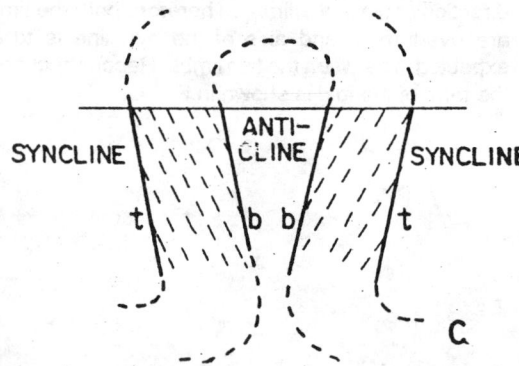

Fig. 192 C. It is an anticlinal fan fold, the axis being located in between the two limbs.

Figs. 192 A,B,C.

features and are not produced by the deformative forces which produce the different structures (like the folds, the faults etc.) in the rocks. However these primary sedimentary features are utilisable to determine the "top and bottom" of the beds. A very brief account of these features is given in the following paragraphs.

OSCILLATION RIPPLES

Distinction between the "original and the print or the cast" ripple is necessary, since the position of the top and the bottom of a bed changes, accordingly. Disposition of these two varieties is shown in Figs. 193 A,B,C. On the top of a bed, the oscillation ripples are *concave upwards* (Fig. 193 B). The next layer will get the print or the cast of these ripples on its bottom portion. These latter ripples will therefore be *convex upwards* as shown in Fig. 193 C. Thus the top and the bottom of a bed can be ascertained by the concavity and the convexity of the ripples. If it is concave upwards, then the top also is in that direction. If it is concave downwards, then the top also is towards the bottom side (this becomes the topsy turvy situation). In short it is to be remembered that *the concavity points to the top, and the convexity points to the bottom of a bed, respectively.* The disposition of the oscillation ripples in respect of the different attitudes of the bed, are shown in Figs. 193 D to K.

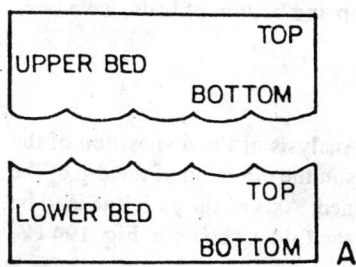

Fig. 193 A. Oscillation ripples on the top part of the lower bed, and bottom part of the immediately overlying bed.

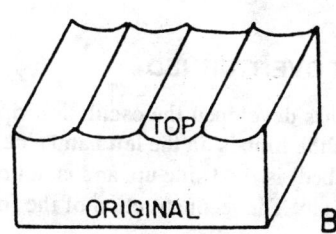

Fig. 193 B. Original oscillation ripple. Note that it is concave upwards or towards the top of the bed.

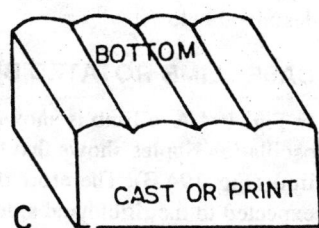

Fig. 193 C. Cast or print of oscillation ripple. Note that it is convex upwards or towards the bottom of the bed.

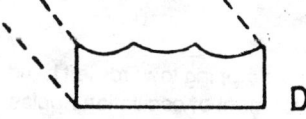

Fig. 193 D. bed horizontal, top towards top of bed.

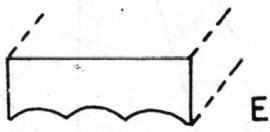

Fig. 193 E. bed horizontal, top towards bottom side.

Fig. 193 F dipping bed, top towards left hand side.

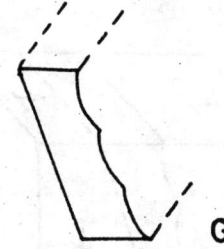

Fig. 193 G. dipping bed, top towards right hand side.

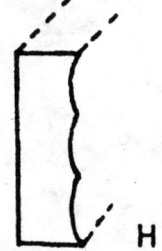

Fig. 193 H. bed vertical, top towards right hand side.

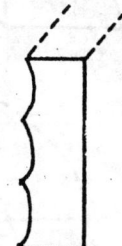

Fig. 193 I. bed vertical, top towards left hand side.

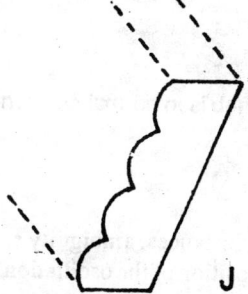

Fig. 193 J. bed dipping left, top towards right hand side.

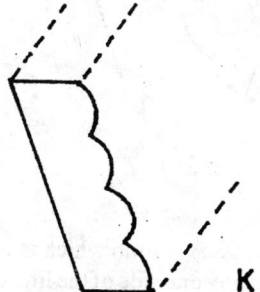

Fig. 193 K. bed dipping right, top towards left hand side.

Figs. 193 A to K. Appearance of oscillation ripples.

Some cases of the application of the oscillation ripples in ascertaining top and bottom of beds, have been described below.

CASE I LIMB ROTATED BUT NOT OVERTURNED

In Fig. 194 A, a limb is shown which has developed the oscillation ripples. Analysis of the disposition of the oscillation ripples shows that the top of the limb is on the left hand side, that is in the direction of the dip of the limb (Fig. 194 B). Therefore the limb (bed) is right side up, and is not overturned. Axis of the anticline is to be expected to the right hand side of the limb. The reconstruction of the form of the fold is shown in Fig. 194 C.

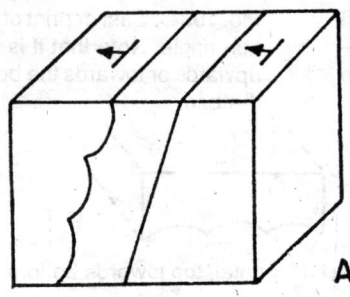

Fig. 194 A. A limb dipping towards left hand side along with the development of oscillation ripples, is shown. Analysis of the ripples is given in Fig. B.

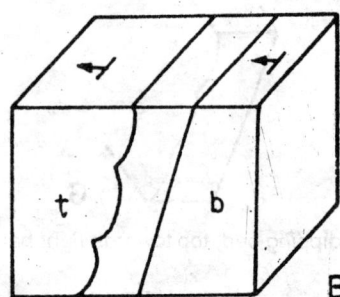

Fig. 194 B. Ripples are concave upwards on the left hand side. Top is also on the same side. Axis of syncline is to be expected on the left hand side of the limb. Reconstruction of the form of the fold is shown in Fig. C.

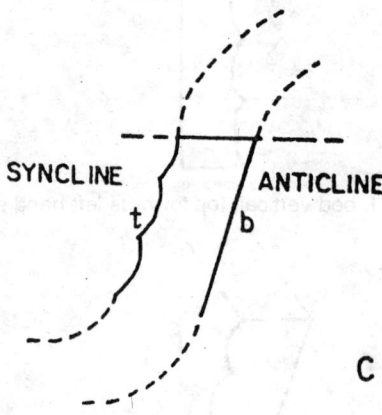

Fig. 194 C. The limb is in normal position. The fold is also normal one.

Figs. 194 A,B,C.

In Fig. 195 A, a limb which is vertical in attitude is shown. In such instances, ambiguity exists because the top may be on any one side of the limb, and therefore the nature of the disposition of the oscillation ripples assumes a great importance. Analysis of the disposition of the oscillation ripples shows that these are concave towards the right hand side of the limb, and hence it marks the top of the limb (bed), (Fig. 195 B). Therefore the axis of the anticline is to be expected on the left hand side of the limb. The limb is therefore rotated to a vertical position but it is not overturned. The reconstruction of the form of the fold is shown in Fig. 195 C.

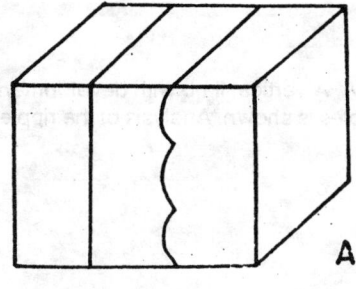

Fig. 195 A. A vertical limb with the development of oscillation ripples is shown. Analysis of ripples is given in Fig. B.

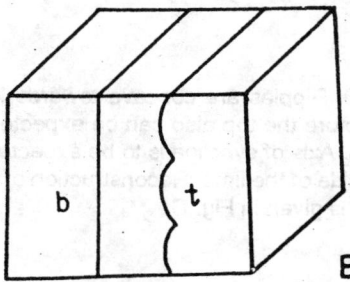

Fig. 195 B. Ripples are concave towards right hand side indicating top of the limb also to be in the same direction. Axis of syncline is to be expected on the right hand side. The limb is rotated but is not overturned. The reconstruction of the form of the fold is shown in Fig. C.

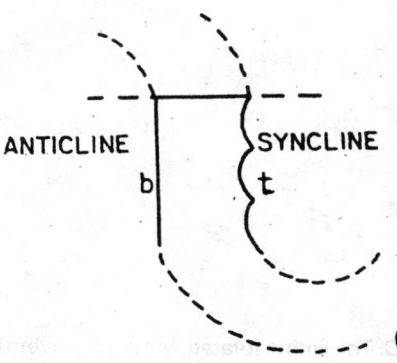

Fig. 195 C. It is a rotated fold, with one limb being vertical.

Figs. 195 A,B,C.

Yet another situation is described in Fig. 196 A, where the limb is again vertical, but the disposition of the oscillation ripples is different. Analysis of the disposition of the oscillation ripples shows that these are concave towards the left hand side of the limb (Fig. 196 B). Therefore the top of the limb (bed) also is to be expected on the same side. The limb is rotated to a vertical position, but it is not overturned. Therefore the axis of the anticline is to be expected on the right hand side of the limb (in the previous case, Fig. 195 C, it was on the left hand side of the limb). The reconstruction of the form of the fold is shown in Fig. 196 C.

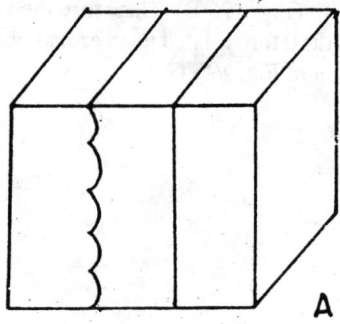

Fig. 196 A. A vertical limb with development of oscillation ripples is shown. Analysis of the ripples is given in Fig. B.

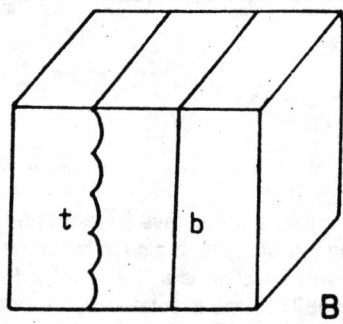

Fig. 196 B. Ripples are concave towards left hand side, therefore the top also can be expected on the same side. Axis of syncline is to be expected on the left hand side of the limb. Reconstruction of the form of the fold is given in Fig. C.

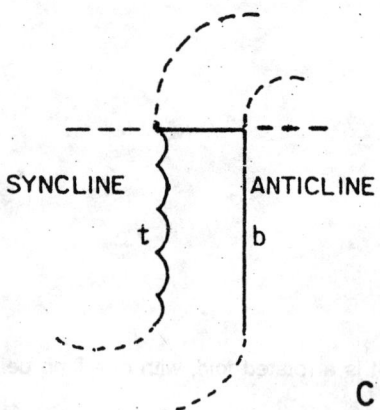

Fig. 196 C. The limb is rotated, but it is not overturned.

Figs. 196 A,B,C.

CASE II ONE LIMB OVERTURNED

In Fig. 197 A, an overturned limb of a fold alongside with the development of the oscillation ripples, is shown. Analysis of the oscillation ripples shows that these are concave towards the left hand side, that is against the direction of the dip of the limb (Fig. 197 B). The limb therefore is rotated and is further overturned. Axis of the anticline is to be expected in the direction of the dip of the limb i.e., to the right hand side of the limb. The reconstruction of the form of the fold is shown in Fig. 197 C.

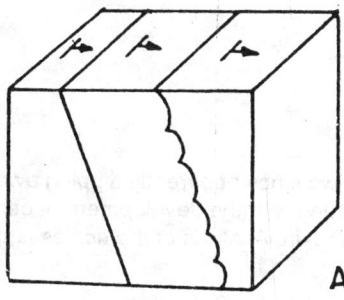

Fig. 197 A. A limb dipping together with the development of oscillation ripples is shown. Analysis of the ripples is given in Fig. B.

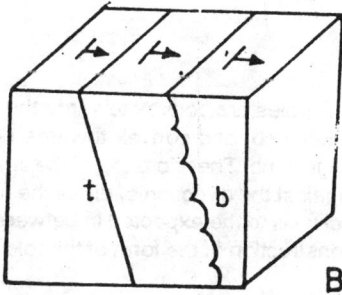

Fig. 197 B. Ripples are convex on the right hand side of the limb. Therefore top of the limb is located against the direction of dip of the limb. The limb is therefore overturned and axis of the syncline is to be expected on the left hand side of the limb. Reconstruction of the form of the fold is shown in Fig. C.

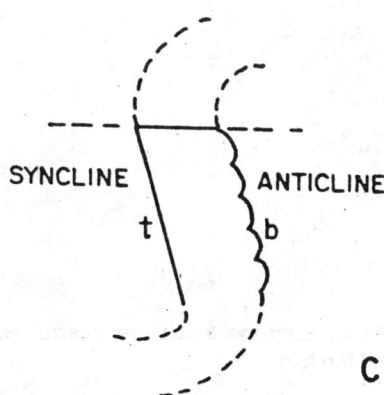

SYNCLINE   ANTICLINE

Fig. 197 C. The limb is overturned.

Figs. 197 A,B,C.

CASE III BOTH LIMBS OVERTURNED

In such cases "fan folds" are produced. In Fig. 198 A, two limbs alongside with the development of the oscillation ripples are shown. Analysis of the disposition of the oscillation ripples of the two limbs of the fold, indicates that the top of the limb (bed) is in the opposite direction of the dip for both the limbs (Fig. 198 B). Therefore both the limbs are rotated and are further overturned. The fold therefore belongs to the category of a fan fold. Further because the limbs apparently dip away from each other, yet it turns out to be a syncline. Therefore it is designated as a "synclinal fan fold". The reconstruction of the form of the fold is shown in Fig. 198 C.

In Fig. 199 A, yet another case of an overturned fold, alongside with the development of the oscillation ripples, is shown. The two limbs apparently dip towards each other. Analysis of the disposition of the oscillation ripples shows that both the limbs are overturned, because the concavity of the ripples is on the left hand side,

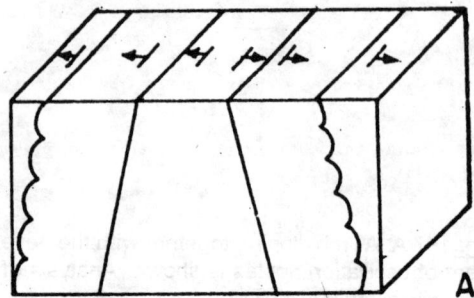

Fig. 198 A. Two limbs apparently dipping away from each other along with the development of oscillation ripples, are shown. Analysis of the ripples is given in Fig. B

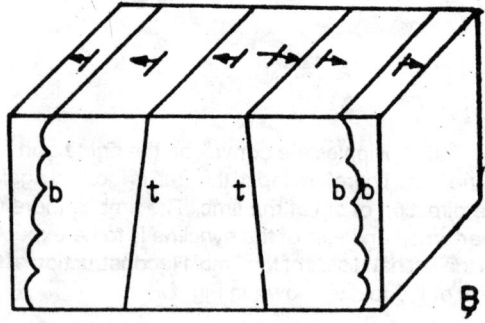

Fig. 198 B. Ripples are convex towards the left hand side of the left limb, and convex towards right hand side of the right limb. Therefore top of the limb is to be expected against the direction of dip of the two limbs. Axis of syncline is to the expected in between the two limbs. Reconstruction of the form of the fold is shown in Fig. C.

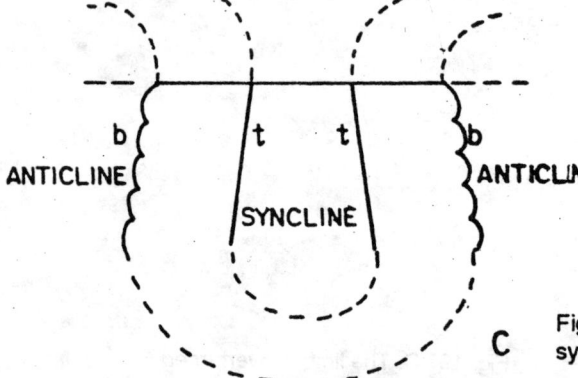

Fig. 198 C. Both the limbs are overturned. It is a synclinal fan fold.

Figs 198 A,B,C.

though the left limb dips to the right hand side. Likewise the right hand side limb dips towards left, but the concavity of the ripples is towards the right hand side (Fig. 199 B). Since both the limbs are overturned, the fold belongs to the category of a "fan fold". From the dip directions, the fold apparently looks like a syncline. It is however inferred that the fold is an "anticlinal fan fold". The reconstruction of the form of the fold is shown in Fig. 199 C.

CURRENT BEDDING

This structure also is encountered in the sedimentary rocks, and it is quite useful in ascertaining the "right side up" position of the beds. The normal disposition of the current bedding is shown in Fig. 200 A. Study of this figure

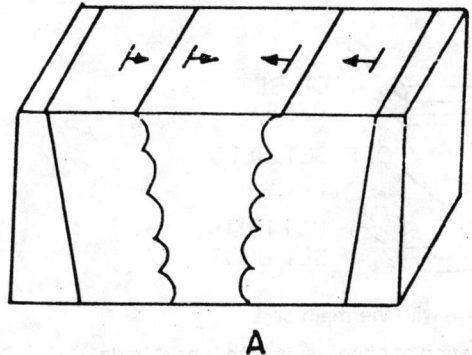

A

Fig. 199 A. Two limbs apparently dipping towards each other along with the development of oscillation ripples are shown. Analysis of the ripples is given in Fig. B.

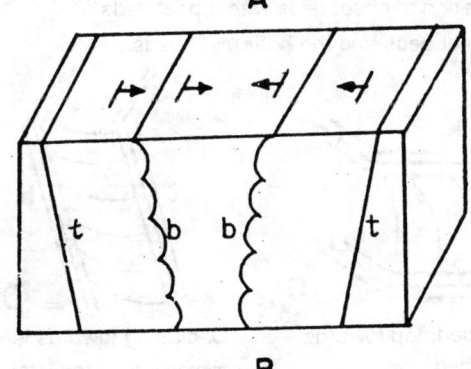

B

Fig. 199 B. Ripples are convex on the right hand side for the left limb, and left hand side for the right limb. Therefore top of the limb is to be expected against the direction of dip of the two limbs. The two limbs are therefore overturned, and the axis of the anticline is to be expected in between the two limbs. Reconstruction of the form of the fold is shown in Fig. C.

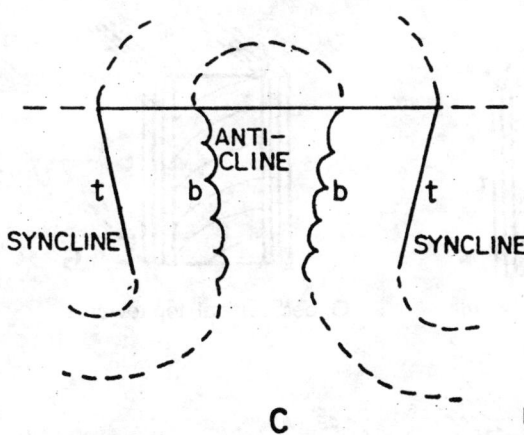

C

Fig. 199 C. It is an anticlinal fan fold.

Figs. 199 A,B,C.

reveals that the disposition of the offset beds with respect to the top set ones is quite different, and thus this aspect can be conveniently and effectively used to fix up the top and bottom of the bed. No ambiguity like the "original and the cast" of the current bedding exists. Therefore this structure can be applied with more confidence. The salient featues of the current bedding are as under;

   (i) top of the bed, or the direction of the younging of the beds is towards the acute truncation of the offset beds;

  (ii) bottom of the bed, or the direction of the beds becoming old, is towards the gentler disposition of the offset beds.

    The dispositions of the current bedding in respect of the different attitudes of the beds, are shown in Figs. 200 B to I.

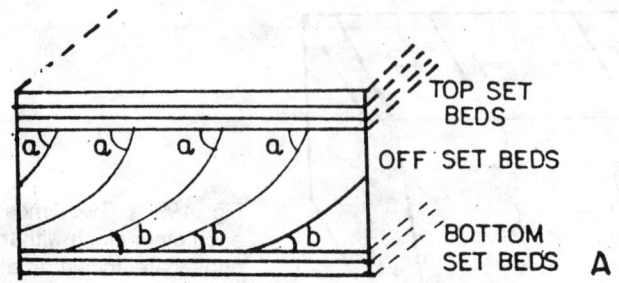

A. current bedding and its relation with the main bed.

a = high angle truncation/abrupt termination of offset beds with topset beds.

b = gentle angle of contact between offset beds and the bottomset beds.

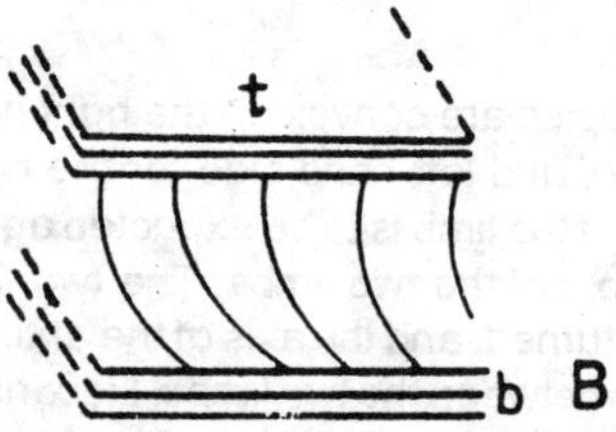

B. horizontal beds, top towards top.

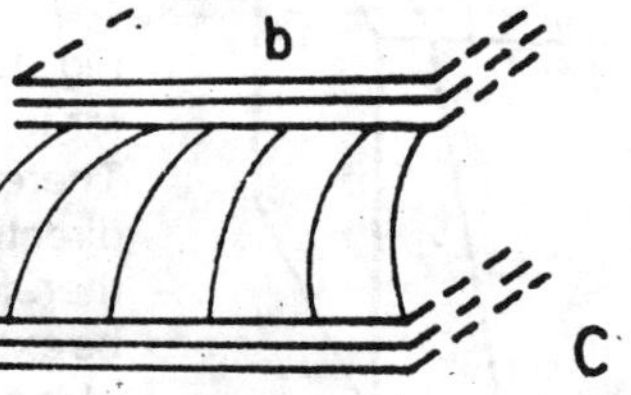

C. horizontal bed, top towards bottom of the bed.

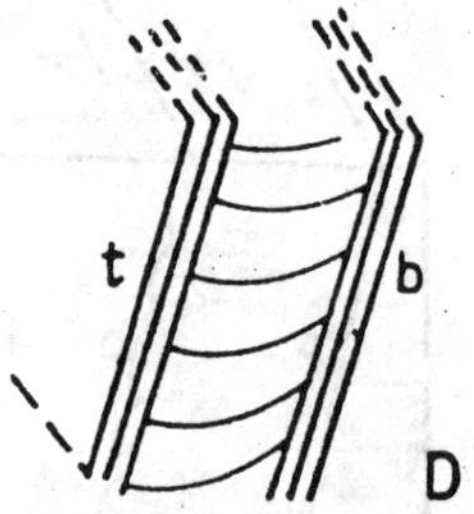

D. dipping towards left, top also towards left.

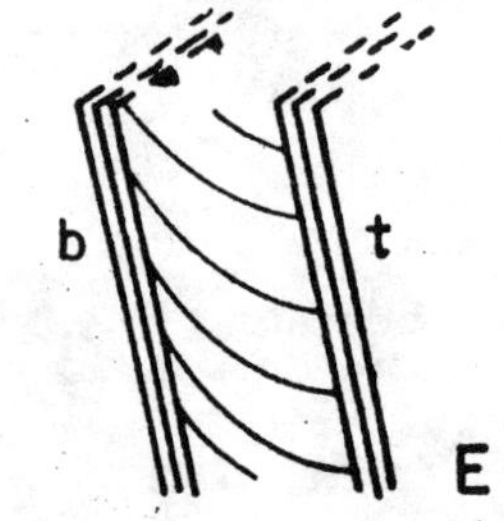

E. dipping towards right, top also towards right.

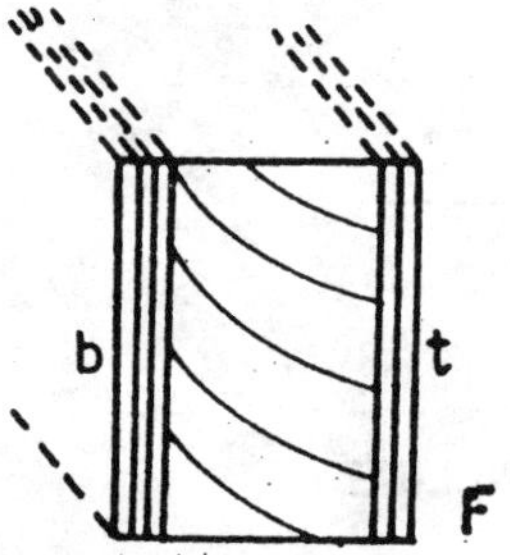

F. bed vertical, top towards right.

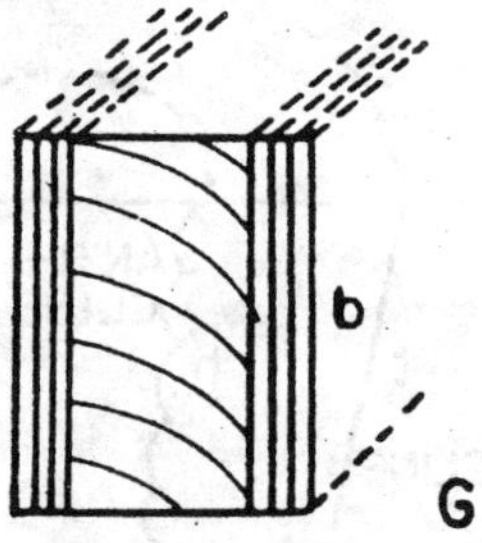

G. bed vertical, top towards

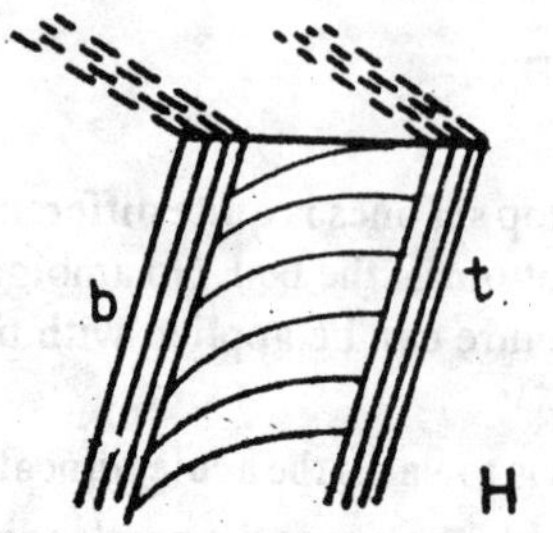

H. dipping towards left, but

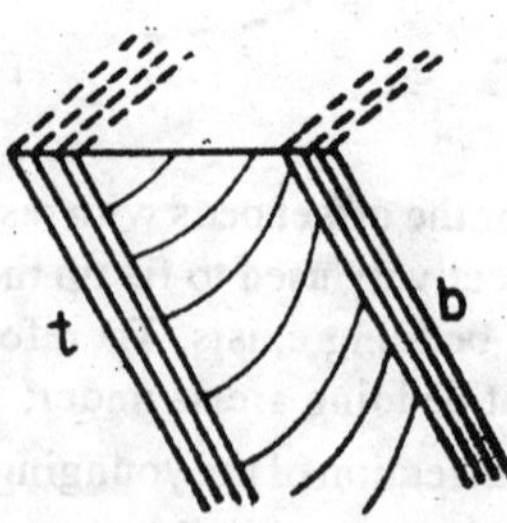

I. dipping towards right, but the top is towards left.

Figs. 200 A to I. Attitude of current bedding.

CASE I LIMBS ROTATED BUT NOT OVERTURNED

In Fig. 201 A, is shown a limb of a fold alongside with the development of the current bedding. The limb is dipping but is not vertical. Analysis of the disposition of the current bedding shows that the direction of acute truncation is on the left hand side of the limb (bed), (Fig. 201 B). Therefore the top of the limb also is to be expected in the same direction. The dip of the limb is also to the left hand side. Hence the limb right side up, and it is not overturned. Axis of the anticline is to be expected on the right hand side of the limb. The reconstruction of the form of the fold is shown in Fig. 201 C.

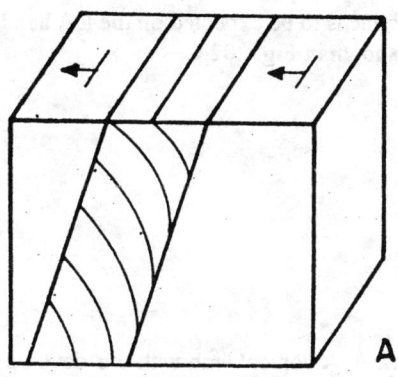

Fig. 201 A. A dipping limb with development of current bedding is shown. Analysis of current bedding is given in Fig. B.

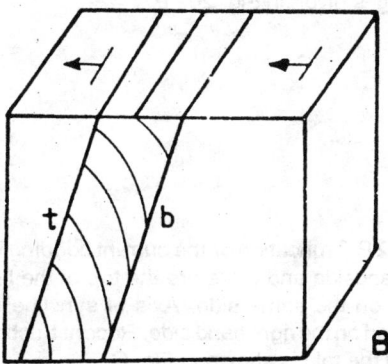

Fig. 201 B. Truncation of current bedding is on the left hand side. The top of the limb is therefore to be expected on the same side. Axis of syncline also is to be expected in the same direction. Reconstruction of the form of the fold is shown in Fig. C.

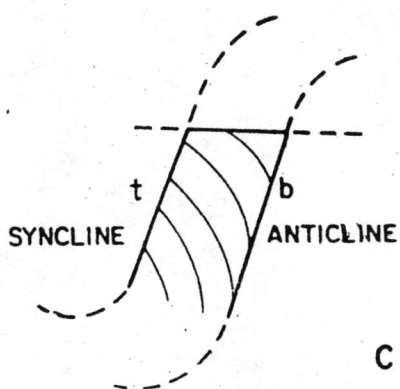

Fig. 201 C. The limb is right side up. The fold is normal one.

Figs. 201 A,B,C.

CASE II LIMB ROTATED TO VERTICALITY

In Figs. 202 A and 203 A are shown limbs of a fold alongside with the development of the current bedding. The limb is vertical and therefore the disposition of the current bedding will help to determine the top of the bed (limb). Two possibilities arise; the top is on the righthand side (Fig. 202 A) or it is on the left hand side (Fig. 203 A).

TOP ON THE RIGHT HAND SIDE (Figs. 202 A,B,C)

The disposition of the current bedding shows that the offset beds are acutely truncated on the right hand side of the limb, indicating thereby that the top is in the same direction (Fig. 202 B). Now because the limb is vertical, it is therefore inferred that it is rotated, but is not overturned. Anticline is to be expected on the left hand side of the limb of the fold. The reconstruction of the form of the fold is shown in Fig. 202 C.

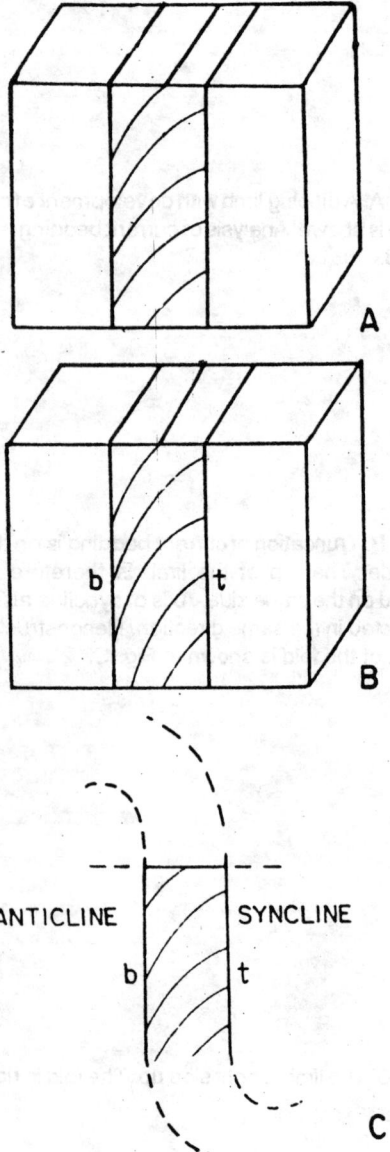

Fig. 202 A. A vertical limb with the development of current bedding is shown. Analysis of the current bedding is given in Fig. 202 B.

Fig. 202 B. Truncation of the current bedding is on the right handside and therefore the top of the limb also will be on the same side. Axis of syncline is to be expected on the right hand side. Reconstruction of the form of the fold is shown in Fig. C.

ANTICLINE   SYNCLINE

Fig. 200 C. The limb is rotated but is not overturned.

Figs. 202 A,B,C.

TOP ON THE LEFT HAND SIDE (Figs. 203 A,B,C)

Analysis of the disposition of the offset beds shows that the acute truncation is on the left hand side of the limb (Fig. 203 B). The top is therefore on the same side. As the limb is vertical, it has therefore undergone rotation, but it is not overturned. Anticline is to be expected on the right hand side of the limb. The reconstruction of the form of the fold is shown in Fig. 203 C.

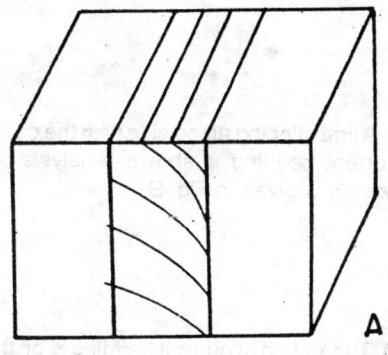

Fig. 203 A. A vertical limb with the development of current bedding is shown. Analysis of the current bedding is shown in Fig. B.

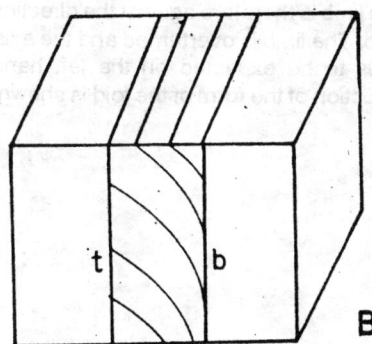

Fig. 203 B. Truncation of the current bedding is on the left hand side, and therefore the top of the limb will be on the left hand side. Axis of syncline is to be expected on left hand side of the limb. Reconstruction of the form fold is shown in Fig. C.

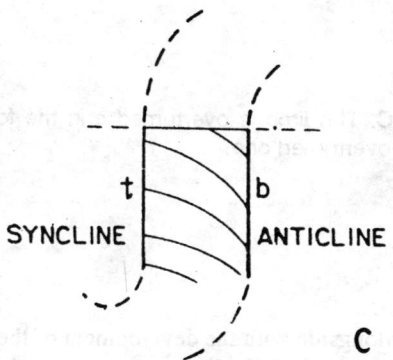

Fig. 203 C. The limb is rotated but is not overturned.

Figs. 203 A,B,C.

CASE III LIMB OR LIMBS OVERTURNED

Folds may develop with one or both the limbs rotated and overturned. In such cases the disposition of the current bedding gains importance to the maxium. In Fig. 204 A, only one limb is shown alongside with the development of the current bedding. Analysis of the disposition of the current bedding indicates that the top of the limb is in the opposite direction of the dip of the limb (bed), (Fig. 204 B). Therefore the bed (limb) is rotated and is further overturned. Hence the axis of the anticline is to be expected in the direction of the dip of the limb i.e., on the right hand side of the limb. The reconstruction of the form of the fold is shown in Fig. 204 C.

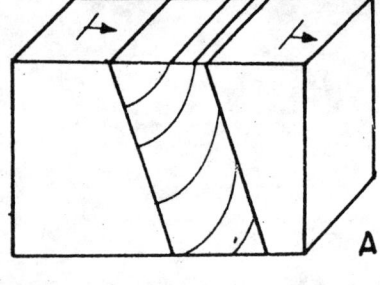

Fig. 204 A. A limb dipping along side with the development of current bedding is shown. Analysis of the current bedding is given in Fig. B.

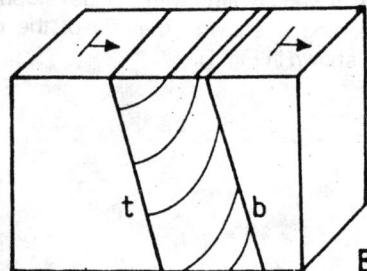

Fig. 204 B. Truncation of current bedding is on the left hand side while the limb dips to the right hand side. Top of the limb is therefore against the direction of dip of the limb. The limb is overturned and the axis of the syncline is to be expected on the left hand side. Reconstruction of the form of the fold is shown in Fig. C.

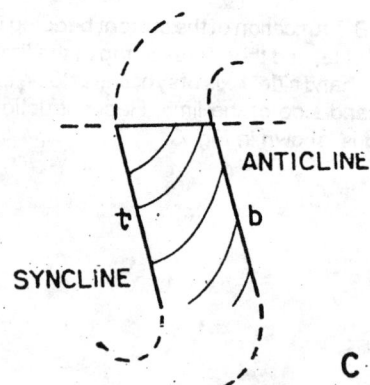

Fig. 204 C. The limb is overturned and the fold is therefore overturned one.

Figs. 204 A,B,C.

BOTH LIMBS OVERTURNED

In Fig. 205 A, two limbs apparently dipping towards each other, alongside with the development of the current bedding, is shown. Analysis of the disposition of the offset beds shows that both the limbs are overturned, because the direction of the dip of the limbs points to the bottom of the limb (bed), and not the top (Fig. 205 B). Since both the limbs are overturned, the fold therefore turns out to be a fan fold. Further it belongs to an "anticlinal fan fold". The reconstruction of the form the fold is shown in Fig. 205 C.

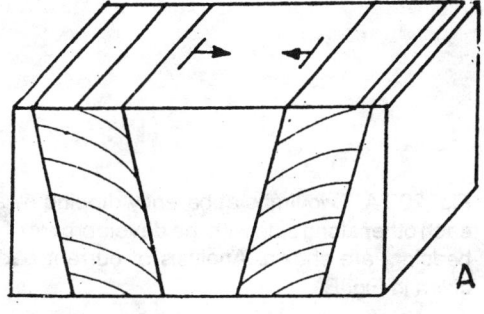

Fig. 205 A. Two limbs apparently dipping towards each other along side with the development of current bedding, are shown. Analysis of current bedding is given in Fig. B.

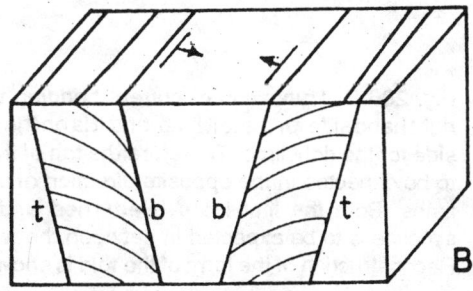

Fig. 205 B. Truncation of current bedding is on the left hand side for the left limb, and it is on the right hand side for the right limb. Therefore top of the limb is to be expected in the opposite direction of dip of the limbs. The two limbs are hence overturned. Axis of anticline is to be expected in between the two limbs. Reconstruction of the form of the fold is shown in Fig. C.

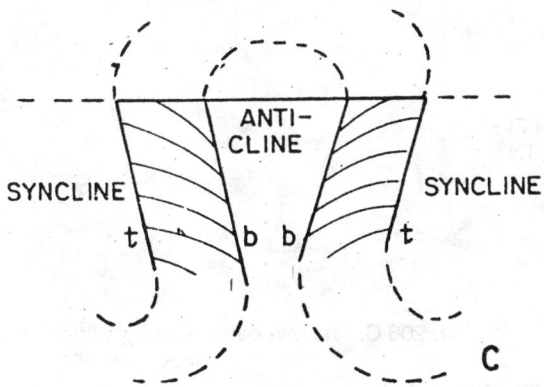

Fig. 205 C. The fold belongs to anticlinal fan fold.

Figs. 205. A,B,C.

In Fig. 206 A, two limbs apparently dipping away from each other alongside with the development of the current bedding, is shown. Analysis of the disposition of the current bedding shows that both the limbs are overturned, because the truncation of the offset beds is in the opposite direction of the dip of the limbs (Fig. 206 B). Therefore the fold belongs to the category of a fan fold. Further, the limbs apparently dip away from each other, but the top of the limb (bed) is in the opposite direction i.e., the limbs were originally dipping towards each other. Hence the fold is a "synclinal fan fold". The reconstruction of the form of the fold is shown in Fig. 206 C.

GRADED BEDDING

This structure also helps to ascertain the "right side up" position of the beds. Akin to the current bedding, graded bedding also does not have any ambiguity of application, because there is no consideration like the original and

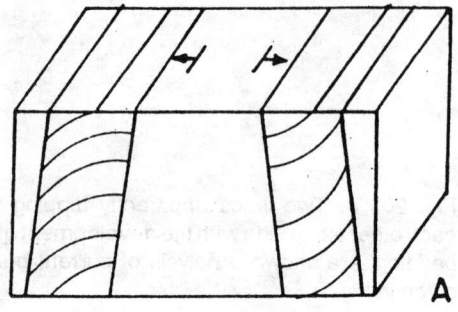

Fig. 206 A. Two limbs apparently dipping away from each other along side with the development of current bedding, are shown. Analysis of current bedding is given in Fig. B.

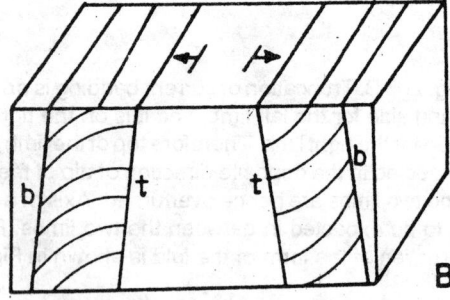

Fig. 206 B. Truncation of current bedding is on the right hand side for the left limb, and it is on the left hand side for the right limb. Therefore the top of the limb is to be expected in the opposite direction of dip of the limbs. Both the limbs are overturned and axis of syncline is to be expected in between the two limbs. Reconstruction of the form of the fold is shown in Fig. C.

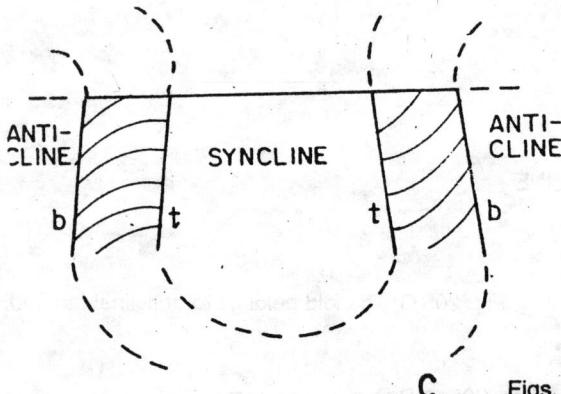

Fig. 206 C. The fold belongs to synclinal fan fold.

Figs. 206 A,B,C.

the cast or the print of the graded bedding. The disposition of the graded bedding is such that the grain size of the sediments goes on becoming fine towards the top of the succession, or in the direction of the "younging" of the beds. A typical graded bedding is shown in Fig. 207 A. In a sedimentary succession, there may be several horizons that have developed the graded bedding. All of them will indicate only one direction of "younging" of the beds. The disposition of the graded bedding in respect of the different attitudes of the beds, is shown in Figs. 207 B to I.

The application of the graded bedding in establishing the form of the fold will be described in the following paragraphs.

CASE I LIMB ROTATED, BUT IS NOT VERTICAL OR OVERTURNED

In Fig. 208 A, a limb of a fold which is dipping and has developed graded bedding, is shown. Analysis of the

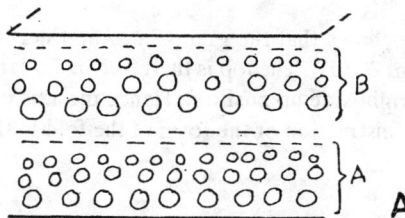

A. Grains become fine towards top of the beds A and B. Therefore in graded bedding, the grain size becomes finer towards top, meaning direction of younger beds.

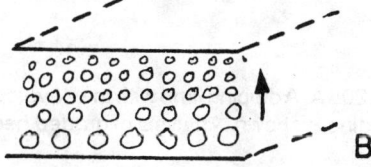

B. bed horizontal, top towards top.

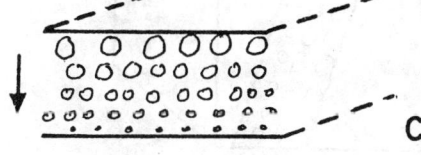

G. bed horizontal, top towards the bottom of the bed.

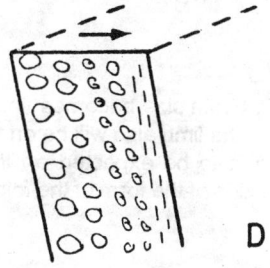

D. dipping towards right, top also to the right side.

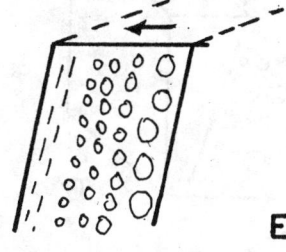

E. dipping towards left, top also towards left side.

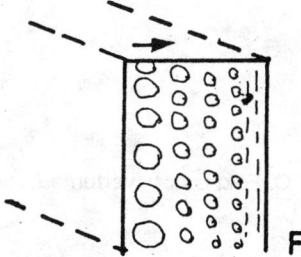

F. bed vertical, top to the right side.

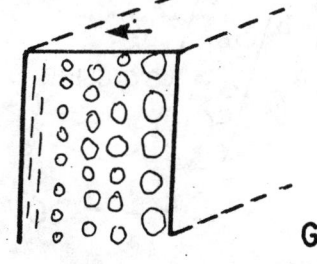

G. bed vertical, top to the left side.

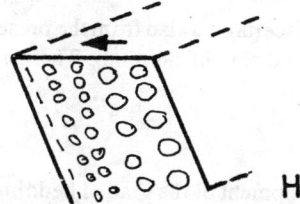

H. dipping towards, right side, top

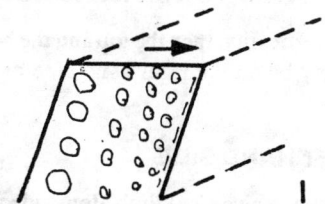

I. dipping towards left side, top towards

Fig. 207. A-I

disposition of the graded bedding shows that the grains become finer towards the right hand side i.e., in the direction of the dip of the limb (Fig. 208 B). The top is therefore in the same direction as that of the dip direction of the limb. The limb is thus in a right side up position. Hence the axis of the anticline is to be expected on the left hand side of the limb. The reconstruction of the form of the fold is shown in Fig. 208 C.

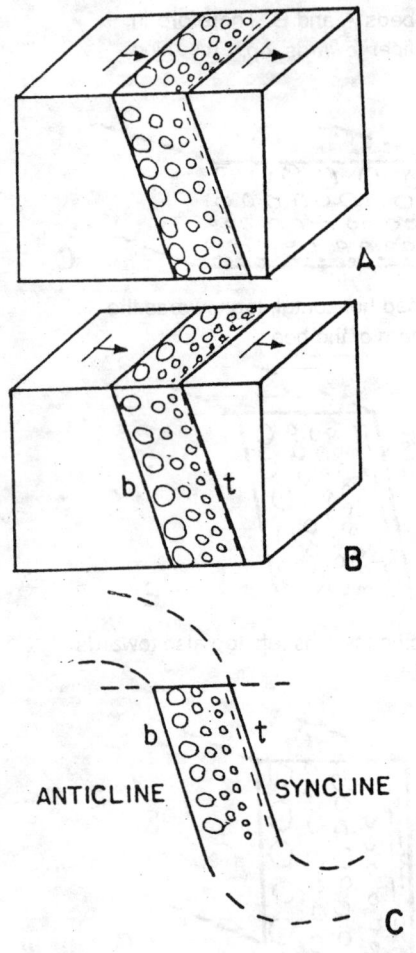

Fig. 208 A. A dipping limb with the development of graded bedding is shown. Analysis of graded bedding is given in Fig. B.

Fig. 208 B. Grain size becomes finer on the right hand side. Top of the limb also will be on the same side. Axis of syncline is to be expected on the right hand side. Reconstruction of the form of the fold is shown in Fig. C.

Fig. 208 C. Fold is not overturned one.

Figs. 208 A,B,C.

CASE II LIMB ROTATED TO VERTICAL POSITION

When a limb attains verticality, then the top and the bottom can be ascertained also from the presence of the graded bedding. The top may be on the right hand side or on the left hand side of the limb. These two situations are described below.

TOP ON THE LEFT HAND SIDE

In Fig. 209 A, is shown in a vertical limb alongside with the development of the graded bedding. Analysis of the graded bedding shows that the grain size becomes finer on the left hand side of the limb (Fig. 209 B). Therefore the top of the limb is to be expected on the left hand side of the limb. Axis of the anticline is thus to be expected on the right hand side of the limb. The reconstruction of the form of the fold is shown in Fig. 209 C.

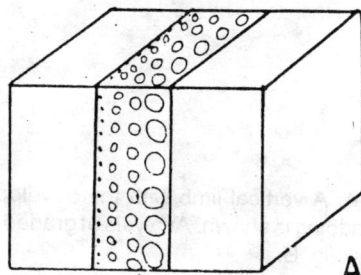

Fig. 209 A. A vertical limb with the development of graded bedding is shown. Analysis of the graded bedding is given in Fig. B.

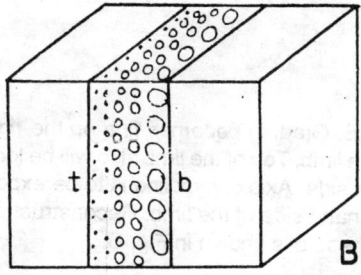

Fig. 209 B. Grain size becomes finer on the left hand side. The top of the limb also will be on the same side. Axis of syncline is to be expected on the left hand side of the limb. Reconstruction of the form of the fold is shown in Fig. C.

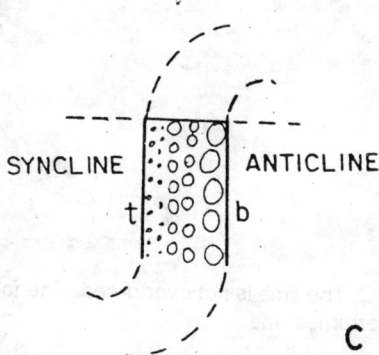

Fig. 209 C. The limb is not overturned. The fold likewise is not overturned one.

Figs. 209 A,B,C.

TOP ON THE RIGHT HAND SIDE

In Fig. 210 A, is shown a vertical limb which has developed graded bedding. Analysis of the disposition of the graded bedding reveals that the grain size becomes finer on the right hand side of the limb (Fig. 210 B). Therefore the right hand side marks the top of the limb (bed). Anticline is therefore to be expected on the left hand side of the limb. The reconstruction of the form of the fold is shown in Fig. 210 C.

CASE III LIMB OR LIMBS OVERTURNED

Rotation beyond verticality gives rise overturning. One or both the limbs might get overturned. These situations are described below.

ONE LIMB OVERTURNED

In Fig. 211 A, a limb dipping to the left hand side together with the development of the graded bedding is shown. Analysis of the disposition of the graded bedding shows that the grain size becomes finer to the right hand side,

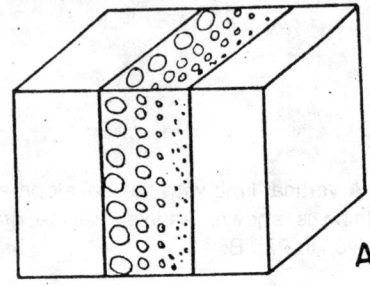

Fig. 210 A. A vertical limb with the development of graded bedding is shown. Analysis of graded bedding is given in Fig. B.

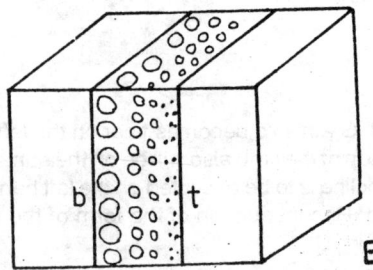

Fig. 210 B. Grading becomes fine on the right hand side of the limb. Top of the limb also will be located on the same side. Axis of syncline is to be expected on the right hand side of the limb. Reconstruction of the form of the fold is shown in Fig. C.

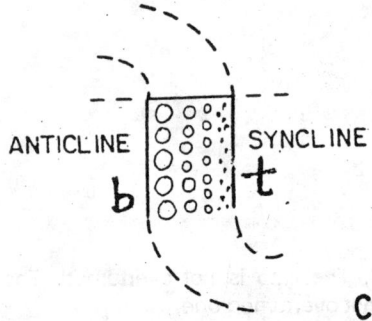

Fig. 210 C. The limb is not overturned. The fold also is not overturned one.

Figs. 210 A,B,C.

while the limb is dipping to the left hand side (Fig. 211 B). The limb is therefore overturned and hence in the direction of the dip of the limb, an anticline is to be expected. The reconstruction of the form of the fold is shown in Fig. 211 C.

BOTH THE LIMBS OVERTURNED

In Fig. 212 A, two limbs which are apparently dipping towards each other, alongside with the development of the graded bedding, are shown. Analysis of the disposition of the graded bedding shows that both the limbs are overturned, because in the direction of the dip of the limbs, coarser grain size is encountered (Fig. 212 B). The fold therefore belongs to the category of a fan fold. Further, the limbs dip apparently towards each other, but the limbs are overturned. Therefore the fold is further classifiable as an "antialine fan fold". The reconstruction of the form of the fold is shown is Fig. 212 C.

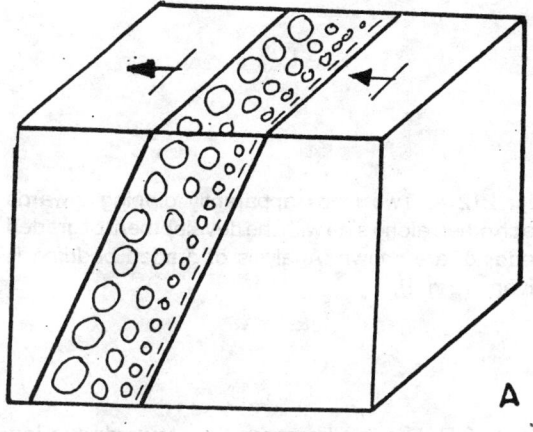

Fig. 211 A. A limb dipping towards left hand side, along side with the development of graded bedding, is shown. Analysis of graded bedding is given in Fig. B.

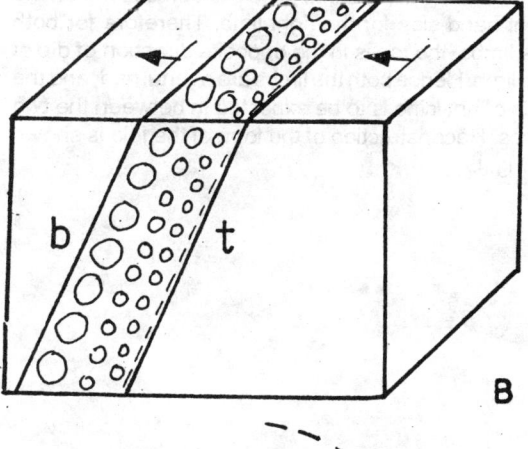

Fig. 211 B. Grading becomes finer on the right hand side while the limb dips towards left hand side. Therefore the limb is overturned, and the axis of the syncline is to be expected on the right hand side of the limb. Reconstruction of the form of the fold is shown in Fig. C.

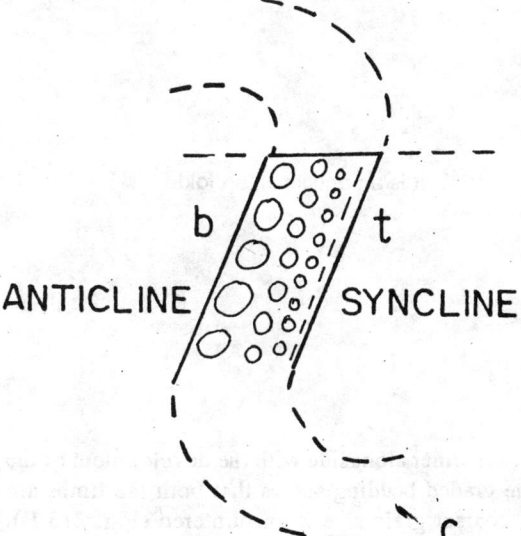

ANTICLINE   SYNCLINE

Fig. 211 C. Fold is overturned.

Figs. 211 A,B,C.

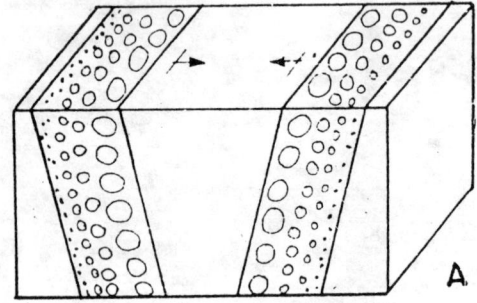

Fig. 212 A. Two limbs apparently dipping towards each other, along side with the development of graded bedding, are shown. Analysis of graded bedding is given in Fig. B.

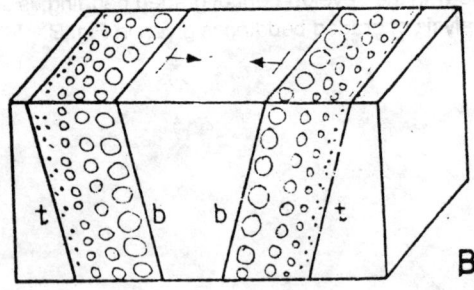

Fig. 212 B. Grading becomes finer towards the left hand side for the left limb, and it becomes finer on the right hand side for the right limb. Therefore for both the limbs, the top is in the opposite direction of dip of the limb. Hence both the limbs are overturned, and the axis of antilcine is to be expected in between the two limbs. Reconstruction of the form of the fold is shown in Fig. C.

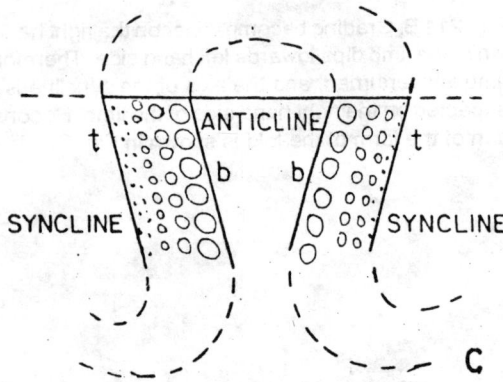

Fig. 212 C. It is an anticlinal fan fold

Figs. 212 A,B,C.

In Fig. 213 A, two limbs apparently dipping away from each other alongside with the development of the graded bedding, are shown. Analysis of the disposition of the graded bedding shows that both the limbs are overturned, because in the direction of the dip of the limbs, coarser grain size is encountered (Fig. 213 B). Therefore the fold belongs to the category of a fan fold. Further because the limbs apparently dip away from each other, the fold gets classified as a "synclinal fan fold". The reconstruction of the form of the fold is shown in Fig. 213 C.

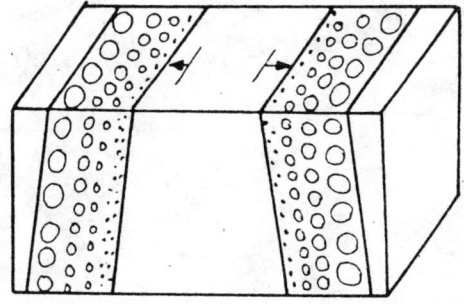

Fig. 213 A. Two limbs apparently dipping away from each other along side with the development of graded bedding are shown. Analysis of the graded bedding is given in Fig. B.

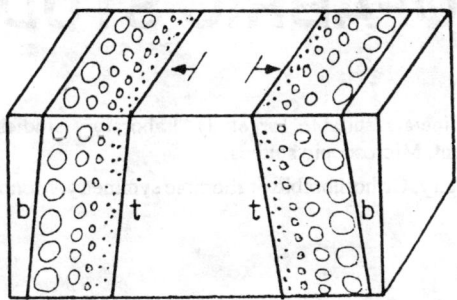

Fig. 213. B. Grains become finer on the right hand side for the left limb, and on the left hand side for the right limb. Therefore top of the limb is expected in the opposite direction of dip of the two limbs. As both the limbs are overturned, axis of syncline is to be expected in between the two limbs. Reconstruction of the form of the fold is shown in Fig. C.

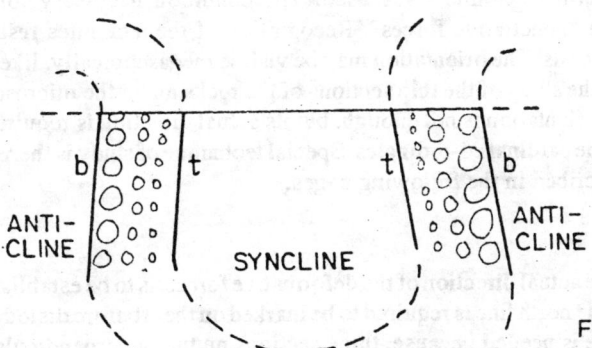

Fig. 213 C. The fold is a synclinal fan fold.

Figs. 213 A,B,C.

There are other structures by which the top and the bottom of the beds can be established, like the local unconformity, the rain prints, the contemporaneous deformation, the columnar joints and so on. But these are not very frequently developed and hence are not described in this book.

8

PETROFABRIC STUDIES

Field technique, Selection of rock samples and minerals suitable for study, Laboratory studies, Conventions regarding labelling direction of movement, Microscopic studies.

Classification of tectonites, Axial or spheroidal symmetry, Orthorhombic or rhombic symmetry, Monoclinic symmetry, Triclinic symmetry.

Figures 214 to 221

INTRODUCTION

While classifying the different structures, the term "tectonite" was used. The condition necessary for the development of the tectonites is the action of the "penetrative forces". Recognition of the tectonites rests on ascertaining the existence of orientation of the minerals. The orientation may be visible megascopically, like that seen in the schists, the foliated rocks and so on, or the study of the thin sections of the rocks under the microscope may be necessary. The mere establishment of the orientation is not enough, but its actual direction is required to be determined with respect to the north and south, the cardinal co-ordinates. Special technique of study is therefore required to be followed, and the same will be described in the following pages.

FIELD TECHNIQUE

Oriented samples of rocks are necessary because the actual direction of the deformative forces, is to be established. Oriented samples only means that the direction of the north line is required to be marked on them before dislodging them from their outcrop. A large and fresh sample is needed because, three sections mutually perpendicular to each other are required to be prepared from it. Very coarse grained rocks have two drawbacks, namely, grains of minerals included within the size of the thin section will be very few, and while preparing the thin sections, grains may come off from the slide. It is therefore necessary to break off a large chunk of rock from the outcrop, and then trim it, and put it back to its original position, as much as possible. The north line is then drawn on it with the help of a clinometer or a brunton compass.

SELECTION OF ROCK SAMPLES AND MINERALS SUITABLE FOR STUDY

Petrofabric studies are generally undertaken on the rocks which are suspected to have undergone some deformation. In the neighbourhood of faults, samples are therefore collected such that some are close to the fault plane, and others are away from it. Like wise, samples are collected close to and away from the axis of the folds. Metamorphic rocks like the schists, the gneisses, the granulites, the quartzites and so on, are normally subjected to the petrofabric analysis, and hence samples are usually collected from these rocks. But when no indications of deformation are apparently noticeable, even then samples may be collected. However in such cases, the selection of the samples becomes a random process. Petrofabric studies may be conducted on any rock, however some of them are more favourable like the sandstone, the quartzite, the schist, the acidic igneous rocks and so on. This is so because, though the penetrative forces affect any rock, the effect of it is recognisable on such favourable minerals like the quartz, the micas, minerals with accicular habit, and so on. Quartz is the most suitable mineral, because it is optically uniaxial, and therefore the orientation or otherwise of the optic axis, is determinable. Quartz also is present in many rocks - igneous, sedimentary or metamorphic rocks. Mica (biotite) is likewise very suitable, because it has only one set of cleavages. Thus in thin section, the direction of orientation of the cleavages, can be measured. Elongated minerals like the needles of hornblende, plates of staurolite, kyanite and so on, may be used. There is yet another consideration in the selection of the minerals. The mineral must be present in large quantity in the rock to be studied. Percentage of the orientated grains is important in recognising the existence and the intensity of the penetrative forces. Obviously values greater than 60 or 70 percent are acceptable to infer that the minerals are oriented. Therefore, quartz and biotite mica are invariably chosen because the former mineral is present in large quantity and in most of the rocks, and it is the only mineral occurring in the sandstones and the quartzites.

LABORATORY STUDIES

Samples collected in the field are subjected to the preparation of three thin sections from each sample. As already noted, the sections should be taken along directions that are mutually perpendicular to each other. There are definite guide lines for selecting the three directions in which the sections are to be taken. If the rock be showing foliation, lineation, schistocity, banding or any other linear or a planar structure, then one section is taken perpendicular to such a structure. The second one is taken perpendicular to the first direction, while the third is taken perpendicular to the plane containing the earlier two directions. This has been shown in Fig. 214. In the absence of any structure developed in the rock to be studied, then three directions (sections) are taken at random, but even these directions between themselves should be again mutually perpendicular to each other.

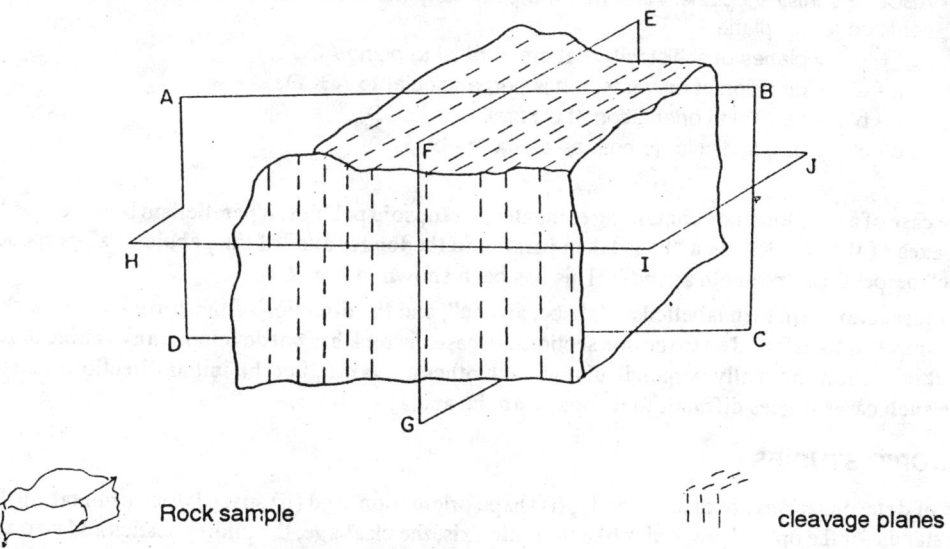

Rock sample

cleavage planes

ABCD = vertical section perpendicular to schistose planes

EFG = vertical section parallel to schistose planes.

HIJ = horizontal section perpendicular to schistose planes.

Note that ABCD, EFG and HIJ planes are mutually perpendicular to each other.

Fig. 214. Planes along which thin sections are to be prepared.

CONVENTIONS REGARDING LABELLING DIRECTIONS OF MOVEMENT

The penetrative forces rotate the minerals and arrange them into parallel positions. Therefore this direction of the forces is conveniently called as the "a" direction or axis. The direction of orientation (parallel dispositions) is called "b", this being perpendicular to "a". "c" is then taken perpendicular to the plane containing "a and b". Therefore in a schist, the strike of the schistocity becomes "b", "a" is taken perpendicular to the plane of schistocity, while "c" is perpendicular to "a and b". This has been shown in Fig. 215.

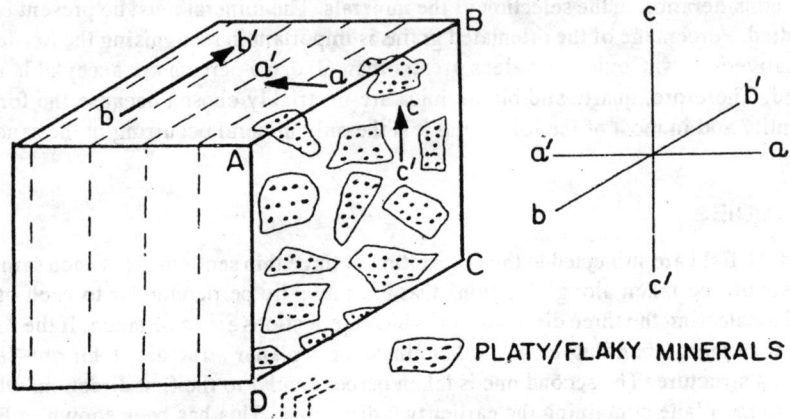

PLATY/FLAKY MINERALS

Fig. 215. Convention in labelling directions of movements.
ABCD = schistose plane. Note that the platy/flaky minerals with their flat surfaces are confined to this plane.

           = planes of schistosity that are parallel to plane ABCD.
a - a'   = direction of movement. It is perpendicular to ABCD.
b - b'   = parallel to orientation. It is parallel to ABCD.
c - c'   = perpendicular to both a - a' and b - b'.

      In the case of a conglomerate containing elongated or ellipsoid pebbles, if parallelism be developed amongst the longer axes of the pebbles, then "b" is taken parallel to the longer axes of the pebbles, "a" perpendicular to "b", and "c" perpendicular to both a and b. This has been shown in Fig. 216.

      The thin sections are then labelled as "ab, bc, and ac", and the direction of the north line as marked on the main rock sample, is transferred on to the thin sections. In case the rock has not developed any visible deformation, then three thin sections mutually perpendicular to each other are taken, but the initial directions chosen, are at random. In such cases it goes difficult to recognise ab, bc and ac sections.

MICROSCOPIC STUDIES

Two kinds of determinations are made namely, (i) shape orientation, and (ii) space lattice orientation. The latter parameter stands for the optical properties like the optic axis, the cleavage, the gliding planes and so on. Uniaxial minerals only are useful, because in all the grains studied, the same element (optic axis) gets considered, which is not possible if the mineral were to be biaxial one. Like wise, minerals with only one set of cleavages are useful. As a consequence of this, quartz and biotite mica are considered for the determinations of the optic axis, and the cleavages, respectively. The thin section is mounted on a 5 axes Universal Stage, and the readings for the directions of the optic axis, cleavages, longer axis of elongated or ellipsoidal minerals, pebbles etc., are taken, from the several grains available in the three sections, separately. From these readings, point diagrams are constructed, and these are further evaluated by applying the "Mellis or the Circle "method of evaluation (see Chapter 2 for further details, also refer Turner et. al. 1963, pp. 61-64).

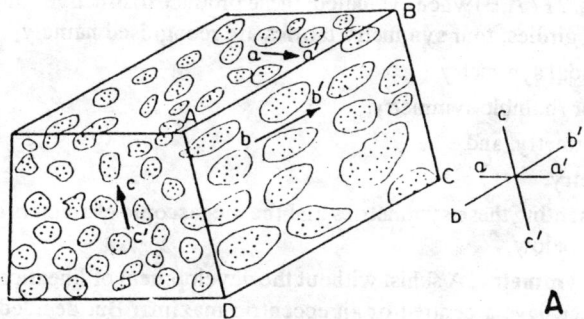

Fig. 216. A conglomerate bed containing elongated/ellipsoidal pebbles.

ABCD = A plane showing parallely oriented longer axes of ellipsoidal pebbles.
Note that there is no schistosity developed, because platy/flaky minerals are not present,
but orientation is developed.
a - a' = perpendicular to longer axes of elongated pebbles.
b - b' = parallel to the longer axes of elongated pebbles.
c - c' = perpendicular to both a - a' and b - b'.

CLASSIFICATION OF TECTONITES

As mentioned earlier, two varieties of tectonites are distinguished namely, S and B. In the former variety, planar structures produce the orientation, while in the latter variety, the linear structures produce the orientation. Shapes of the grains, if elongated or ellipsoidal in shape, may also cause the orientation. In such cases, planar structure may not be produced. Laths of felspars, elongated or ellipsoidal pebbles of the conglomerates belong to this category. The B - tectonites are recognised only on the optical properties (space lattice) of the grains. Such tectonites are classified on the basis of the fabric symmetry noticed in the stereograms. The symmetry is judged on the consideration of the "maxima" and the "girdles". A "maxima" is said to be developed when the plotted points cluster together into an isolated unit (Fig. 217 A). A girdle is said to be developed when the plotted points are concentrated into a linear fashion (Fig. 217 B).

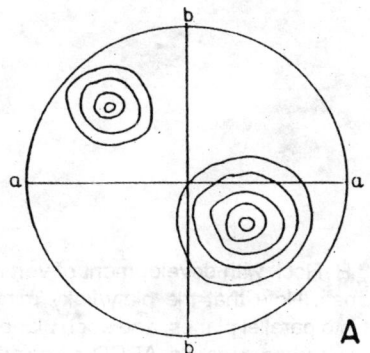

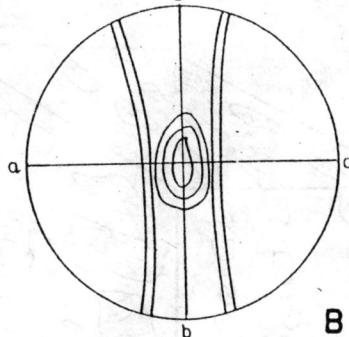

Fig. 217 A. Fabric symmetry with development of two isolated maxima.
Fig. 217 B. Fabric symmetry with development of a girdle and an elongated maxima.

Such stereograms (Figs. 217 A,B) when evaluated, these produce distinctive contour patterns. Based on the development of maxima and girdles, four symmetry fabrics are recognised namely,

    (i) axial or spheroidal symmetry,

    (ii) orthorhombic or rhombic symmetry,

    (iii) monoclinic symmetry, and

    (iv) triclinic symmetry.

The stereograms representing these symmetries and the megascopic characters of the rocks in which these are developed, are described below.

(a) **Axial or spheroidal symmetry:** A schist without the development of lineation, produces this symmetry. The stereograms either have a centred or an eccentric maxima. But decidedly a girdle is not formed. Depending upon the attitude of the plane of schisticity, the position of the maxima changes. If the schistose plane be vertical, then the maxima will be located at the centre; if it is having an intermediate dip amount, then it will lie inbetween the centre and the periphery of the circle; and if the dip be low, then the maxima will be located close to the periphery of the circle. A rock possessing lineation alone also can produce axial symmetry. The stereograms and the rocks corresponding to them are shown in Figs. 218 A to I.

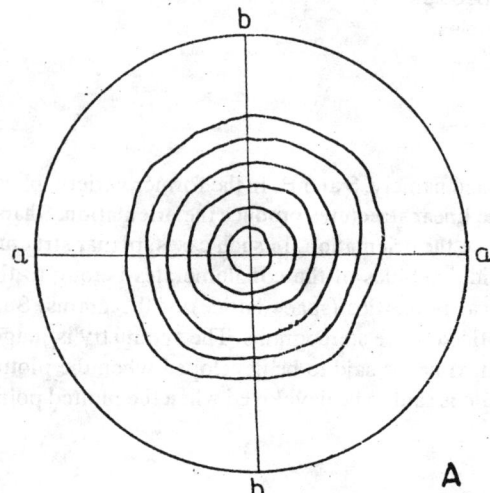

Fig. 218 A. Fabric symmetry with centrally located maxima. Note that girdle is not developed. Rock with vertical schistosity or lineation may produce such a symmetry - fabric. It is called as axial or spheroidal symmetry.

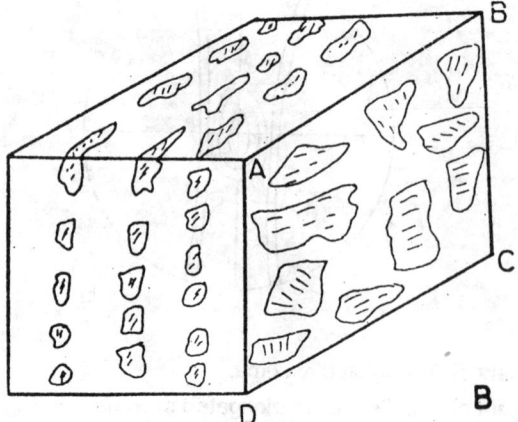

Fig. 218 B. Rock with development of vertical schistose planes. Note that the platy/flaky minerals are brought into parallel planes, and each plane contains flat but unoriented minerals. ABCD is one such schistose plane to which others are parallel. The disposition of minerals produces only one maxima, and that too located at the centre of the stereogram (Fig. 218A) because the schistosity is vertical.

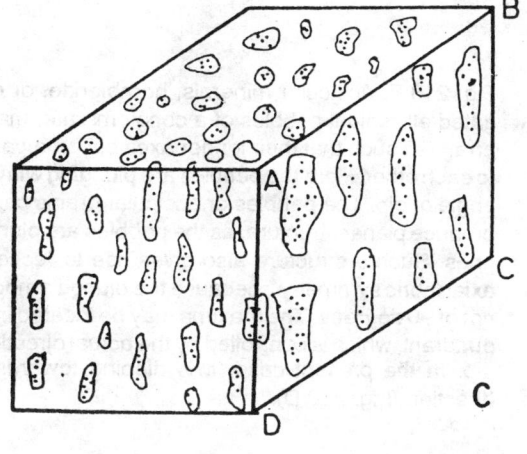

Fig. 218 C. A rock with the development of lineation alone due to accicular, elongated or ellipsoidal minerals, is shown. These minerals need not be confined to any plane, but if these do so, even then only one maxima will be produced. This is so because the plunge of lineation is 90 degrees. It will be at the centre of the stereogram. ABCD is one such plane in which a few minerals are present and whose longer axes are parallel to each other.

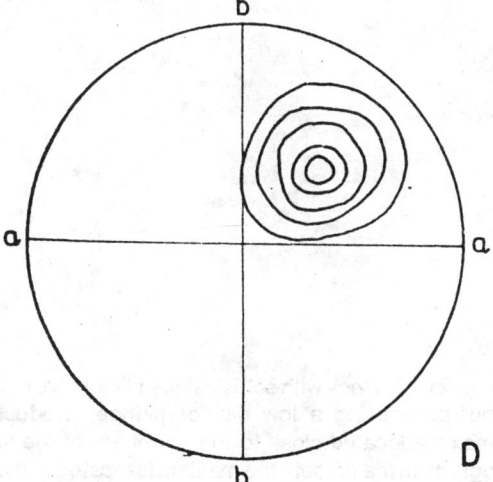

Fig. 218 D. A rock with the development of schistocity or lineation, or both, and dipping due northeast direction, produces an eccentric maxima. It may be due to SE, SW or due NW, as per the geographic location of the sample in the crust of the earth. Such eccentric maxima belongs to "axial or spheroidal" symmetry fabric.

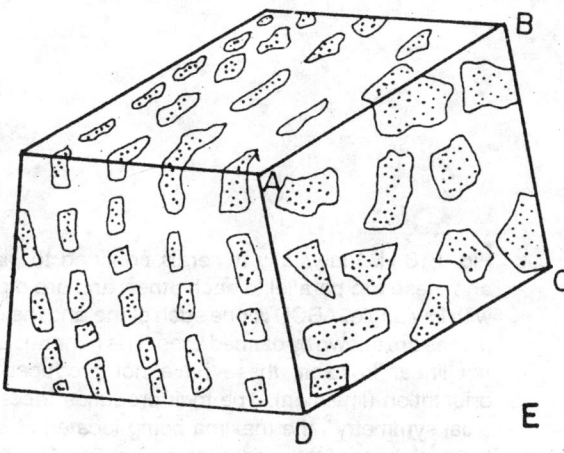

Fig. 218 E. Platy/flaky minerals like biotite, chlorite producing schistosity without lineation. ABCD is one such schistose plane which is dipping one. Other planes are parallel to it. Note that in the plane ABCD, the platy/flaky minerals do not have any common direction of orientation, but their flat surfaces are confined to several planes. As the schistose plane is dipping one, it produces "eccentric maxima", and it will be located in the NE quadrant, because the schistose plane also dips in that direction.

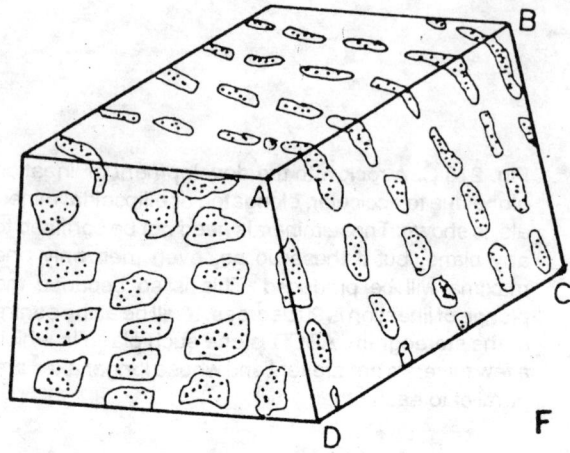

Fig. 218 F. Accicular minerals, hornbiende, or elongated/ellipsoidal pebbles of a conglomerate, may be arranged such that their longer axes become parallel to each other. Here the pebbles are plunging with high angle of dip. The pebbles or accicular grains can not produce planar structure, as the pebbles are plunging ones. Such a structure also gives rise to "eccentric axial fabric symmetry", because the plunge amount is not of 90 degrees. The maxima may be located in any quadrant, which is controlled by the actual direction of dip. In the present case, it is dipping towards NE direction (Fig. 218 D).

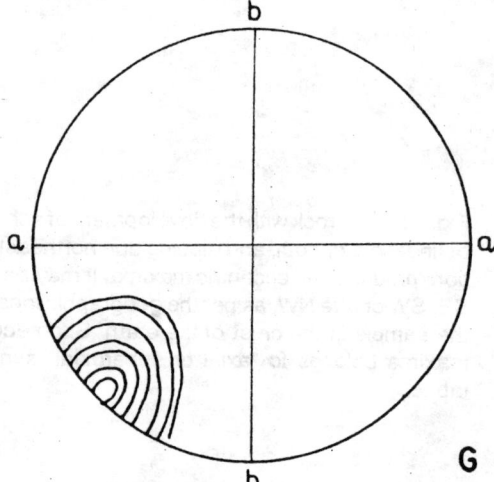

Fig. 218 G. Rock with schistosity, or lineation, or both, but possessing a low dip, or plunge, produces a maxima located close to the periphery of the stereogram. In the present the maxima is located in the SW quadrant, and it is also "eccentric axial fabric symmetry"

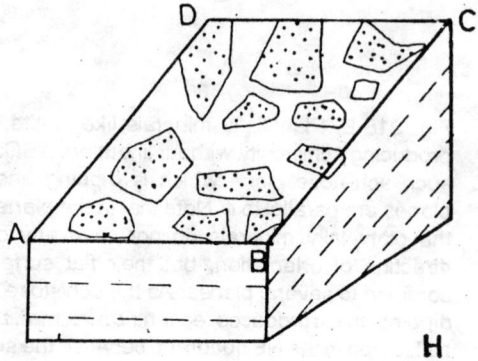

Fig. 218 H. Platy/flaky minerals confined to planes and these are parallel to each other, and are dipping with low angle. ABCD is one such plane and the other planes are shown by dashed lines. The minerals being not linear in habit, these have not produced any orientation (lineation). The rock produces "eccentric axial symmetry", the maxima being located close to the periphery of the stereogram (Fig. 218 G).

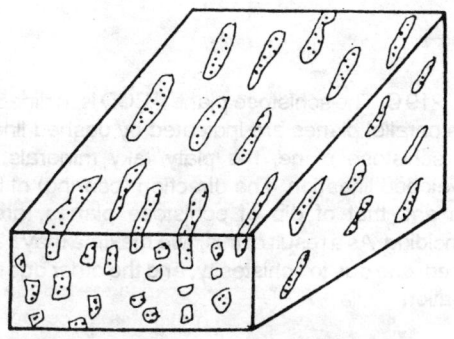

Fig. 218 I. Accicular minerals, elongated or ellipsoidal grains can produce lineation. The plunge of mineral grains is low. The grains being not flat, no schistosity can be produced. Such rocks develop "eccentric axial symmetry" the maxima being located very close to the periphery of the stereogram, because the dip of the plunge angle is low. In the present case, it is developed towards the SW quadrant (Fig. 218 G).

(b) Orthorhombic or rhombic symmetry: This is characterised by the development of a girdle and two maxima. In this case two planes of symmetry are present. A schist with a distinctly developed lineation, produces such a symmetry. Schistocity produces one maxima, while the lineation produces the other. These features are shown in Figs. 219 A to D.

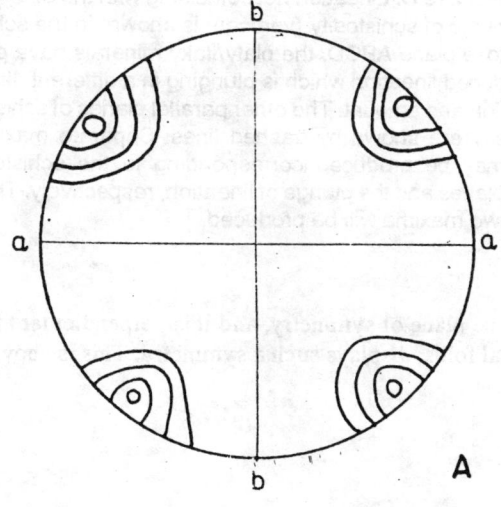

Fig. 219 A. 300 axes of quartz in mylonitised quartzite from Moine thrust. Here both a and b are planes of symmetry (after Turner and Weiss, 1963).

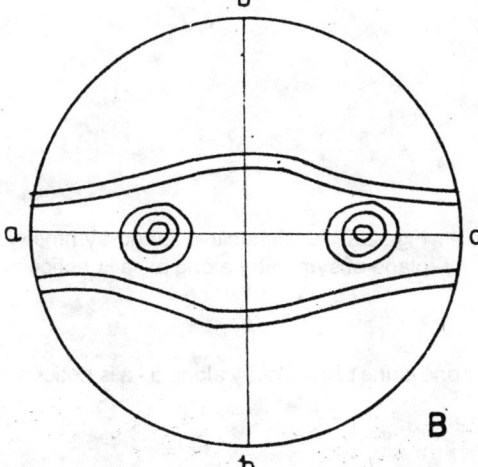

Fig. 219 B. Fabric symmetry with two maxima and a girdle (after Billings 1960).

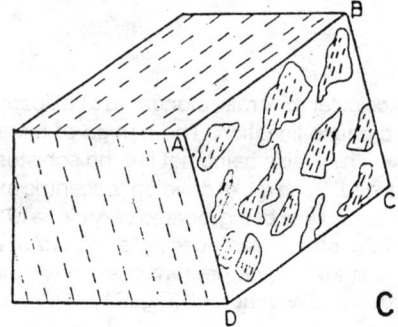

Fig. 219 C. The schistose plane ABCD is inclined one. The parallel planes are indicated by dashed lines. In the schistose plane, the platy/flaky minerals have developed lineation. The direction (bearing) of lineation and that of dip of schistose planes, are not coinciding. As a result of this, two maxima may be produced, one due to schistosity, and the other due to the lineation

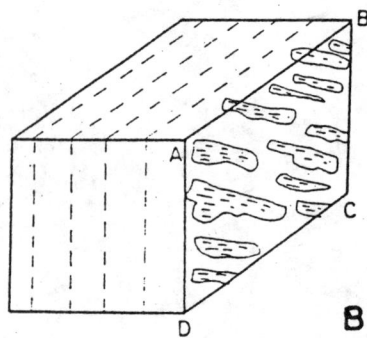

Fig. 219 D. Lineation not coinciding with the direction of dip of schistosity (vertical), is shown. In the schistose plane ABCD, the platy/flaky minerals have produced lineation which is plunging in a different direction and amount. The other parallel planes of schistosity are shown by dashed lines. Separate maxima may be produced corresponding to the schistose planes and the plunge of lineation, respectively. Thus two maxima will be produced.

(c) **Monoclinic symmetry:** In this case, there is only one plane of symmetry, and it is perpendicular to the lineation. A schistose rock thrown into asymmetrical folds, displays such a symmetry. This is shown in Figs. 220 A,B.

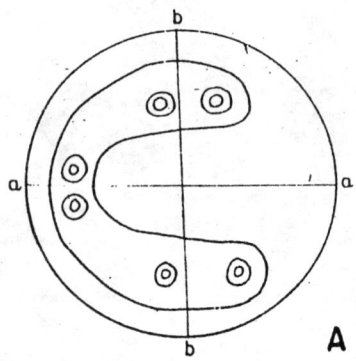

Fig. 220 A. Monoclinic fabric symmetry. Only one plane of symmetry along a - a is noticed.

Fig. 220 A. Monoclinic fabric symmetry. Only one plane of symmetry along a - a is noticed.

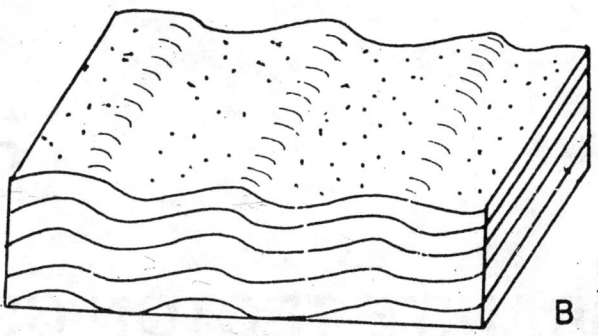

Fig. 220 B. A schistose rock thrown into asymmetrical folds, produces monoclinic fabric symmetry (after Billings, 1960).

(d) **Triclinic symmetry:** In this case even the plane of symmetry is destroyed. A schist which is folded and possesses lineation gives rise to such a symmetry. This is shown in Figs. 221 A,B.

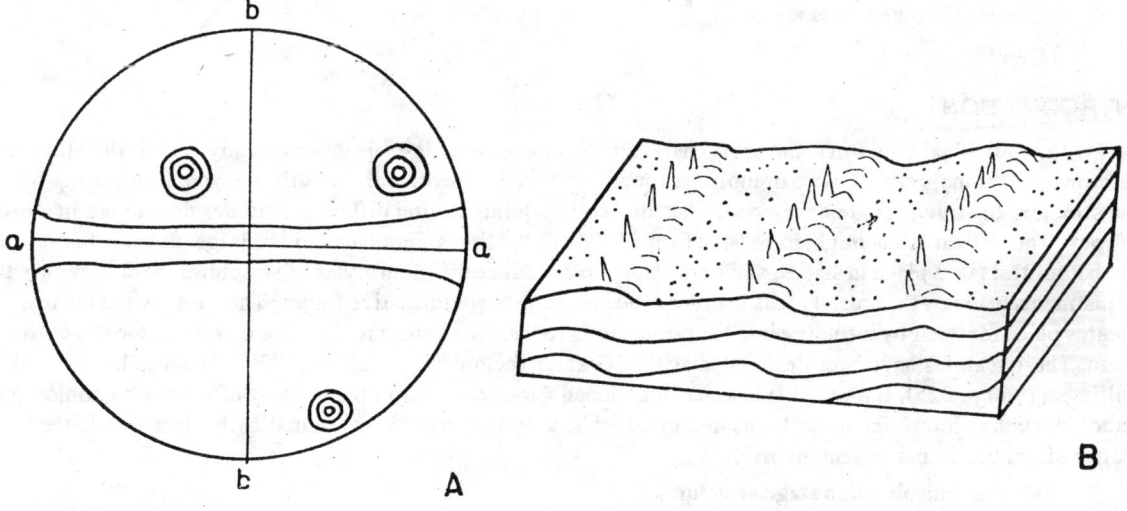

Fig. 221 A. Triclinic fabric symmetry. Note that there is no plane of symmetry at all.

Fig. 221 B. A schistose rock thrown into assymmetrical folds along with development of lineation produces a triclinic fabric symmetry (after Billings 1960).

Tectonites are also classified into (i) planar, and (ii) linear varieties, corresponding to schists, and rocks in which lineation alone is developed, respectively. Schistose rocks, foliated rocks etc., wherein planar structures are present, such rocks develop S-tectonites. Augen gneisses, hornblende gneisses and other similar rocks wherein accicular minerals dominate, or rocks containing ellipsoidal pebbles like the conglomerates, lava flows possessing trachytic structure, may develop lineation, and such rocks give rise to B - Tectonites. In such tectonites, two maxima are seen to develop. Depending upon the angular relation between the two maxima, B ∧ B and B ⊥ B tectonites are recognised.

9

GRANITE TECTONICS

Introduction, Pre-Consolidation stage structures. Post-Comsolidation stage structures. Structures produced in the country rocks.
Figures - 222 and 223.

INTRODUCTION

So far the individual structures and their characteristics have been described. As already noted, the structures seldom occur alone, generally two or more are found together. However there is a difference of time during which the structure or structures are developed. Thus one notices joints having different attitudes developed in a mass of rock. The several joints might have been formed at different times. Emplacement of large plutonic bodies like the batholiths, takes place in stages, and each stage may produce different styles of structures. Structures are the adjustments made by the crustal rocks in order to accomodate the deformative forces. Therefore when the magma creates place for itself by intruding into the pre-existing rocks, the country rocks also produce some structures in them. These features have been designated as the "Granite Tectonics" by Cloos (1939). This has been noted by Billings (1960 p. 325). Balk (1937) also has mentioned Cloos (op.cit.) in the context of granite tectonics. The tectonics such as joints, fractures, faults, marginal schistocity and so on, are systematically classified in the three stages of granite emplacement, namely.

    (a) pre-consolidation stage structures,

    (b) post-consolidation stage structures, and

    (c) structures developed in the intruded country rocks.

Thus there is an evolutionary process followed in the development of the structures, namely,

    (i) intrusion of the magma,

    (ii) movement and consolidation of it, when some structures are developed, and

    (iii) either during the time of the intrusion of the magma or after wards, the adjacent country rocks develop structures in them.

Obviously there will be a systematic relation between these various structures produced under the different stages and hence under different "structural environments". As large batholithic instrusions register the tectonics more clearly, and as granites form the largest igneous bodies, it is logically designated as "Granite Tectonics". In the following paragraphs, the several structural features associated with the stages of intrusion will be described.

PRE-CONSOLIDATION STAGE STRUCTURES

During this stage, greater part of the magma is in the molten state, and it is forcibly rising upwards during its process of emplacement. However some minerals might have been already crystallised which belong to the high temperature affiliation. The magma is also likely to contain some xenoliths that have been acquired from the walls of the country rocks, through the process of "magmatic stoping". These xenoliths as well as the early formed minerals will flow at a rate slower than the rest of the magma. Likewise, a differential rate of flowage will be set in the magma located in the vicinity of the country rocks, because, the walls prevent the upward movement of the magma. As a result of this, the early formed minerals and the xenoliths (if any), get oriented at an angle to the walls of the country rocks. However in the more central parts of the magma chamber, minerals or the xenoliths are not at all oriented, because there is no opposition offered to the movement of the magma. Thus the marginal parts of the magma chamber, develop foliation (if platy, flaky minerals be present), or lineation (if linear, accicular minerals, or elongated, ellipsoidal xenoliths be present). The granite therefore develops a gneissic foliation and or lineation, all along its border with the country rocks. The disposition of the foliation is such that it is steeper on the sides and is nearly flat towards the roof portion of the pluton. In a cross section therefore, such plutons give rise to a domical structure (Fig. 222) owing to the orientation of the suitable minerals, xenoliths and so on.

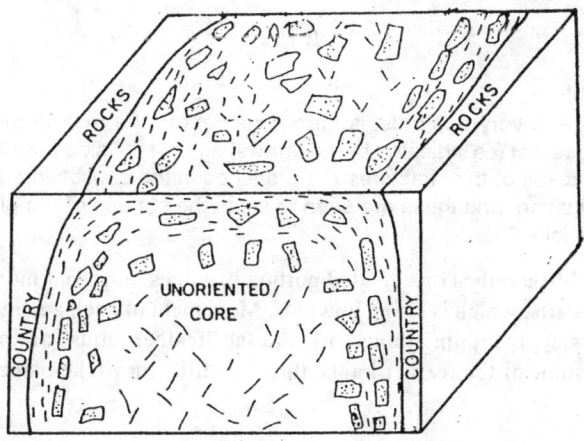

Fig. 222. Structural features of large plutonic intrusives.

Note that the intrusive has developed orientation of the xenoliths and early formed minerals, along the contact of it with the country rocks. Foliation is produced by the platy/flaky minerals, and the dip of foliation produces a domical structure. The central part of the pluton is strucreless and is therefore called as "unoriented core". The foliation and orientation together produces a "domical or arch" structure, when a vertical section is taken from one end of the pluton to the other. This is appreciable in the ABCD face of the block diagram (Fig. 222).

Alongside with the flowage of the magma, consolidation also takes place, and this proceeds from the periphery of the pluton towards the more interior parts. Thus a solid crustal layer of the consolidated part of the magma envelopes the still molten portion of the magma beneath it, which is still keeping on rising upwards into the crust of the earth. Such a consolidated layer which is thin in the initial stage, it experiences tension, and joints are produced in it. Such joints are nearly flat and are seen to cross the consolidated part of the magma (Fig. 223 A).

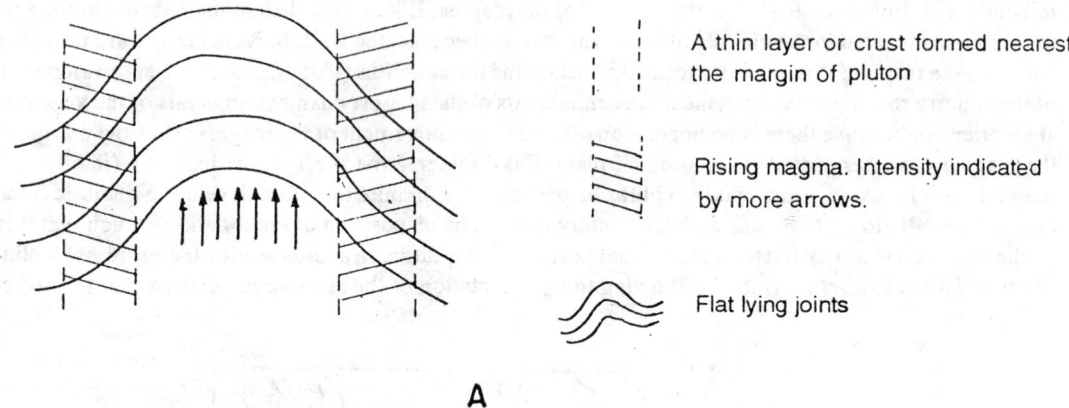

A thin layer or crust formed nearest the margin of pluton

Rising magma. Intensity indicated by more arrows.

Flat lying joints

A

Fig. 223 A. Stage A.

During this stage, a very thin crust is formed nearest to the margin of the pluton. The magma rises quite fast (as indicated by the arrows) against the sides (walls) of the magma chamber. As a result of this, fractures are produced in the consolidated part, which are having low inclination, and these are seen to cross the surface of the pluton. These are called as "cross joints".

As cooling advances further, the consolidated portion becomes more and more thick, but there still exists some magma at the deeper parts, which is rising upwards. Movement of such a magma is however considerably slow. Under this situation, steeply dipping joints called as the "feather joints" are produced in the consolidated part of the magma. Along some of the feather joints, thrust faulting may take place (Fig. 223 B).

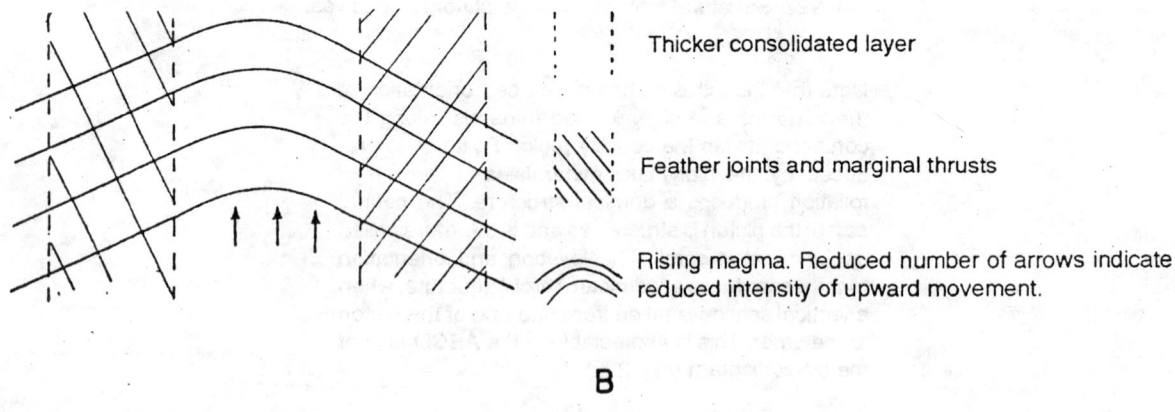

Thicker consolidated layer

Feather joints and marginal thrusts

Rising magma. Reduced number of arrows indicate reduced intensity of upward movement.

B

Fig. 223 B. Stage B.

In this stage, a very thick crust of the consolidated rock is formed at the sides as well as at the top of the magma which is still pushing upwards. The intensity of upward movement is much reduced. Under such a situation, fractures and joints that are steeply dipping into the magma chamber, are formed. Because the magma is still pushing upwards, some of the fractures/joints are utilised as fault planes. This gives rise to "upthrusts". The high angle joints are termed "feather joints".

POST-CONSOLIDATION STAGE STRUCTURES

At the final stage when most part of the magma has consolidated, the remaining volume of the magma pushes the thick crust of the granitic rocks, and gives rise to the joints that are nearly perpendicular to the dome shaped, consolidated, thich blanket of the granitic rocks. These joints are called as the "fan joints", because these are observed to "fan out" in the pluton. Such joints are nearly vertical in attitude. These features are shown in Fig. 223 C.

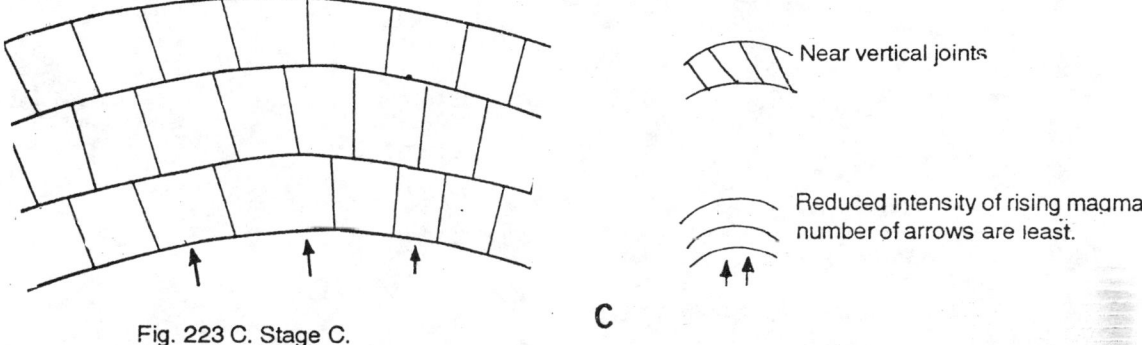

Near vertical joints

Reduced intensity of rising magma number of arrows are least.

C

Fig. 223 C. Stage C.

This is the last stage in the emplacement of large sized plutons. Almost the entire magma has consolidated leaving behind a very small volume which is still rising upwards. Because a very thick crust of rock is formed, nearly vertical joints are developed. These are observed to "fan" out, on the roof part of the magmatic body (chamber). These joints simulate the tension fractures developed at the crest of the anticlines. These joints are called as "fan joints".

Thus in the "Pre consolidation stage" of the magma emplacement, two kinds of structures are produced namely,

 (i) plastic structures when the marginal orientation and foliation are produced, and

 (ii) rupture structures, when flat lying joints, feather joints, marginal faults, are produced. The "domal or arch" structure" is also produced during this stage

In the "Post consolidation stage" of the magma emplacement, only rupture structures like the fan joints are produced. In this fashion a large sized granitic pluton acquires the various structures during the two stages of consolidation of the magma.

STRUCTURES PRODUCED IN THE COUNTRY ROCKS

As a result of the forceful intrusion of the large sized igneous bodies, the country rocks also make adjustments in them. It is a sympathetic reaction. While describing the causes of folding, it has been already noted that the compressive forces necessary for the folding, may be produced by the intrusions of magmas. Depending upon the size of the intrusions, the country rocks develop corresponding structures. Resistant competent rocks are ruptured, fractured or faulted nearest to the intrusion. This leads to the brecciation of the marginal rocks. Flowage takes place, if the confining pressure be available, to give rise to augen like structures. Autoclastic conglomerates may be developed. The intensity of folding will be very complex and intense, nearest the intrusion. The laccolithic form of the igneous intrusion is entirely due to the pre-existing bedded rocks, which get "up arched" due to the intrusion of the magma. If the bedded rocks were not to be present, then the "qua qua versal dip" can not be produced in the country rocks due to the igneous intrusion alone.

Granite tectonics only demonstrates that the geological events occur in pairs. The magma is looking for room (this forms one geological event), and the space is created by pushing up and or aside the country rocks (this forms the second geological event) when the structures are produced in the country rocks. Likewise, when faulting takes place (this forms one geological event), then the rocks develop shears, joints, folds etc. (this forms of the second geological event). Numerous such cases can be cited.

BIBLIOGRAPHY

Balk, R. (1937); Structural behaviour of igneous rocks. Geol. Soc. Am. Mem. 5

Banerji, A.K. (1962); Cross folding, migmatisation and ore localisation along part of Singhbhum shear zone south of Tatangar, Bihar. Eco. Geol. 57, pp. 50-71.

Bhimsen, K. (1989); Geological studies of the rocks of Ajara-Gadhinglaj area, Kolhapur district, Maharashtra state, India. Unpub. Ph. D. thesis, Kar. Univ. Dharwad.

Billings. M.P. (1960); Structural Geology. 2nd Edition, Asia Publishing House, Bombay.

CH. Sudarsana, Raju, S. Abbas Fazeli, and M. Basavachary (1973); Some mechanical properties of rocks of Hyderabad region. Jour. Ind. Acad. GeoSc. Vol. 16, No. 2, pp. 53-60.

Chavadi, V.C. (1974); Geology of the mafic and other associated rocks of Savantvadi, Ratnagiri district, Maharashtra. Unpub. Ph. D. thesis, kar. Univ. Dharwad.

Cloos, H. (1939); Hebung, spalting, vulkanismus. Geol. Rund. Band. 30. Zwisschheft 4 A, pp. 406-527.

Deendar, D.I. (1982); Geology of the Vengurla area, Ratnagiri district, Maharashtra state. Unpub. Ph. D. thesis, Kar. Univ. Dharwad.

Desai, H.D. (1991); Geology of the rocks at and around Jambunath hills, Bellary district, Karnataka state. Unpub. Ph. D. thesis, Kar. Univ. Dharwad.

Durg, N.L. (1966); A structural study of the phyllites near Dandeli, Mysore state. Quart. Jour. Geol. Min. Met. Soc. Ind. XLI, No. 3, pp. 153-155.

Foote, R.B. (1874); The auriferous rocks of Dambal hills, Dharwar district, Rec. Geol. Surv. Ind. Vol. 7, Pt. 4.

Foote, R.B. (1876); Geological features of Southern Marhatta country and adjacent districts. Mem. Geol. Surv. Ind. Vol. XII, pp. 37-164.

Foote, R.B. (1882) : Notes on a traverse across some gold fields of Mysore. Rec. Geot Surv. Ind. Vol. 15, pp. 191-202.

Fote, R.B. (1886) : Notes on the geology of Bellary and Anantpur districts. Rec. Geol. Surv. Ind. Vol. 19. part 2, pp. 97-111.

Foote, R.B. (1888) : Dharwar system, the chief auriferous rock series in South India. Rec Geol. Surv. Ind Vol. 21, pp. 40-56.

Foote, R.B. (1895); Dharwar-Shimoga band of Dharwars. Geol. Surv. Ind. Vol. 25, pp. 75-76.

Ghaisas, K.R. (1960); Structure of some crushed Kaladgi rocks near Saundatti. Jour. Univ. Poona, No. 2, pp. 35-40.

Gokhale, N.W. (1954); The Geology of the Dharwar district. M.Sc. Unpub. Dissertation, Benares Hindu Univ. Banaras.

Gokhale, N.W. (1964); A structural study of the Quartz phyllite and granite occurrence near fluorite-galena mine of Patka, Velence hills, Hungary. Acta Geologica. Tomus VIII, Fasciculi 1-4, pp. 337-345.

Gokhale, N.W. (1970); Structural studies on the granites and the associated schists of the Velence Mountains, Hungary, and the granite emplacement. Acta Geologica Tomus. 14. pp. 5-22.

Gokhale, N.W., Gothe, N.N., and Koppad, V.B. (1971); Thrust faulting amd related structures in the Bannikoppa-Bagewadi conglomerates of Gadag schist belt. Jour. Geol. Soc. Ind. Vol. 12, pp. 157-163.

Gokhale, N.W. (1977); Sinistral faulting and other structures in sandstones of Saundatti, Belgaum district, Karnataka Ind. Min. Vol. 18, pp. 73-78.

Gokhale, N.W., Gourashettar, V.K., and Puranik, S.C. (1983); Peculiar jointing in banded hematite quartzites of Nagavi, Gadag taluka, Dharwad district, Karnataka state. Prof. Kelkar Memorial volume, Ind. Soc. Earth Scientists, pp. 25-28.

Gokhale, N.W. and Hegde, G.V. (1987); Fold mullions and certain other structures in the rocks of Halgatti village, Belgaum district, Karnataka state. Jour. Sc. Kar. Univ. Vol. XXXII. pp. 65-69.

Gokhale, N.W. and Hegde, G.V. (1987); Giant, abnormal cataclasites of Timmapur, Nilgund and Kardigud, Belgaum-Bijapur districts, Karnatak state. Curr. Sc. Vol. 56, No. 22, pp. 1161-1164.

Gokhale, N.W. and Bhimsen, K. (1987); Possible structural control on the meandering pattern of Hiranyakeshi river, Kolhapur district, Maharashtra state. Abs. No. 55, Ind. Sc. Cong, 74th Sesson, p. 41.

Gokhale, N.W. (1987); Manual of Geological Maps. CBS Publishers and Distributors, Delhi-32.

Gokhale, N.W. and Pujar, G.S (1989); Bedding plane fault in the Kaladgi rocks, Basidoni, Belgaum district, Karnataka state. Curr. Sc. Vol. 58, No. 19 pp. 1088-1089.

Gokhale, N.W. and Waghamare, B.P. (1989); K - Ar. ages on three basic intersecting dykes from Gadag schist belt Karnataka. Jour. Geol. Soc. Ind. Vol. 34, pp. 663-664.

Gokhale, N.W. and Joshi, V.S. (1990); Comparative structural studies of four outliers of Kaladgi formations and their mode of upliftment. Bull. Ind. Geol. Assoc. 23 (1), pp. 13-16.

Gokhale, N.W., Waghamare, B.P., Bhimsen, K., Pujar, G.S. and Godbole Sushama; Axial rotational strike and dip faults in the quartz arenites, south of Jamkhandi, Bijapur district, Karnataka state (submitted for publication).

Gothe, N.N. (1973); Geology of the granitic and the associated rocks of Mundargi-Hadagalli area, Karnataka state. Unpub. Ph. D. thesis, Kar. Univ. Dharwad.

Guha, S.K., Goswamy, P.D., Verma, M.M. Agrawal, S.P., Padale, T.G. and Marwadi, S.C. (1970); Recent seismic disturbance in the Shivajingar lake area of the Koyna Hydrolectric Project, Maharashtra, India. Cent. Water Power Res. Station, Poona. Jan.

Hegde, G.V. (1984); Structural and sedimentological studies of the rocks of the Mudakavi-Lakhmapur area, Belgaum-Bijapur districts. Karnataka state. Unpub. Ph. D. thesis, Kar. Univ. Dharwad.

Hegde, V.N. (1984); Geology of the Gangavati-Kamalapuram area, Raichur-Bellary districts, Karnataka state. Unpub. Ph. D. thesis, Kar. Univ. Dharwad.

Heim, A. and Gansser (1939); Central Himalayas. Denkschr Schweiz, Naturf Gesell 73 (1), quoted by Krishnan, M.S. (1968).

Holmes, A. (1981); Principles of physical geology. ELBS Publication, 3rd Edition

Koppad, V.B. (1976); Geology of the area around Belhatti and Bannikoppa, shirhatti taluka Dharwad district, Karnataka state, India Unpub. Ph. D. thesis, Kar. Univ. Dharwad.

Kouhsari, A.H. (1986); Geological studies of bauxite deposits of Shrivardhan area, Kolabla district, Maharashtra state, India. Unpub. Ph. D. thesis, Kar. Univ. Dharwad.

Krishnan, M.S. (1968); Geology of India and Burma. Higginbothams (P) Ltd. Madras.

Maclaren, J.M. (1906); Notes on auriferous tracts in Southern India. Rec. Geol. Surv. Ind. Vol. XXXIV, pp. 90-115.

Muralidharan, D. (1991); Geohydrology of Aurepalle watershed in semiarid India. Unpub. Ph. D. thesis, Kar. Univ. Dharwad.

Patil, H.G., Deendar, D.I., and Gokhale, N.W. (1984); Mini Columnar joints in basalts of Kodali village, Maharashtra. Curr. Sc. Vol. 53 No. 20, p. 1089.

Patil, M.R. (1989); Geology of the area at and around Jamkhandi, Bijapur district, Karnataka state. Unpub. M.Sc. Dissertation, Kar. Univ. Dharwad.

Pujar, G.S. (1989); Geology of the area east of Manoli, Belgaum district, Karnataka state. Unpub. Ph. D. thesis, Kar. Univ. Dharwad.

Puranik. S.C. (1979); Iron formations of Nagavi-Doni area, Dharwad district, Karnataka state, India. Unpub. Ph. D. thesis, Kar, Univ. Dharwad.

Puranik. S.C., Gokhale, N.W., Gourashettar, V.K. (1982); Rotational faulting in Kaladgi formations near Nargund, Karnataka, India GeoViews, Vol. 1, No. 1, pp. 42-43.

Savanur, R.V. (1966); A study of the fault patterns in some of the blocks of South Karanapura and Ramgarh coal fields. Jour. Min. Met. Fuels. pp. 280-285.

Turner, F.J., and Weiss, E. (1963); Structural analysis of metamorphic tectonites. New York. pp. 61-64.

Wilson, C.J.L. (1973); A prograde microfabric in deformed quartzite sequence, Mount Isa, Australia. Tectonophysics. Vol. 19, pp. 33-81.

Y. Janaradan Rao and Ch. Sudarshan Raju (1964); structural studies in the PreCambrian rocks of the area northeast of Khammam, A.P. Proc. Sem. Peninsular Geology, pp. 115-126.

Y. Janardan Rao and Ch. Sudarsan Raju (1968); Tectonic history of the area east of Khammam, A.P. Jour. Geol. Soc. Ind. Vol. 9 No. 2 pp. 138-145.